Materials Issues for Tunable RF and Microwave Devices

MATERIALS RESEARCH SOCIETY
SYMPOSIUM PROCEEDINGS VOLUME 603

Materials Issues for Tunable RF and Microwave Devices

Symposium held November 30–December 2, 1999, Boston, Massachusetts, U.S.A.

EDITORS:

Quanxi Jia
Los Alamos National Laboratory
Los Alamos, New Mexico, U.S.A.

Félix A. Miranda
NASA Glenn Research Center
Cleveland, Ohio, U.S.A.

Daniel E. Oates
Lincoln Laboratory
Massachusetts Institute of Technology
Lexington, Massachusetts, U.S.A.

Xiaoxing Xi
The Pennsylvania State University
University Park, Pennsylvania, U.S.A.

Materials Research Society
Warrendale, Pennsylvania

CAMBRIDGE UNIVERSITY PRESS
Cambridge, New York, Melbourne, Madrid, Cape Town,
Singapore, São Paulo, Delhi, Mexico City

Cambridge University Press
32 Avenue of the Americas, New York NY 10013-2473, USA

Published in the United States of America by Cambridge University Press, New York

www.cambridge.org
Information on this title: www.cambridge.org/9781107413238

Materials Research Society
506 Keystone Drive, Warrendale, PA 15086
http://www.mrs.org

First published 2000
First paperback edition 2013

Single article reprints from this publication are available through
University Microfilms Inc., 300 North Zeeb Road, Ann Arbor, MI 48106

CODEN: MRSPDH

ISBN 978-1-107-41323-8 Paperback

This work was supported in part by the Office of Naval Research under Grant Number
ONR: N00014-99-1-1047. The United States Government has a royalty-free license
throughout the world in all copyrightable material contained herein.

CONTENTS

FREQUENCY AGILE MATERIALS
FOR ELECTRONICS

ELECTRIC-FIELD TUNING

*Invited Paper

*Invited Paper

HIGH-FREQUENCY APPLICATIONS
OF FERROELECTRICS

FERROELECTRICS

MAGNETIC AND OTHERS

FUNDAMENTALS

*Invited Paper

MATERIALS CHARACTERIZATION

*Invited Paper

PREFACE

This proceedings contains papers presented at Symposium KK, "Materials Issues for Tunable RF and Microwave Devices," held November 30–December 2 at the 1999 MRS Fall Meeting in Boston, Massachusetts. Electric or magnetic tunability of RF and microwave devices is desirable for a variety of civilian and military applications. In recent years, there have been tremendous advances in thin-film processing, in particular the metal-oxide thin films. Consequently, it has been recognized that the integration of nonlinear-dielectric, ferrite, colossal magnetoresistive (CMR), and superconductor materials could revolutionize tunable devices by providing new capabilities while significantly reducing size and cost.

Some of the crucial issues facing the tunable RF and microwave device technology concern the material properties, in particular the loss in thin films of the tunable materials. Extensive efforts are being devoted to understand the tuning and loss mechanisms, improve thin-film processing and characterization, develop new materials, and design novel device concepts. The papers contained in this volume are a reflection of the work currently in progress in this exciting area. They range from electric-field to magnetic-field tuning, from devices to materials, and from fundamental understanding to materials characterization. The authors come from different materials communities: nonlinear dielectric, ferrite, colossal magnetoresistive (CMR), semiconductors, superconductivity, and device engineering, etc. We hope that this volume will serve as an up-to-date reference for the researchers inside and outside the field of tunable RF and microwave devices, and thus contribute to new breakthroughs in the basic and applied research in this rapidly expanding field.

Quanxi Jia
Félix A. Miranda
Daniel E. Oates
Xiaoxing Xi

MATERIALS RESEARCH SOCIETY SYMPOSIUM PROCEEDINGS

MATERIALS RESEARCH SOCIETY SYMPOSIUM PROCEEDINGS

Prior Materials Research Society Symposium Proceedings available by contacting Materials Research Society

Frequency Agile Materials
For Electronics

ANALYSIS AND OPTIMIZATION OF THIN FILM
FERROELECTRIC PHASE SHIFTERS

R. R. ROMANOFSKY*, F. W. VAN KEULS**, J. D. WARNER*, C. H. MUELLER*, S. A.
ALTEROVITZ*, F. A. MIRANDA*, AND A. H. QURESHI***
*NASA Glenn Research Center, Cleveland OH 44135
**Ohio Aerospace Institute, Cleveland OH 44142
***Cleveland State University, Cleveland OH 44101

ABSTRACT

Microwave phase shifters have been fabricated from ($YBa_2Cu_3O_{7-\delta}$ or Au)/$SrTiO_3$ and
Au/$Ba_xSr_{1-x}TiO_3$ films on $LaAlO_3$ and MgO substrates. These coupled microstrip devices rival the
performance of their semiconductor counterparts at Ku- and K-band frequencies. Typical
insertion loss for room temperature ferroelectric phase shifters at K-band is ≈ 5 dB. An
experimental and theoretical investigation of these novel devices explains the role of the
ferroelectric film in overall device performance. A roadmap to the development of a 3 dB
insertion loss phase shifter that would enable a new type of phased array antenna is discussed.

INTRODUCTION

Evolving high data rate communications systems demand greater attention to subtle
aspects of information theory and electromagnetic engineering. As the ratio of signaling
bandwidth to carrier frequency decreases, less familiar phenomenon enter into system
performance. And, new coding techniques are pushing channel capacity ever closer to the
Shannon limit [1]. Some of these effects are expected to become quite pronounced if the trend
toward wide-band scanning phased array antennas and efficient high-speed modulators continues
[2]. For example, in a phased array antenna inter-element spacing, the physical size of the array,
and the steering vector can conspire to introduce pulse distortion from group delay, inter-symbol
interference, and beam squinting [3,4]. And the operating point of the amplifiers can affect the bit
error rate depending on the modulation type and the number of carriers. Naturally one wants the
phased array to operate as efficiently as possible given power limitations and thermal management
problems. This desire necessitates that the power amplifiers operate in a nonlinear region near
saturation. Nonlinear effects cause amplitude-to-amplitude modulation (AM/AM) and amplitude-
to-phase modulation (AM/PM) distortion. The net effect of AM/AM distortion is to alternately
compress and expand the signal constellation. The net effect of AM/PM conversion is a rotation
of the signal constellation [3]. In a receive array, the third order intercept of the low noise
amplifiers largely determines inter-modulation distortion and heat dissipation [5]. Phase shifters
typically follow low noise amplifiers in a receive array and precede power amplifiers in a transmit
array. Since the phase shifter's insertion loss depends on its phase setting and since its switching
action represents some finite time domain response, its potential contribution to bit error rate
degradation cannot generally be ignored. There will always be some effects in any phase shift
keyed (PSK) modulation system, to what degree depends on the steering vector update rate and
data rate.

In 1963 Berry introduced a new class of antennas that utilized an array of elementary
antennas as a reflecting surface [6]. The "reflectarray" has the potential to combine the best
attributes of a gimbaled parabolic reflector, low cost and high efficiency, and a direct radiating

3

phased array, vibration-free beam steering. A key advantage is the elimination of a complex corporate feed network. The reflectarray consists of a two-dimensional aperture characterized by a surface impedance and a primary radiator to illuminate that surface. The ferroelectric phase shifters described in this paper can be integrated with microstrip patch radiators to form a such a phase agile antenna [7]. Because the antenna elements and phase shifters can be defined using a two-step lithography process, the ferroelectric reflectarray holds promise to dramatically reduce manufacturing costs of phased array antennas and alleviates thermal management problems associated with microwave integrated circuit transmit arrays. (Note that while the devices to be described actually operate in the paraelectric regime slightly above the Curie temperature, it has become customary for workers in the field to still refer to the materials as ferroelectrics. This paper adheres to that convention.) A receive reflectarray has been designed at 19 GHz and is pictured in figure 1. The governing design assumption is a phase shifter insertion loss of 3 dB, about 40% better than we have consistently demonstrated from $Ba_xSr_{1-x}TiO_3$ films on $LaAlO_3$ and MgO substrates. A circular aperture was approximated arranging 177 subarrays as shown to improve the aperture efficiency.

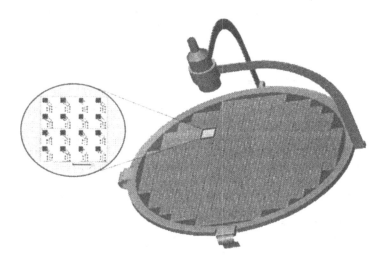

Figure 1. A 2832 element 19 GHz ferroelectric reflectarray concept. The callout shows a 16 element subarray patterned on a 3.1 x 3.1 cm^2, 0.25 mm thick MgO substrate. The array diameter is 48.5 cm. The unit cell area is 0.604 cm^2 and the estimated boresight gain is 39 dB.

The array was designed to scan past a 45 degree angle with an inter-element spacing of 0.52 wavelengths. The callout shows 16 4-section phase shifters coupled to a square patch antenna with a 90 degree hybrid coupler. The phase shifter is terminated abruptly in an open circuit and is used in a reflection mode. One output of the coupler has an additional 45° microstrip extension to feed the orthogonal edges of the patch 90° out of phase. For right hand circular polarization the phase shifter is attached to the left port of the coupler so the vertical edge of the patch receives the reflected energy with a 90° delay. While a triangular grid pattern permits the fewest elements per unit area, it was simpler to fit the phase shifters in a square unit cell. The

4

governing assumption in the design is that a 3 dB insertion loss phase shifter can be consistently reproduced. Indeed the phase shifter performance drives the performance and cost of the entire array. Even with a 3 dB loss device, assuming a receiver noise figure of 2 dB, the system noise temperature exceeds 800 K. Because the phase shifter is inserted between the antenna terminals and the low noise amplifier, it has the same effect as a feed line with equivalent loss in the determination of noise. The array noise does not increase with the number of elements since the noise is non-coherent but the signal at each element is correlated [8]. Assuming an aperture efficiency of 70% and a scan loss that falls off as $\cos\theta^{1.2}$, the gain-to-noise temperature ratio (G/T) of the array is estimated at ≈ 3.1 dB/K. *If the phase shifter loss could be reduced to 2 dB, the number of elements would be cut approximately in half.*

EXPERIMENT

The phase shifters analyzed here hold promise for reflectarray applications because they are compact, low-loss, and can be lithographed on the same surface as the radiating element. The designs used are based on a series of coupled microstriplines of length l interconnected with short sections of nominally 50 Ω microstrip. The maximum coupled voltage occurs when the coupled sections are a quarter wavelength long (i.e. $\beta l=90°$). Bias up to 400 V is applied to the sections via printed bias-tees consisting of a quarter-wave radial stub in series with a very high impedance quarter-wave microstrip. A sketch of the coupled microstrip cross-section is show in figure 2. By concentrating the fields in the odd mode, the phase shift per unit length is maximized and by using the ferroelectic in thin film form the effects of high loss tangent are minimized. Selecting the strip spacing "s" involves a compromise among: minimizing insertion loss, simplifying lithography, and minimizing the tuning voltage.

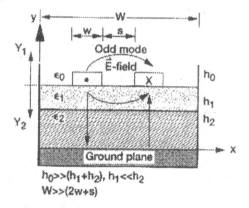

Figure 2. Cross-section of the coupled microstripline phase shifter showing the odd-mode electric field configuration. Y_1 and Y_2 represent the admittance looking in the positive and negative y direction, respectively, from the charge plane. The thickness of the ferroelectric layer is h_1 while the host substrate has thickness h_2.

A key advantage of this technology is the relatively large feature size. Active devices at the frequencies of interest here would necessitate submicron gate length GaAs FETs. The finest feature size associated with the coupled line phase shifters is the electrode separation s, typically

5

≈10 μm. Whereas the GaAs FET performance is largely dictated by transconductance and hence carrier transit time across the gate region, the coupled line phase shifters are static devices. The electrode gap separation determines the degree of electromagnetic coupling and the dc potential required to tune the film. As a rule-of-thumb the cutoff frequency f_c of a MESFET scales roughly with gate size as $f_c = 9.4/l$ where the frequency is in GHz and l is in microns. Hence the ferroelectric phase shifters have much larger feature sizes at a given frequency of operation and consequently much less demanding process requirements. The total phase shift can be increased by cascading coupled line sections. Each section is linked by a 50 Ohm or so microstrip jumper. A photograph of a 4-section phase shifter patterned on 0.5 mm MgO is shown in figure 3.

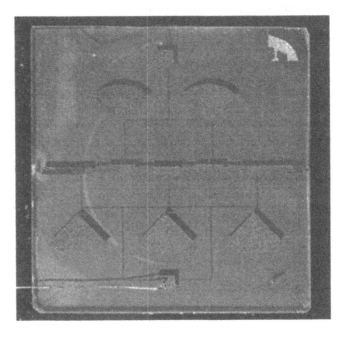

Figure 3. A 4-element $Ba_xSr_{1-x}TiO_3$ phase shifter on 0.5 mm MgO. The circuit measures 1 cm x 1 cm. l=457 μm, s=10 μm, and w=56μm.

These ferroelectric phase shifters are fabricated using standard lithography techniques. The "lift-off" processing recipe is straightforward. Starting with a clean substrate, AX4210 photoresist is spun on at 4000 rpm. This is followed by a soft bake at 75° C for 30 minutes. The photoresist is exposed through a visually translucent iron oxide mask for 30 seconds using a Carl Zeuss mask aligner with 300 nm optics. To facilitate the lift-off process, the wafer is soaked in chlorobenzene for 10 minutes at 25° C and blown dry with N_2. This is followed by a second bake at 90° C for 10 minutes. The wafer is developed for about 2 minutes in 4:1 deionized $H_2O:AZ400K$ developer, rinsed in deionized H_2O for 5 minutes, and blown dry with N_2. Metalization consists of evaporating a 150 Angstrom adhesion layer followed by 1.8 μm of Ag followed by a 500 Angstrom Au cap. The wafer is then soaked in acetone until the metal lifts off. It is advantageous to etch the ferroelectric from all regions except the coupled lines so that the bias tees are insensitive to tuning. A dilute 5% HF solution has been used to etch a $Ba_xSr_{1-x}TiO_3$

rectangular mesa beneath the coupled sections. The tradeoff is that positioning the pattern over a good region of material is tougher.

The performance of these devices is measured using an HP 8510C automatic network analyzer. The device is paced in a simple test fixture with SMA connectors and usually the launchers are attached to the microstrip input and output with silver pint. Measurements are usually done in a vacuum of about 10 mT to prevent dielectric breakdown of the air between the coupled lines. Alternatively, paraffin can be used to coat the lines but occasionally air bubbles trapped inside may contribute to arcing. In the future, a spray coated teflon coating will be used to permit safe operation under ambient conditions. Coupled microstripline FE phase shifters capitalize on the odd mode propagation constant and so have much more phase shift per unit length than simple microstripline while avoiding the need for a coplanar ground. Microwave measurements of coupled microstripline phase shifters of epitaxial $Ba_{0.5}Sr_{0.5}TiO_3$ films grown by combustion chemical vapor deposition (CCVD) on 0.5 mm MgO showed very low loss (estimated $tan\ \delta$ was between 0.03-0.002 at 20 GHz at 23°C) [9]. As shown in figures 4 and 5, the transmission coefficient, S_{21}, at all frequencies exhibits symmetrical curve as a function of bias voltage at 297 K, as expected. The best figure of merit was 53°/dB [79]. It is evident that the hysteresis is negligible.

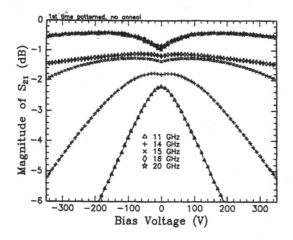

Figure 4. Measured 50 Ω 4-section phase shifter on MgO at 297 K. The $Ba_{0.50}Sr_{0.50}TiO_3$ film was grown by MicroCoating Technologies using CCVD.

The best films have been deposited by laser ablation at temperatures between 650 and 750 C and a dynamic oxygen pressure near 100 mTorr [10,11]. Thicker $Ba_xSr_{1-x}TiO_3$ films (>350 nm) generally exhibited poorer microwave performance despite having a clear theoretical advantage in terms of maximizing tuning as discussed in the next section. Crystalline quality unfortunately degrades as film thickness increases. The degradation is faster in $Ba_xSr_{1-x}TiO_3$ than in $SrTiO_3$.

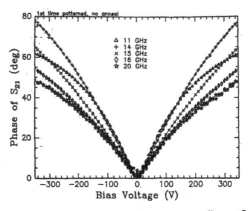

Figure 5. Measured insertion phase corresponding to figure 4.

The use of MgO (ε_r=9.7) allows wider lines for a given impedance compared to LaAlO$_3$ (ε_r=24). Consequently, the conductor loss is lower on MgO. Figures 6 and 7 show measured insertion loss and phase for 0.5 μm Ba$_{0.60}$Sr$_{0.40}$TiO$_3$ film grown on 0.5 mm MgO. Choosing a operating temperature that approaches the Curie temperature from the paraelectric phase usually results in larger phase shifts but correspondingly higher loss.

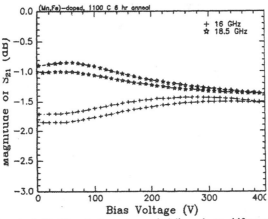

Figure 6. Measured 50 Ohm 4-element coupled line phase shifter at 210 K with Au/Cr electrodes patterned on a 0.5 μm Ba$_{0.60}$Sr$_{0.40}$TiO$_3$ film grown on 0.5 mm MgO. The film was grown by the Naval Research Laboratory. These film were 1% Mn doped and annealed at 1100 C for 6 hours.

The best performance to date has been obtained from YBa$_2$Cu$_3$O$_{7-\delta}$ and laser ablated SrTiO$_3$ films on (100) single crystal LaAlO$_3$ substrates [12]. Data for an 8-section nominally 50 Ω coupled microstrip device at 16 GHz is shown in figure 8. The superconducting film was 350 nm thick and the ferroelectric film was 2.0 µm thick. A figure of merit approaching the goal of 120 °/dB was obtained. It is unclear what role surface effects may have played in the superior

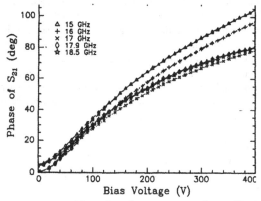

Figure 7. Measured insertion phase corresponding to figure 6.

performance of this particular phase shifter that used $YBa_2Cu_3O_{7-\delta}$ electrodes instead of metal electrodes. Note that while bulk $SrTiO_3$ is an incipient ferroelectric, thin films exhibit a relative dielectric constant maximum between 40 and 80 K.

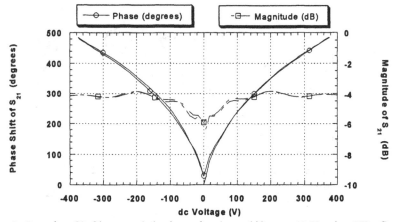

Figure 8. 8-section 50 Ohm coupled microstrip phase shifter at 40 K using $YBa_2Cu_3O_{7-\delta}$ electrodes and laser ablated $SrTiO_3$ films on (100) single crystal 0.25 mm $LaAlO_3$. Hysteresis is unremarkable. l=470 µm, s=7.5 µm, and w=25µm.

Some recent results suggest that after extended voltage cycling an anomalous discontinuity in loss and phase occurs under certain bias and film growth conditions. The origin of this polarization change is not understood at this time. Fatiguing effects have been observed in ferroelectric films for DRAM applications, where occasionally micro-cracks occur to absorb the stress. Such effects are irreversible. The study of this phenomenon is ongoing.

DEVICE MODELING

The multi-layer structure has been analyzed using a computationally efficient variational method to calculate the even and odd mode capacitance [7,13]. If a quasi-TEM type of propagation is assumed the propagation constant and impedance can be completely determined from line capacitance. Since the cascaded coupled line circuit resembles a series of one-pole bandpass filters, as the dc bias increases, the dielectric constant of the BST film decreases, causing the passband to rise in frequency and the *tan δ* of the BST to decrease. The impedance matrix of the cascaded network can be derived by well-known coupled line theory using the superposition of even and odd mode excitation. Then an equivalent S-parameter model can be extracted and used to predict the pass-band characteristics of the phase shifter.

The bandwidth compression from tuning is evident in figure 9 which is data from an 8-section phase shifter on 0.3 mm MgO using a 400 nm $Ba_{0.60}Sr_{0.40}TiO_3$ laser ablated film. The roll-off at the upper end of the frequency range is attributed to bias-tee effects.

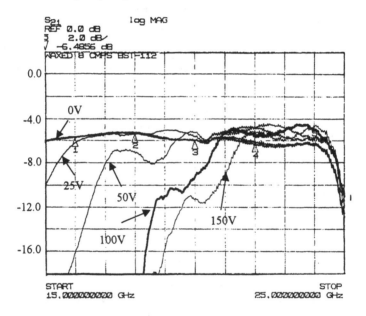

Figure 9. Measured Insertion Loss (including SMA launchers) of an 8-element ≈50 Ω PLD coupled microstripline phase shifter at 290 K as a function of bias voltage. Substrate is 0.3 mm MgO with 400 nm $Ba_{0.60}Sr_{0.40}TiO_3$ film. *l*=350 μm, s=7.5 μm and w=30 μm. Bandwidth compression from the filtering effect is evident. Marker 1, 2, 3, and 4 are at –5.75, -5.38, -6.00, and –6.49 dB, respectively.

In order to gauge the impact of the ferroelectric film on overall performance, Table 1 summarizes several important parameters for a *single* coupled microstrip section on h_2=0.3 mm MgO (ε=9.7) derived using the quasi-TEM analysis where $ε_1$ and h_1 correspond to figure 2. The

10

insertion phase is designated as ϕ_I, the composite dielectric loss as αd, and the characteristic impedance is taken as $Z_o=[Z_{oe}Z_{oo}]^{1/2}$. In all cases the loss tangent of the host substrate was 0.001 and the loss tangent of the ferroelectric film of thickness h_1 was taken as 0.05, 0.028, and 0.005 for ε_1 equal to 2500, 1000, and 500, respectively.

Table 1. Theoretical propagation characteristics of a single coupled microstrip section on 0.3 mm MgO based on the quasi-TEM method described in [7] and [13]. l=350 µm, s=10µm, and w=30µm.

ε_1	$h_1 = 2$ µm			$h_1 = 1$ µm			$h_1 = 0.5$ µm		
	ϕ_I °	αd (Np/m)	Zo (Ω)	ϕ_I °	αd (Np/m)	Zo (Ω)	ϕ_I °	αd (Np/m)	Zo (Ω)
2500	65.9	66.3	29.7	50.5	45.3	37.9	40.0	30.7	46.7
1000	46.6	22.5	40.7	37.4	15.3	49.6	31.2	10.3	58.4
500	37.3	3.0	49.7	31.2	2.2	58.4	27.1	1.6	66.4

The net phase shift is 2.2 times greater for the 2 µm film compared to the 500 nm film. Table 2 summarizes propagation characteristics for a *single* coupled microstrip section on h_2=0.25 mm LaAlO$_3$ (ε=24) derived using the quasi-TEM analysis. The insertion phase is greater because the effective permittivity of the composite structure is substantially greater than that of Table 1. But the correlation between phase shift and film thickness is about the same.

Table 2. Theoretical propagation characteristics of a single coupled microstrip section on 0.25 mm LaAlO$_3$ based on the quasi-TEM method described in [7] and [13]. l=457 µm, s=8µm, and w=25µm.

ε_1	$h_1 = 2$ µm			$h_1 = 1$ µm			$h_1 = 0.5$ µm		
	ϕ_I °	αd (Np/m)	Zo (Ω)	ϕ_I °	αd (Np/m)	Zo (Ω)	ϕ_I °	αd (Np/m)	Zo (Ω)
2500	100.1	70.4	24.4	79.4	47.9	30.1	65.3	32.3	35.9
1000	73.9	23.9	32.1	61.7	16.2	37.8	53.6	10.8	43.0
500	61.4	3.4	37.9	53.5	2.5	43.1	48.2	1.8	47.4

It is clear from the experimental and modeled data that the inherent dielectric loss of epitaxial ferroelectric films isn't necessarily devastating insofar as microwave device performance is concerned. Indeed the loss tangent of a thin dielectric film ($h_1 \leq 2$µm) on a good substrate (tan$\delta \leq 0.001$) can deteriorate substantially (tan$\delta \leq 0.05$) before the insertion loss of the structures presented here is compromised. Even when the bulk of the substrate has relatively high loss, useful devices can be rendered for certain applications. Microwave designers generally quote a loss tangent of 10^{-4} as desirable. But as can be seen from figure 10, a loss tangent as poor as 10^{-2} translates into an insertion loss of only 1 dB at 18 GHz for a 1 cm long 50 Ω microstrip line on a homogeneous substrate.

Figures 11-12 represent simulated S-parameter data comparing the effect of loss tangent, using IE3D software, for the 8-section ferroelectric phase shifter corresponding to figure 9. IE3D uses a method of moments based solution and has been found to be very accurate when compared to experiment for these multi-layer structures. In both graphs the tanδ of the MgO was 10^{-4}.

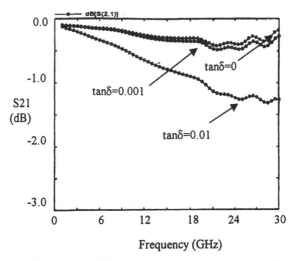

Figure 10. Simulated effect (using IE3D) of substrate loss tangent on insertion loss of a 1 cm long 50 Ω microstrip line terminated in a matched source and load on 0.25 mm LaAlO₃. The 2 μm thick metal strip and ground plane had an electrical conductivity of 4.9 x 10⁷ S/m.

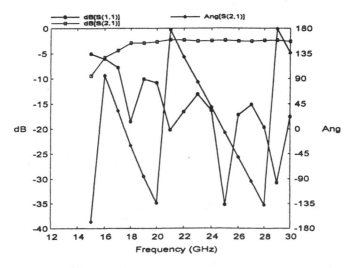

Figure 11. Modeled S-parameters for an 8-section coupled microstrip ferroelectric phase shifter on 0.3 mm MgO with 400 nm $Ba_{0.60}Sr_{0.40}TiO_3$ film. l=350 μm, s=7.5 μm and w=30 μm. ε_1=2500, tanδ=0.05, and σ=4.9 x 10⁷ S/m for the 2.5 μm metal layers.

The difference in S21 was about 1.25 dB when the effect of tanδ is considered, in this particular case. The loss due to impedance mismatch is about 0.5 dB or less for the zero field condition but can degrade to 1 dB or more if the ferroelectric is tuned to ε_1=300. Conductor loss is evidently about 1 dB and radiation is considered negligible since the lines are intentionally operated in the

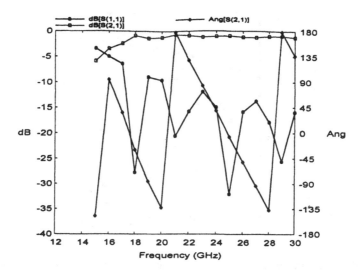

Figure 12. Modeled S-parameters for an 8-section coupled microstrip ferroelectric phase shifter on 0.3 mm MgO with 400 nm $Ba_{0.60}Sr_{0.40}TiO_3$ film. l=350 μm, s=7.5 μm and w=30 μm. ε_1=2500, tanδ=0.005, and σ=4.9 x 10^7 S/m for the 2.5 μm metal layers.

odd mode. In this mode the currents in the coupled lines are equal but opposite so the radiated fields tend to cancel. The overall loss is ideally 3.5 to 4.0 dB when the phase shifters are embedded between a 50 Ω source and load.

It is fundamentally important to ensure that a good conductor with the proper thickness is used for the electrodes and ground plane. Figure 13 illustrates the difference between the measured resistivity of a thin gold film and bulk gold. Since the conductor loss is proportional to surface resistance, there can be a dramatic impact on loss if the conductor isn't at least several skin depths thick. And the proper value for resistivity must be used. For example, for a single section coupled line on 0.3 mm MgO and using the thin film resistivity values from figure 13, the insertion loss at 19 GHz and 290 K for 1.0 μm and 2.5 μm thick gold electrodes is 0.63 dB and 0.48 dB, respectively. Hence an excess loss of 1 dB per 8-section phase shifter can easily occur from improper lithography. Some of the experimental phase shifter data provided earlier may have suffered from this effect.

CONCLUSIONS

Coupled microstripline ferroelectric phase shifters have already demonstrated performance superior to their semiconductor counterparts at microwave frequencies. Typical de-embedded insertion loss for these novel ≈360° phase shifters is about 5 dB. But in order to realize the full technical and economic benefits of a new type of phased array antenna, called the ferroelectric reflectarray, a 3 dB insertion loss ferroelectric phase shifter must be produced on a consistent basis. Incorporating a microstrip matching network between the coupled sections and the terminations can potentially produce a 0.5 to 1 dB improvement. Lower characteristic impedance

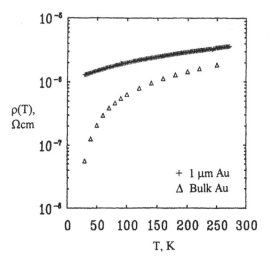

Figure 13. Comparison between resistivity of a thin evaporated Au film on Al2O3 and bulk Au as a function of temperature.

devices could reduce conductor loss even further. And the deposition of dense, high conductivity electrodes that are at least ≈3 skin depths thick can result is further improvement. If the tanδ of the ferroelectric film could be maintained at 0.005 or less, the films contribution to total loss would be essentially negligible except for the mismatch it introduces as it is tuned. But even a tanδ of <0.05 is manageable. It is most desirable to maintain the crystalline properties of the ferroelectric films to a thickness of at least 1 μm. Relatively thick films increase the phase shift per coupled section while the conductor loss stays more or less constant. Consequently the number of sections required to produce a full 360° phase shift can be reduced.

REFERENCES

1. C. Berrou, A. Glavieux and P. Thitmajshima, ICC 1993, pp. 1064-1070.
2. J.M. Budinger et al., IEEE Int'l. Conf. Comm., Montreal, Canada, June 1997.
3. C.A. Jensen, J.D. Terry, and M. Vanderaar, *The Implications of Encoder/Modulator/Phased Array Designs for Future Broadband LEO Communications*, October, 1997.
4. S. Ohmori, S. Taira, and M. Austin, J. Comm. Res. Lab., Tokyo, Japan, **38** (2), (1991) 217.
5. D. Collier, Microwave Systems News, (1990) 37.
6. D.G. Berry, R.G. Malech, and W.A. Kennedy, IEEE Trans. Ant. Prop., **11** (1963) 645.
7. R.R. Romanofsky, PhD Thesis, Cleveland State University, 1999.
8. M.P. Delisio, R.M. Weikle, and D.B. Rutledge, IEEE rans. MTT, **46** (11) 1949.
9. W-Y, Lin et al, Ferroelectric Workshop, Puerto Rico, May 1999.
10. F.W. VanKeuls et al., IEEE MTT Symp. Digest, (1999) 737.
11. F.W. VanKeuls et al., Microwave & Optical Tech. Lett., **20** (1) 53.
12. F.W. VanKeuls et al., Appl. Phys. Lett., **71** (21) 3075.
13. R.R. Romanofsky and A.H. Qureshi, Intermag 2000, Toronto, Canada, 2000.

30 GHz ELECTRONICALLY STEERABLE ANTENNAS USING $Ba_xSr_{1-x}TiO_3$-BASED ROOM-TEMPERATURE PHASE SHIFTERS

C. M. Carlson
Department of Physics, University of Colorado, Boulder, CO 80309

T. V. Rivkin, P. A. Parilla, J. D. Perkins, D. S. Ginley
National Renewable Energy Laboratory, Golden, CO 80401

A. B. Kozyrev, V. N. Oshadchy, A. S. Pavlov, A. Golovkov, M. Sugak, D. Kalinikos
St. Petersburg Electrotechnical University, St. Petersburg, Russia 197376

L. C. Sengupta, L. Chiu, X. Zhang, Y. Zhu, S. Sengupta
Paratek Microwave, Inc., Columbia, MD 21045

ABSTRACT

We report the performance of 16-element phased array antennas operating at 30 GHz and ambient temperature. These antennas use $Ba_xSr_{1-x}TiO_3$(BST)-based phase shifters to produce the beam steering. Ferroelectric phase shifters offer advantages over current semiconductor and ferrite devices including faster switching speeds and lower costs. Also, ferroelectric phase shifters offer higher power handling capability than semiconductor devices and also have high radiation resistance. We made phase shifters from laser-ablated epitaxial BST films as well as from polycrystalline BST-oxide composite films. Although neither the devices nor the materials themselves are fully optimized, phase shifters have shown > 360° of phase shift with < 350 V DC bias ($E < 9$ V/μm) and ~8 dB insertion loss. With ferroelectric phase shifters incorporated, antennas show radiation patterns with central-lobe half-power widths of ~13° and side lobe intensities down by more than 10 dB. Using the phase shifters, the central lobe can be shifted, or "steered," by ±18° in either direction. These results demonstrate a first step toward a prototype steerable antenna for 20–30 GHz satellite communications as well as other applications.

INTRODUCTION

Presently, phased array antennas are increasingly being used for long-range (X-band, ~8–12 GHz) and short range (Ka–W band, ~26–110 GHz) communications and radar systems. Depending on the requirements of the application, anywhere from a few (< 10) up to thousands of individual phase shifters are needed to produce electronic steering of the radiation pattern. Currently, there are two principal means of electronically controlling the relative phase of microwave signals. These use phase shifters that are based on either semiconducting diode or ferrite technologies. Both these technologies have been under development since about 1960, but neither has established a clear superiority over the other. The choice between these two dissimilar approaches depends on many factors including usable frequency range and other application-constrained parameters.

A third option for phase shifter technology currently under investigation is based on ferroelectric (FE) materials. Ferroelectric phase shifters have marked advantages over both semiconductor and ferrite devices in terms of switching speed and cost. They also have high power handling capability and high radiation resistance in comparison to semiconducting phase shifters. Table I further illustrates some of the strengths and weaknesses of the three competing technologies.

15

Table I. Comparison of FE with current phase shifter technologies.

	Ferroelectric	Semi-Cond.	Ferrite
Price	Cheap	Expensive	Very Expensive
Power Handling	Good (>1 W)	Poor ~200 mW	Excellent
Switching Speed	Very Fast (< 10^{-10} sec)	Fast (at low power)	Slow (inductance)
Radiation Resistance	Excellent	Poor	Excellent
Power Consumption	Low (few 10^2 V, no current)	Very Low (few V, no current)	High (large current)

In this paper, we report a theoretical and experimental investigation of ferroelectric-based 30 GHz phase shifters and their incorporation into an electronically steerable antenna module.

FERROELECTRIC MATERIALS

Since we are targeting ambient-temperature applications, we employ $Ba_xSr_{1-x}TiO_3$(BST)-based ferroelectric materials. Specifically, we have focused on making films of pure BST (x=0.4) as well as BST-based oxide composite films. The pure BST films are epitaxially grown on MgO (001) substrates using pulsed laser deposition (PLD) as described elsewhere [1]. As many have found, the main problem with such films is controlling the loss [1]. One way to control the loss, as well as to tailor other dielectric properties, is to use composite materials of various compositions [2]. BST-composite films are also deposited on MgO (001) substrates using a proprietary process.

As mentioned above, the pure BST films are epitaxial and show high structural quality. Figure 1 shows typical θ/2θ and pole figure x-ray diffraction scans for a 500 nm-thick film.

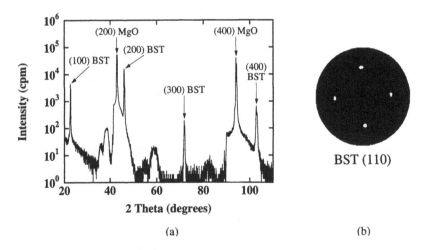

(a)   (b)

Fig. 1. (a) θ/2θ and (b) pole figure x-ray diffraction scans for laser-ablated BST/MgO film showing single epitaxial orientation.

The θ/2θ scan (Fig. 1a) shows only ($h00$) peaks indicating a single out-of-plane texture, and the pole figure (Fig. 1b) confirms the in-plane orientation of the epitaxial film. Atomic force microscopy (not shown) indicates a smooth surface morphology with ~100 nm round grains and a surface roughness of < 10 Å RMS. The BST-composite films, by contrast, are polycrystalline with large grains as shown in Fig. 2.

Fig. 2. Scanning electron microscope image of BST-composite film surface.

Dielectric characterization of the epitaxial BST films has been done at ~2 GHz using a modified ring resonator technique [1,3]. These films show loss tangents (tanδ) of ~0.06 and tunabilities ($\kappa = \varepsilon(E=0)/\varepsilon(E\neq0)$) of ~2.5–3.0 under an applied field of $E \sim 7$ V/μm, where $\varepsilon(E)$ is the film's dielectric constant. BST-composite films were characterized at 30 GHz using a technique described elsewhere [4]. These films have tanδ ~ 0.04 and κ ~ 1.8 for $E \sim 20$ V/μm.

PHASE SHIFTERS

Phase Shifter Fabrication

For these experiments, we employ a grounded coplanar waveguide (CPW) design for the phase shifter circuit. As shown in Fig. 3, it consists of an active segment (L) between input and output impedance matching segments. The device size is 12x6 mm^2. The grounded CPW has a layered structure consisting of a BST-based ferroelectric film prepared on a 0.5 mm–thick single crystal MgO (001) substrate. This device shifts the phase of transmitted radiation by varying the

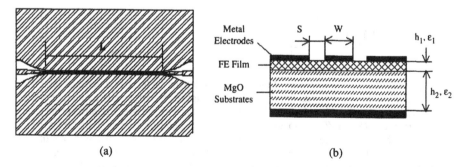

(a)   (b)

Fig. 3. Grounded CPW phase shifter (a) layout and (b) cross-section.

propagation speed of the CPW line. This variation is a consequence of the change in dielectric constant of the FE layer under a DC controlling voltage applied between the central and side electrodes.

These devices were fabricated by a photolithographic lift-off technique [5]. The lift-off method produces straight, non-ragged, metal edges and leaves no metal residue in the gaps, which can increase the probability of electrical breakdown during DC biasing. The surface of the FE film is not affected by lift-off as it is by the chemical etching technique. Furthermore, the lift-off technique can handle the thick metal structures (500 Å Au / 4 μm Ag / 500 Å Au / 100 Å Ti / FE / MgO) that are necessary to minimize loss in the metal due to skin depth effects.

Theory of Phase Shifter Design

The multilayer coplanar structures that make up our CPW phase shifters can be analyzed with standard theoretical methods [6,7]. The figure of merit for phase shifters ($F = \Delta\phi/\alpha$) depends on the phase shift ($\Delta\phi$) and on the total line losses $\alpha = \alpha_d + \alpha_c$, where α_d and α_c are the attenuation due to losses in the dielectric and conductor layers respectively. Analytical expressions for the phase shift and losses can be obtained using a conformal mapping transformation with the approximation that the dielectric constant of the FE film is much greater than that of the substrate ($\varepsilon_1 \gg \varepsilon_2$). Equations 1-3 show these expressions.

$$\Delta\phi\left[\frac{rad}{m}\right] = \frac{2\pi}{\lambda_0}\left(\sqrt{\varepsilon_{eff}(0)} - \sqrt{\varepsilon_{eff}(U_b)}\right) = \frac{\pi}{\lambda_0}\cdot\sqrt{F_1}\cdot\sqrt{\varepsilon_1(U_b)}\left(\sqrt{\kappa}-1\right) \tag{1}$$

$$\alpha_d\left[\frac{dB}{m}\right] = A\cdot\omega\cdot\sqrt[4]{\varepsilon_1(0)\cdot\varepsilon_1(U_b)}\cdot\sqrt{\tan\delta_1(0)\cdot\tan\delta_1(U_b)}\cdot\sqrt{F_1} \tag{2}$$

$$\alpha_c\left[\frac{dB}{m}\right] = B\cdot R_s\cdot\sqrt[4]{\varepsilon_1(0)\cdot\varepsilon_1(U_b)}\cdot\frac{\sqrt{F_1}}{K(k_0')\cdot W} \tag{3}$$

In Eqs. 1-3, λ_0 is the radiation wavelength in free space, $\kappa = \varepsilon_1(0)/\varepsilon_1(U_b)$ is the tunability of the FE film under a DC bias U_b. $K(k_0')$ is the complete elliptic integral of the first kind and is a function of the CPW device dimensions S and W. $\tan\delta(0)$ and $\tan\delta(U_b)$ are loss tangents of the FE film under zero and non-zero applied bias respectively. R_s is the microwave surface resistance of the metal conductors. A and B are constant coefficients, and F_1 is a fill factor from the conformal map. Based on Eqs. 1-3, a rough estimate of the figure of merit for our CPW phase shifters is

$$F\left[\frac{rad}{dB}\right]_{\substack{\varepsilon_1 \gg \varepsilon_2 \\ \tan\delta_1 \gg \tan\delta_2}} = \frac{\frac{\pi}{\lambda_0}\left(\sqrt{\kappa}-1\right)}{\sqrt[4]{\kappa}\left(A\omega\cdot\sqrt{\tan\delta_1(0)\cdot\tan\delta_1(U_b)} + \frac{B\cdot R_s}{K(k_0')\cdot W}\right)}. \tag{4}$$

According to Eq. 4, the figure of merit does not depend on the actual value of the FE film's dielectric constant (ε_1) or its thickness (h_1). The figure of merit is only a function of the film's tunability (κ) and the loss from both the film ($\tan\delta$) and electrodes. Variation of the electrode geometry (S and W) has no effect on the FE film's contribution to F (first term in denominator), but can dramatically increase the conductor losses (second term in denominator) for small widths of the central electrode (W) or gap (S). This effect is illustrated in Fig. 4, which shows an abrupt increase in metal loss (α_c) as the gap width (S) decreases below 10–20 μm.

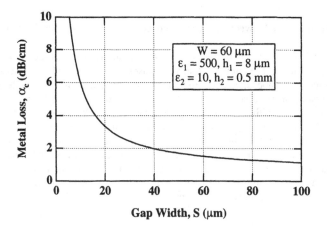

Fig. 4. Numerical calculation of losses in metal electrodes as a function
of the gap width S for CPW phase shifter ($f = 30$ GHz).

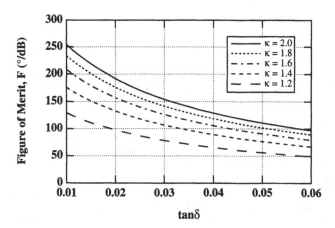

Fig. 5. Figure of merit vs. loss tangent for several values
of tunability ($f = 30$ GHz, $R_s = 0.04$ Ω).

The dependence of the figure of merit (F) on $\tan\delta$ as calculated from Eq. 4 is shown in Fig. 5 for several values of the tunability (κ). According to this calculation, values of the figure of merit of about 100 °/dB, which is more than adequate for many applications, should be possible by using FE films with $\tan\delta \sim 0.04$ and $\kappa \sim 1.5$. As discussed above, our current FE films satisfy both these requirements.

A quantity which, according to Eq. 4, can be viewed as a figure of merit for the FE film itself in regard to CPW phase shifter applications is

$$F_{FE} = \frac{\sqrt{\kappa} - 1}{\sqrt[4]{\kappa} \cdot \sqrt{\tan\delta_1(0) \cdot \tan\delta_1(U_b)}}.$$

(5)

This provides a comparison between FE samples concerning their suitability for phase shifter devices.

From the calculation of metal loss as a function of gap width (S) as shown in Fig. 4, it seems that the figure of merit (F) monotonically increases with S for fixed tunability of the FE film, while in practice this does not appear to be the case. The reason is that in order for the tunability κ to remain constant, we must apply a constant electric field, typically of about 15–20 V/μm. As the gap width increases, we must apply larger voltages to keep the field constant ($E \sim U_b/S$). In practical situations, U_b is limited by the controlling circuits. This limit for U_b leads to a decrease of the applied electric field as the gap width increases and consequently to the decrease of tunability. Therefore, the retention of maximum values of κ is impossible for large gap sizes, and κ effectively becomes a function of the gap width. In order to account for this effect in our analysis of the figure of merit, we must include the dependence of κ on applied voltage. This dependence can be approximated as

$$\kappa(U_b) = \frac{\kappa_{max} \cdot \left[1 + \left(\frac{U_b}{U_0} \right)^2 \right]}{\kappa_{max} + \left(\frac{U_b}{U_0} \right)^2}, \qquad (6)$$

where κ_{max} is the maximum tunability corresponding to an applied electric field of $E \sim 20$ V/μm, and U_0 is a phenomenological parameter of the FE varactor which depends on the gap width and FE film properties including thickness. The true relationship between U_0 and the gap width is complicated and must be obtained for a given film type by measuring varactors with a variety of gap sizes. For the purpose of rough estimation, we assume that $U_0 = \gamma S$, where γ is a constant coefficient that depends on the properties of the film and its thickness.

Using the approximation in Eq. 6, we have numerically calculated (without requiring $\varepsilon_1 \gg \varepsilon_2$) the dependence of the figure of merit (F) on the gap size assuming $\kappa_{max} = 2$, $\tan\delta = 0.04$, and U_b is limited to 600 V. Figures 6 and 7 show this calculation for several values of the FE film thickness and dielectric constant respectively. As is clear from Figs. 6 and 7, these calculations allow us to find the optimal phase shifter parameters for a given applied voltage limit and the known parameters of the FE film. For the stated constraints, the calculations indicate that maximum values of the figure of merit correspond to a range of gap sizes ($S = 40$–80 μm) and should only weakly depend on the FE film's thickness and dielectric constant.

Measured Phase Shifter Performance

Here we present the measured phase shifter performance of two devices; one based on a BST film and the other on a BST-composite film. Microwave S-parameter measurements were made at 31.34 GHz using an HP 8722D network analyzer. Figure 8 shows the variation of phase shift and insertion loss with applied bias voltage. The maximum phase shift is ~250° with an average insertion loss of ~ 9 dB. This corresponds to a figure of merit of ~28 °/dB. The performance of this device was limited by a low applied voltage. A larger bias would produce more phase shift with no increase in loss, thereby leading to an improved figure of merit. If we assume that this would produce 360° of phase shift with the current losses, the figure of merit would then be ~35–40 °/dB.

20

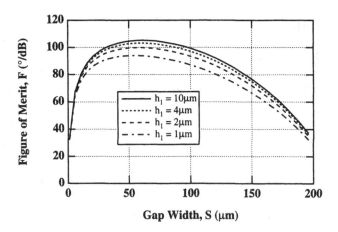

Fig. 6. Figure of merit vs. gap width for several values of the FE film thickness, h_1 ($f = 30$ GHz, $W = 60$ μm, $R_s = 0.04$ Ω, $\varepsilon_1 = 1100$, $\kappa_{max} = 2$, tanδ $= 0.04$, and maximum $U_b = 600$ V).

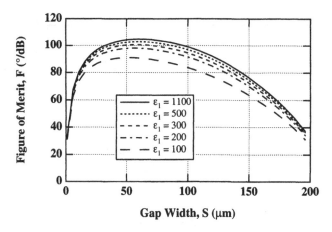

Fig. 7. Figure of merit vs. gap width for several values of the FE film permittivity, ε_1 ($f = 30$ GHz, $W = 60$ μm, $R_s = 0.04$ Ω, $h_1 = 10$ μm, $\kappa_{max} = 2$, tanδ $= 0.04$, and maximum $U_b = 600$ V).

The phase shift and insertion loss for a device based on a BST-composite film are given in Fig. 9. This device is capable of a continuous phase shift between 0 and 360° with a maximum insertion loss of 8 dB. This corresponds to a figure of merit of ~45 °/dB. While the performance of these phase shifters is respectable, their optimization is ongoing, both from the device design and the materials optimization perspectives. Our best measured result to date for the figure of merit is ~65 °/dB. This was measured for a BST-composite device that showed only 4.0 dB of insertion loss. We expect that further optimization will lead to performances of ~80 °/dB in the near future.

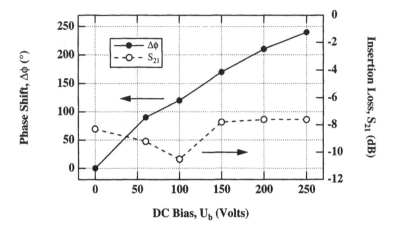

Fig. 8. Phase shift and insertion loss vs. applied bias voltage for a phase shifter made from a PLD-grown BST FE film ($f = 31.34$ GHz, $S = 40$ μm, $W = 60$ μm).

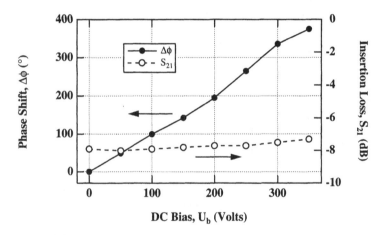

Fig. 9. Phase shift and insertion loss vs. applied bias voltage for a phase shifter made from a BST-composite FE film ($f = 31.34$ GHz, $S = 40$ μm, $W = 60$ μm).

STEERABLE ANTENNAS

Antenna Design

For the purposes of demonstration, we have made 30 GHz antennas using a very simple design. The antenna module consists of 16 microstrip (patch) radiating elements and 4 phase shifters. Each phase shifter controls the relative phase of a 4-element sub-array of radiators. The module is shown in Fig. 10 both in schematic form and as a photograph. The microstrip substrate is 0.25 mm thick, with a dielectric constant of ~3 and loss tangent of ~0.0013.

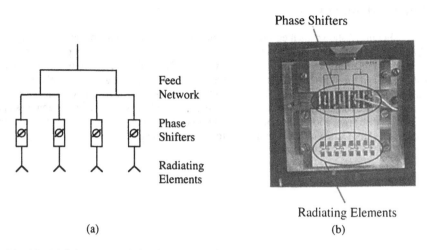

Phase Shifters

Feed
Network

Phase
Shifters

Radiating
Elements

Radiating Elements

(a)   (b)

Fig. 10. (a) Schematic and (b) photograph of 16-element antenna with 4 phase shifters

To confirm that this antenna performs as expected, we measured its radiation pattern before the incorporation of the phase shifters. Figure 11 shows both the simulated and measured radiation patterns, which correspond well. The measured frequency bandwidth corresponding to a VSWR (voltage standing wave ratio) < 2 is $\Delta f/f \sim 5\%$. The half-power (3 dB) width of the central radiation lobe is $\sim 13°$, and the level of the side lobes are less than 11 dB. These also correspond well with the simulation.

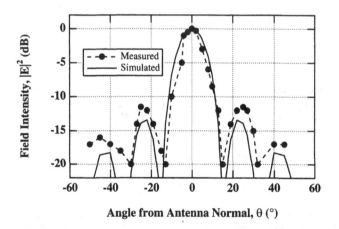

Angle from Antenna Normal, θ (°)

Fig. 11. Simulated and measured radiation patterns of a 30 GHz 16-element
antenna before the incorporation of FE phase shifters.

After incorporating the FE phase shifters, the minimal value of VSWR is ~ 1.5 at the antenna input and occurs at a frequency of ~ 29.6 GHz. All subsequent data are taken at this frequency.

Antenna Performance

Figure 12 shows the measured radiation pattern of the antenna for several sets of phase shifter bias values. As is clear from Fig. 12, changing the bias values of the phase shifters causes the central radiation lobe to shift or "steer" by ±18°. The half-power width of the central lobe is still ~ 13°, and the side lobe intensity is still low. As indicated in Fig. 11, these phase shifters are only subjected to a maximum DC bias of 250 V (E ~ 6 V/μm). Above this voltage, signs of current leakage (break down) begin. This limits the total phase shift available using these phase shifters as discussed above. The best phase shifters (also discussed above) have not yet been incorporated into antennas and should lead to better steering performance.

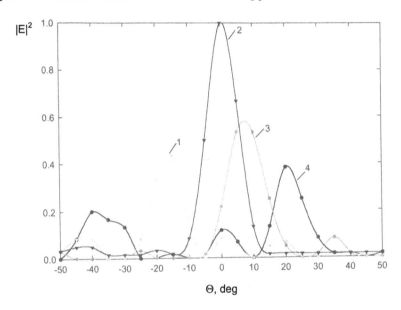

Fig. 12. Measured radiation patterns of 30 GHz 16-element steerable antenna incorporating four BST-composite FE phase shifters. Patterns 1–4 are for the following phase shifter bias values:

    (1) $U_1 = 19$ V, $U_2 = 195$ V, $U_3 = 196$ V, $U_4 = 91$ V
    (2) $U_1 = 6$ V, $U_2 = 20$ V, $U_3 = 74$ V, $U_4 = 44$ V
    (3) $U_1 = 196$ V, $U_2 = 0$ V, $U_3 = 107$ V, $U_4 = 162$ V
    (4) $U_1 = 245$ V, $U_2 = 98$ V, $U_3 = 245$ V, $U_4 = 31$ V

CONCLUSIONS

We have fabricated and characterized 30 GHz coplanar waveguide phase shifters using both epitaxial BST and polycrystalline BST-composite FE films. Numerical simulations of the CPW structure have indicated that the CPW phase shifter figure of merit depends weakly on the absolute value of the dielectric constant and thickness of the FE film, but is a strong function of its tunability and losses. Continuous phase shifts of > 360° with a maximum insertion loss of 8 dB were measured for a Ka-band FE phase shifter at 30 GHz. The lowest insertion loss observed to date is 4.0 dB. Further improvements are expected with more optimization.

Using four of these phase shifters, we have constructed a prototype 30 GHz antenna. We demonstrated that these phase shifters allow the central radiation lobe of this antenna to be "steered" by ±18° without significantly affecting its 3 dB width. We expect improved results when we incorporate our best phase shifters into the antenna, which has not yet been done. This simple demonstration indicates the potential of FE-based phase shifters to compete with existing technologies.

ACKNOWLEDGMENTS

This work was supported by the U. S. Department of Energy (DOE) under contract no. DE-AC36-99G010337.

REFERENCES

1. C. M. Carlson, T. V. Rivkin, P. A. Parilla, J. D. Perkins, D. S. Ginley, A. B. Kozyrev, V. N. Oshadchy, and A. S. Pavlov, submitted to Appl. Phys. Lett. (1999).

2. L. C. Sengupta and S. Sengupta, Materials Research Innovations **2**, p. 278 (1999).

3. D. Galt, T. Rivkina, and M. W. Cromar in *Ferroelectric Thin Films VI*, edited by R. E. Treece, R. E. Jones, C. M. Foster, S. B. Desu, and I. K. Yoo (Mater. Res. Soc. Proc. **493**, Pittsburgh, PA, 1998) pp. 341–356.

4. A. B. Kozyrev, V. N. Keis, G. Keopf, R. Yandrofski, O. I. Soldatenkov, K. A. Dudin, and D. P. Dovhan, Microelectronics Engineering **29**, p. 257 (1995).

5. D. Brambley, B. Martin, and P. D. Prewett, Advanced Materials for Optics and Electronics **4**, p. 55 (1994).

6. K. C. Gupta, et. al., *Microstrip Lines and Slotlines*, Art House, 1996.

7. M. N. Malishev and I. G. Mironenko, Izvestiya vuzov Radioelektronika **26**, p. 31 (1983). (In Russian)

ELECTRODYNAMIC PROPERTIES OF
SINGLE-CRYSTAL AND THIN-FILM STRONTIUM TITANATE, AND
THIN-FILM BARIUM STRONTIUM TITANATE

A.T. FINDIKOGLU *, Q.X. JIA *, C. KWON **, B.J. GIBBONS *, K.Ø. RASMUSSEN *,
Y. FAN *, D.W. REAGOR *, A.R. BISHOP *
*Los Alamos National Laboratory, Los Alamos, NM 87545, findik@lanl.gov
**Physics Department, California State University, Long Beach, CA 90840

ABSTRACT

We have used a coplanar waveguide structure to study broadband electrodynamic properties of single-crystal and thin-film strontium titanate (STO), and thin-film barium strontium titanate (BSTO). We have implemented low-frequency capacitance (100 Hz - 1 MHz), swept-frequency transmittance (45 MHz - 4 GHz), and time-domain transmittance (dc - several GHz) measurements to determine effective refractive index (or, dielectric constant), and dissipation factor (or, loss tangent) as a function of dc bias (up to 4×10^6 V/m) and temperature (20 - 300 K). The STO samples used superconducting electrodes and were designed to operate at cryogenic temperatures, whereas BSTO samples used normal conducting electrodes and exhibited optimal performance around room temperature. By using nearly identical electrode geometries for all devices, we were able to conduct a direct comparative study among them, and investigate not only single-crystal vs thin-film, but also cryogenic vs room-temperature applications.

INTRODUCTION

Strontium titanate, $SrTiO_3$ (STO), and barium strontium titanate, $Ba_xSr_{1-x}TiO_3$ (BSTO), with x close to 0.5, are some of the most widely studied materials in condensed matter physics [1,2]. STO and SBTO are also important from a technological point of view: their large dielectric constant and large dielectric breakdown field make them a potential candidate for storage capacitor cells in next generation dynamic random access memories [3], and the large dielectric nonlinearity they exhibit (at cryogenic temperatures for STO, and around room temperature for BSTO) is a desirable property for various applications such as tunable filters and phased array antennas [4,5].

Recently, there has been an increased interest in the microwave applications of nonlinear dielectrics such as STO and BSTO [6]. To better assess their high-frequency application potential, we have conducted a comprehensive study of their broadband (dc - 10^{10} Hz) electrodynamic characteristics using prototype coplanar waveguide (CPW) devices. In addition, by implementing nearly identical electrode structure for each device, we have been able to do a direct comparative study between single-crystal and thin-film STO [7,8], and also between cryogenic (using STO with superconducting electrodes) and room-temperature (using BSTO with normal conducting electrodes) devices.

EXPERIMENT

Coplanar Waveguide Device Structure

In this study, we have used prototype devices with 8-cm-long meandering CPW electrodes on 1 cm x 1cm chips. The CPW electrodes have approximately 20-μm-wide centerlines, and 10- to 20-μm-wide gaps. A schematic top view and generic cross-section with signal and bias configuration of the devices is shown in Fig. 1. Cryogenic devices use 0.4-μm-thick superconducting $YBa_2Cu_3O_{7-\delta}$ (YBCO) electrodes with 0.7-μm-thick STO thin films on 0.5-mm-thick (100) $LaAlO_3$ (LAO) substrates, or 1-mm-thick single-crystal (100) STO substrates, whereas room-temperature devices use 0.5-μm-thick Au electrodes with 0.7-μm-thick BSTO thin-films on 1-mm-thick (100) MgO substrates. All YBCO, STO, and BSTO films are pulsed-laser deposited, and patterned using standard photolithography. Au and Ti adhesion layers were rf-sputtered. Schematic cross-sections of the devices are shown in Fig. 2.

27

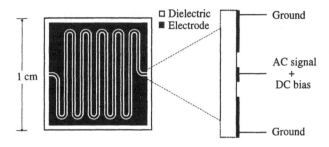

Figure 1. Schematic of generic CPW device with signal and bias configuration.

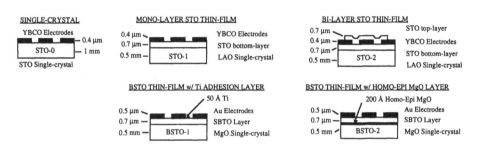

Figure 2. Schematic cross-sections of devices used in this study.

Electrical Measurements and Circuit Model

We have implemented three separate electrical measurement techniques to characterize our devices: low-frequency capacitance, swept-frequency, and time-domain measurements. For low-frequency characterization (100 Hz - 1 MHz), we use a standard LCR meter to measure the capacitance and dissipation factor between the centerline and the groundplanes of the CPW as a function of temperature and dc bias (applied between the centerline and the grounplanes). A maximum dc bias of 40 V corresponds to an electric field of 4x10^6 V/m at the surface of the dielectric film. The swept-frequency measurements (45 MHz - 4 GHz) use a network analyzer to measure reflection (S11), and transmission (S21) characteristics of the CPWs as a function of temperature and bias. The time-domain measurements, on the other hand, use step-pulse or impulse excitations, measure time-domain reflection (TDR) and time-domain-transmission (TDT) coefficients, and, by Fourier analysis, extract broadband (dc-several GHz) electrodynamic characteristics of the CPWs as a function of temperature, dc bias, and impulse amplitude. Figure 3 shows an electrical circuit diagram that we use to model the high-frequency response of the CPWs [7].

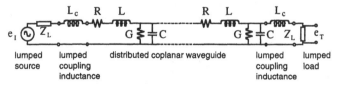

Figure 3. Electrical circuit model.

The electrical circuit model comprises coupling inductance L_c and input/output impedance Z_L for the external circuitry; and distributed element equivalents of series resistance R, series inductance L, shunt conductance G, and shunt capacitance C per unit length of the CPW. In their most general form, the inductance and impedance of the lumped elements are constants; and those of the distributed circuit elements depend on angular frequency ω, temperature T, bias voltage v_{dc}, and signal voltage v_s. The characteristic impedance Z_o and propagation function γ of the coplanar waveguide are given by [9]

$$Z_o = \sqrt{(R + i\omega L)/(G + i\omega C)} \qquad (1)$$

and

$$\gamma = \sqrt{(R + i\omega L)(G + i\omega C)} = \alpha + i\frac{\omega}{c}n \quad , \qquad (2)$$

where c is the speed of light in vacuum, α is the attenuation constant, and n is the effective refractive index.

RESULTS

<u>Single-crystal vs Thin-film Strontium Titanate</u>

Figure 4 shows low-frequency capacitance C and dissipation factor vs temperature for three cryogenic CPW devices; single-crystal (STO-0), monolayer thin-film (STO-1) and bilayer thin-film (STO-2). The dissipation factor is equivalent to effective loss tangent $tan\delta$ if the losses are dominated by dielectric losses. The single crystal device shows much larger capacitance value especially at low temperatures. The corresponding relative dielectric constant of single crystal STO is about 10000 whereas it is about 1500 for thin-film STO samples around 20 K.

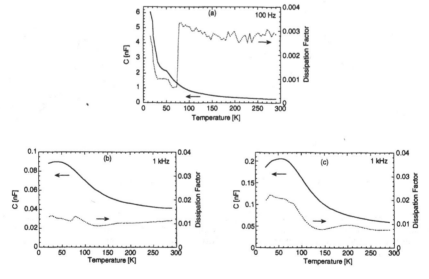

Figure 4. Capacitance and dissipation factor vs temperature for (a) STO-0, (b) STO-1, and (c) STO-2.

The single-crystal STO sample shows about an order of magnitude smaller dissipation than thin-film samples. We note that the inverse of RC time-constant of the single-crystal sample was very close to the measurement frequency when the YBCO electrodes were in normal state. Thus, the high values measured above the transition temperature (~85 K) could have major contribution from series resistive effects. But, below about 80 K, we expect the dominant dissipation to be due to dielectric losses in the parallel channel.

Figure 5 shows percentage change in capacitance $\delta C/C$ and dissipation factor vs bias for these devices at 20 and 60 K. We have used two-loop sweeps (i.e., the voltage was ramped from minimum to maximum value twice continuously) for all bias dependent capacitance measurements. For the single-crystal sample, we have used a smaller range of bias because single-crystals show much smaller dielectric breakdown values than thin-films. After several high-frequency measurements, this particular single-crystal sample showed a breakdown around +10 V bias which was reversible. The monolayer and bilayer samples show virtually no temperature dependence between 20 and 60 K, whereas the bias-dependent effects are significantly reduced with temperature in the single-crystal sample. At 60 K, bilayer thin-film sample shows larger capacitance tunability than the single-crystal sample. Still, low-frequency dissipation factor in thin film samples is more than an order of magnitude larger than that in single-crystal sample.

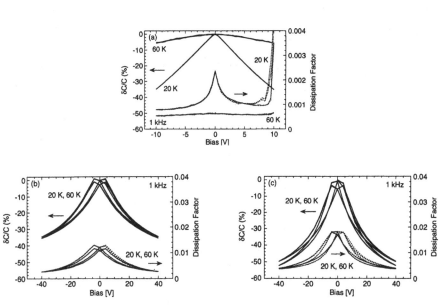

Figure 5. Capacitance and dissipation factor vs voltage bias for (a) STO-0, (b) STO-1, and (c) STO-2.

Figure 6 shows results of swept-frequency measurements. The transmittance is given by the steady-state ratio of transmitted voltage to input voltage for a continuous signal that is swept in frequency. If we use an approximate limit form of Eq. (2), the complex propagation function γ is given by [8]

$$\gamma = i\omega n/c + \frac{1}{2}\omega n \tan \delta/c \quad ,$$ (3)

where $\tan\delta$ is the effective loss tangent of the waveguide medium. Fig. 6 shows fits to experimental data assuming purely inductive Z_c, and frequency-independent Z_o, Z_L, n, and $\tan\delta$. Basically, the separation of peaks (i.e., resonances), initial height of peaks, and decay of peaks with frequency are determined by n, Z, and $\tan\delta$, respectively. The results of the fits are shown in Table I. We note that the high-frequency losses (~100 MHz - 1 GHz) in single crystal sample seems to be at least an order of magnitude higher than those measured at low-frequencies (~1 kHz). Furthermore, these losses increase with bias at high-frequencies, whereas they decrease at low-frequencies. In the case of thin-film samples both the magnitude and the bias-dependence of the loss tangent at high-frequencies seem to be comparable to low-frequency results.

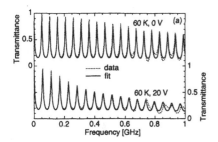 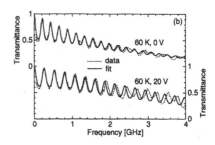

Figure 6. Transmittance vs frequency at 60 K for (a) STO-0, (b) STO-2.

TABLE I. Broadband-frequency (~10^8-10^9 Hz) averages for effective refractive index (n) and its fractional change per dc voltage bias ($\Delta n/\Delta Vn$), and effective loss tangent ($\tan\delta$) and its fractional change per dc voltage bias ($\Delta \tan\delta/\Delta V \tan\delta$) at 17 and 60 K.

Type	n @60 K	$\Delta n/nV$ @60 K	$\tan\delta$ @60 K	$\Delta \tan\delta/\tan\delta V$ @60 K
single-crystal (STO-0)	37.0	-0.4%	0.0035	+16%
mono-layer film (STO-1)	6.5	-0.4%	0.021	-2%
bi-layer film (STO-2)	8.8	-0.8%	0.033	-3%

Figure 7 shows the the results of time-domain measurements (where a Gaussian-like impulse with about 0.05 ns pulse-width has been transmitted through the CPW, and the frequency dependent n and $\tan\delta$ have been extracted by Fourier analysis). These results are consistent with swept-frequency results. Furthermore, they indicate that n and $\tan\delta$ are virtually frequency-independent at microwave frequencies. The ramifications of such bias-dependent small-signal characteristics for large-signal behavior are illustrated by Fig. 8. Here, the input is a 40-V Gaussian-like impulse with about 0.4 ns pulse-width for both the single-crystal and bilayer thin-film device. The transmitted impulse has similar risetime and pulse shape (a shock-like front) for both CPWs at zero bias (the amplitudes are different due to impedance mismatch effects). But under bias, the bi-layer-film CPW shows improved pulse-shaping effects due to reduced microwave loss, whereas the single-crystal CPW performance (dominated by combined effects of nonlinearity and loss) degrades due to increased loss. The solid lines are the results of simulations. A detailed analysis of the large-signal behavior is published elsewhere [7]. Such steep pulse fronts (or trailing edges) could be used, for example, for triggering in electronics.

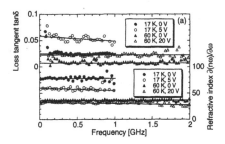 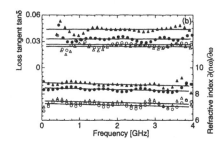

Figure 7. Refractive index and loss tangent vs frequency at 17 and 60 K under 0, 5, and 20 V bias for (a) STO-0, and (b) STO-2 (symbols are data, lines are linear approximations).

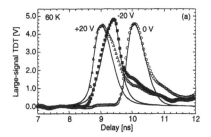 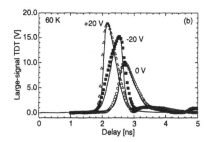

Figure 8. Large-signal impulse transmission at 60 K under 0, -20, and 20 V bias for (a) STO-0, and (b) STO-1 (symbols are data, lines are simulations).

Cryogenic Strontium Titanate vs Room-temperature Barium Strontium Titanate

As shown in Fig. 2, the room-temperature devices, BSTO-1 and BSTO-2, use Au electrodes on 0.7-μm-thick BSTO films grown on MgO substrates. These devices are representative of room-temperature devices since they show their largest dielectric tunability with minimal hysteresis around room-temperature. Figure 9 shows that the capacitance and dissipation factors exhibited by these samples are comparable to those of cryogenic thin-film devices (STO-1 and STO-2) (see Fig. 4). However, BSTO samples, unlike STO samples, show increasing C with temperature. As shown in Fig. 10, BSTO samples exhibit their largest tunability around room-temperature. Unlike STO samples, BSTO samples demonstrate significant tunability with large hysteresis (indicative of remnant polarization) at the temperature where the capacitance has the lowest value, i.e., at 20 K. These results are consistent with measurements on bulk and single-crystals: bulk single-crystal STO is an incipient ferroelectric which enters a quantum paralectric phase around 4 K and shows monotonically weakening nonlinear behavior (i.e., paraelectricity) with temperature, whereas bulk BSTO has a ferroelectric transition around room temperature, below which it remains ferroelectric. The BSTO-1 and BSTO-2 samples at room-temperature show very similar low-frequency characteristics to STO samples at 20 K.

32

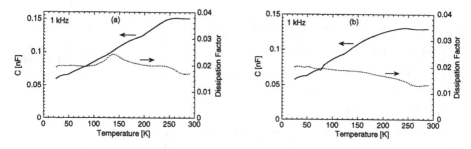

Figure 9. Capacitance and dissipation factor vs temperature for (a) BSTO-1, and (b) BSTO-2.

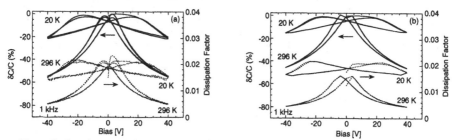

Figure 10. Capacitance and dissipation factor vs voltage bias for (a) BSTO-1, and (b) BSTO-2.

However, the high-frequency characteristics of BSTO samples are quite different from those of STO samples. As shown in Fig. 11, unlike STO samples, the transmittance of the BSTO-2 sample falls precipitously with frequency (BSTO-1 shows very similar characteristics).

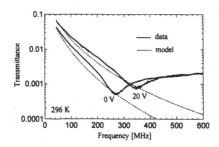

Figure 11. Transmittance vs frequency at 296 K for BSTO-2.

The reason for such high dissipation is the finite resistance of Au electrodes. A closer examination of Eq. (2) yields that in the case of STO samples, once electrodes become superconducting, R becomes negligible with respect to ωL up to at least several GHz, and thus the dominant contribution to attenuation comes from the ratio of $G/\omega C$ (i.e., $\tan\delta$). On the other hand, in the case of BSTO samples, the normal conducting electrodes exhibit finite resistance with R comparable to or larger than ωL at a wide range of frequencies, thus the attenuation is dominated by conductor losses. The dotted lines are the predictions of a simple model which use directly measured R value of 7.3 Ω/mm and C value of 1.68 pF/mm, and calculated L value of 393.4 pH/mm in Eq. (2). The R is approximated by the dc resistance of the centerline (a detailed calculation of high-frequency R is given in ref. [7]). The C was measured with an LCR meter at 1 kHz, and L was calculated using a CPW model [7]. Under bias, the model assumes that only the C changes.

Effect of Anneal in Oxygen

For our STO samples, oxygen anneal at 400 °C for 10 h has been a standard post-patterning process to anneal out some of the damage due to patterning. We have used the same process for our BSTO samples. Figure 12, shows capacitance vs temperature before and after anneal for BSTO-1 and BSTO-2 samples. BSTO-1 characteristics change appreciably with oxygen anneal - both the capacitance and dissipation at high-temperatures increases, with the dissipation showing prominent and shifted peaks after anneal. We are currently pursuing a microstructural investigation to determine whether these effects are caused by possible reaction between the Ti adhesion layer and the BSTO film. BSTO-2 sample shows virtually no change in capacitance or dissipation factor with anneal. BSTO-2 sample shows also no change in tunability, whereas BSTO-1 shows slightly larger tunability in capacitance and much larger tunability in dissipation at 296 K after anneal, as shown in Fig. 13.

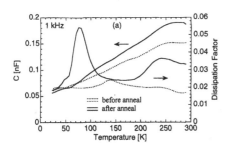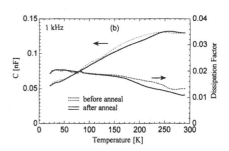

Figure 12. Capacitance and dissipation factor vs temperature before and after anneal for (a) BSTO-1, and (b) BSTO-2.

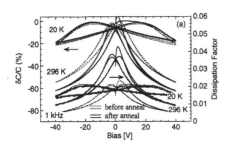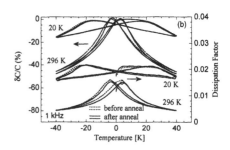

Figure 13. Capacitance and dissipation factor vs bias at 20 and 296 K before and after anneal for (a) BSTO-1, and (b) BSTO-2.

With anneal, the resistance of Au electrodes in both BSTO samples decreases by about a factor of 2, which lead to reduced attenuation as shown in Fig. 14. After anneal, x-ray Θ-2Θ scans show sharper Au (111) peaks, indicative of increased grain size.

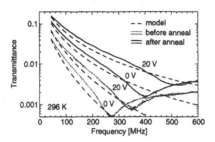

Figure 14. Transmittance vs frequency at 296 K before and after anneal for BSTO-2.

DISCUSSION

The decrease of dielectric constant (or, refractive index) with bias in both single-crystal STO and thin-film STO and BSTO is well described by the hardening of the transverse optic mode of the lowest frequency, also called the soft phonon mode. The dielectric constant in films is reduced and has different temperature dependence, most likely due to increased stress and defect density in the films.

The experimetal results on the dissipative effects in single-crystal STO are difficult to reconcile in a simple model - the cryogenic dissipation increases by more than an order of magnitude from 1 kHz to about 100 MHz, and remains essentially unchanged from about 100 MHz at least up to 1 GHz. Also, whereas the dissipation decreases with bias at low-frequencies, it increases at microwave frequencies.

The STO films show very similar dielectric constant and loss tangent values from 100 Hz up to about 4 GHz. Also, our experimental observation of weak temperature dependence, and reduction with bias, of loss is consistent with a loss mechanism based on transformation of microwave photons into acoustic oscillations at charged defects. But, this mechanism also predicts linearly increasing loss with frequency, whereas our experiments show essentially frequency independent loss.

No comprehensive model exists for dielectric losses in STO. Some possible mechanisms for microwave dielectric losses have recently been considered by Vendik, et al [10]. These loss mechanisms include the fundamental loss connected with multi-phonon scattering of the soft phonon mode, and the transformation of electromagnetic oscillations into acoustic oscillations due to residual ferroelectric polarization or charged defects. The fundamental loss mechanism predicts increasing loss with decreasing temperature below about 60 K and with increasing bias. All three mechanisms also predict linear increase of loss tangent tanδ with frequency up to 100 GHz, in contrast to the frequency-independent behavior we have observed at high-frequencies. The Vendik model relies on the dielectric response of a system with a single degree of freedom [11]. However, this description is inadequate to describe the response of a system with a distribution of relaxation times, i.e., with many degrees of freedom [12]. A general emprical model that accommodates a distribution of relaxation times has been shown to lead to, under certain conditions, a frequency-independent loss tangent behavior in agreement with experiments on many strongly-interacting dielectric systems (i.e., systems with strong interactions among spin, charge, or lattice degrees-of-freedom) [13]. Our observation here of a frequency-independent loss tangent implies that STO might need a more general treatment than that based on a weakly-interacting system.

35

The BSTO samples at room temperature show very similar low-frequency characteristics to those of STO samples at 20 K. However, at high-frequencies, the losses due to normal conducting electrodes dominate the response. For electronics applications, the requirement to use relatively low bias voltage levels (~10 V) would in turn lead to relatively small gap widths (~10 μm) to retain tunability. This requirement combined with the requirement to retain a convenient characteristic impedance level (not much less than 50 Ω) in microwave circuitry would necessitate centerline widths of the order of 10 μm and thus series resistance R values of the order of Ω/mm or higher. Thus, such devices will be intrinsically very lossy at microwave frequencies irrespective of the loss tangent of the nonlinear dielectric medium.

CONCLUSIONS

As presented above, the comprehensive electrical characterization we have implemented in this study has allowed us to determine broadband electrodynamic characteristics of various electrically-tunable CPW devices based on nonlinear dielectrics. Also, by implementing identical electrode geometry for all devices, we were able to make a direct comparison among them in terms of not only electrodynamic properties, but also practical implications.

REFERENCES

[1] R. A. Cowley, Phil. Trans. R. Soc. Lond. A 354, 2799 (1996).

[2] R. Vacher, J. Pelous, B. Hennion, G. Coddens, E. Cortens, K. A. Muller, Europhys. Lett. 17, 45 (1992).

[3] P. Y. Lesaicherre, H. Yamaguchi, Y. Miyasaka, H. Watanabe, H. Ono, M. Yoshida, Integr. Ferroelectr. 8, 201 (1995).

[4] A. T. Findikoglu, Q. X. Jia, X. D. Wu, G. J. Chen, T. Venkatesan, D. W. Reagor, Appl. Phys. Lett. 68, 1651 (1996).

[5] F. A. Miranda, R. R. Romanofsky, F. W. VanKeuls, C. H. Mueller, R. E. Treece, T. V. Rivkin, Integr. Ferroelectr. 17, 231 (1997).

[6] M. J. Lancaster, J. Powell, A. Porch, Supercond. Sci. Tech. 11, 1323 (1998).

[7] A. T. Findikoglu, D. W. Reagor, K. O. Rasmussen, A. R. Bishop, N. Gronbech-Jensen, Q. X. Jia, Y. Fan, C. Kwon, L. A. Ostrovsky, J. Appl. Phys. 86, 1558 (1999).

[8] A. T. Findikoglu, Q. X. Jia, C. Kwon, D. W. Reagor, G. Kaduchak, K. O. Rasmussen, A. R. Bishop, Appl. Phys. Lett. (in press).

[9] R. L. Kautz, J. Appl. Phys. 49, 308 (1977).

[10] O. G. Vendik, L. T. Ter-Martirosyan, and S. P. Zubko, J. Appl. Phys. 84, 993 (1998).

[11] P. Debye, *Polar Molecules* (Chemical catalog Co., New York, 1929).

[12] R. M. Hill, and A. K. Jonscher, Contemp. Phys. 24, 75 (1983).

[13] S. Havriliak, and S. Negami, J. Poly. Sci. C14, 99 (1966).

FIRST DEMONSTRATION OF A PERIODICALLY LOADED LINE PHASE SHIFTER USING BST CAPACITORS

Amit S. Nagra, Troy R. Taylor *, Padmini Periaswamy, James Speck *, Robert A. York

ECE Department, University of California, Santa Barbara, CA 93106
* Materials Department, University of California, Santa Barbara, CA 93106

ABSTRACT

Periodically loaded line phase shifter circuits using voltage tunable BaSrTiO$_3$ (BST) parallel plate capacitors have been demonstrated at X-band. The first such phase shifter circuit was capable of 100° of phase shift with an insertion loss of 7.6 dB at 10 GHz. Subsequently, the monolithic fabrication procedure was refined resulting in an improved phase shifter circuit with 200° of phase shift and an insertion loss of 6.2 dB at 10 GHz. In addition to promising loss performance (32°/dB) at 10 GHz, the circuits reported here have several desirable features such as moderate control voltages (20 V), room temperature operation, and compatibility with monolithic fabrication techniques.

INTRODUCTION

The cost of phase shifters is a major component of the cost of modern phased array antennas. Thus it is of paramount importance to reduce the cost of phase shifters to ensure widespread acceptance of phased arrays in military/civilian applications. Ferroelectric thin film based phase shifters promise to be low cost because of two factors- 1) the ferroelectric material can be deposited relatively inexpensively using RF sputtering/MOCVD 2) the films can be processed using low cost, high volume monolithic fabrication techniques. Apart from cost, the use of ferroelectric thin films in phase shifter circuits also has potential performance advantages such as low insertion loss, high power handling capability and low DC power requirements.

Several groups [1-4] are investigating the possibility of implementing phase shifter circuits using barium strontium titanate (BST) which has an electric field tunable dielectric constant. In these circuits the ferroelectric material (BST) either forms the entire microwave substrate [1,2] on which the conductors are deposited (thick films/bulk crystals) or a fraction of the substrate with thin BST films sandwiched between the substrate and the conductors [3,4]. These circuits rely on the principle that the phase velocity of the waves propagating on these structures can be altered by changing the permittivity of the ferroelectric layer. This approach has several limitations including high conductor losses, inefficient use of the BST tunability and high control voltages. The approach used by us was to periodically load a coplanar wave guide transmission line with voltage tunable ferroelectric (BST) capacitors. The phase velocity of the periodically loaded line depends on the values of the BST capacitors and thus could be changed by applying bias to the BST capacitors. Parallel plate capacitors were employed here since they utilize the tunability of the BST film effectively and require much lower control voltages than interdigitated capacitors. Also, the use of discrete BST capacitors made it easy to control the amount of capacitive loading due to the ferroelectric film and thus allowed the structure to be optimized for good loss performance [5].

37

Mat. Res. Soc. Symp. Proc. Vol. 603 © 2000 Materials Research Society

THEORY

The schematic of the proposed phase shifter circuit is shown in figure 1. The phase shifter basically consists of a high impedance transmission line that is periodically loaded with thin film BST capacitors with spacing L_{sect}. For frequencies much below the Bragg frequency, this structure behaves like a synthetic transmission line [5] with modified propagation velocity and characteristic impedance. Since the capacitance value of the loading BST capacitors can be varied by applying bias, it is possible to change the phase velocity and impedance of the line. For phase shifter applications, the value of the loading capacitors is chosen such that the impedance variation is small but by cascading the correct number of sections the phase shift can be made as large as desired.

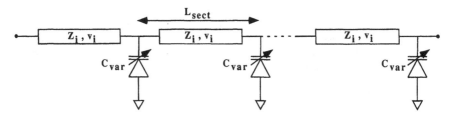

Figure 1: Circuit schematic for a periodically loaded line phase shifter

The phase shifter circuits presented here were designed using the same techniques as [5] and consisted of a CPW transmission line of characteristic impedance 73 Ω on a high resistivity silicon substrate (40 KΩ-cm). A thin insulating layer of silicon nitride was deposited between the CPW metal and the silicon substrate to prevent DC leakage. The CPW line was periodically loaded with BST capacitors whose zero bias capacitance was 270 fF. The BST (150 nm thick) for the tunable capacitors was deposited by RF magnetron sputtering. Platinum was used as the top and bottom electrode for the BST capacitors. The spacing between BST capacitors was chosen to be 580 μm resulting in a Bragg frequency of 25 GHz. A total of 16 periods was used in the phase shifter circuit resulting in an overall length of 9.3 mm. Standard monolithic circuit fabrication techniques were used for the fabrication of the periodically loaded line phase shifters.

RESULTS

RF measurements were made on a HP 8722D network analyzer that was calibrated using on-wafer standards. The two-port s-parameters of the phase shifter circuit were recorded up to 12 GHz. Figure 2a shows the differential phase shift (with respect to the zero bias insertion phase) as a function of frequency for several bias values. As expected for a variable velocity transmission line, the circuit produced a phase shift that varied linearly with frequency (for frequencies well below the Bragg frequency). The circuit was capable of continuous 0-100° phase shift at 10 GHz with any desired resolution. The maximum insertion loss at 10 GHz occurred at zero bias and was 7.6 dB (see figure 2b). The return loss was better than 12 dB over all phase states as shown in figure 2c.

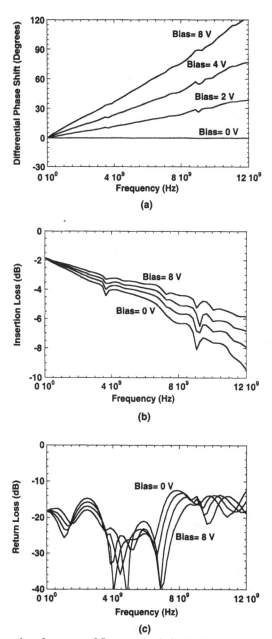

Figure 2: Measured performance of first ever periodically loaded line phase shifter with BST capacitors- a) Differential phase shift b) Insertion loss c) Return loss

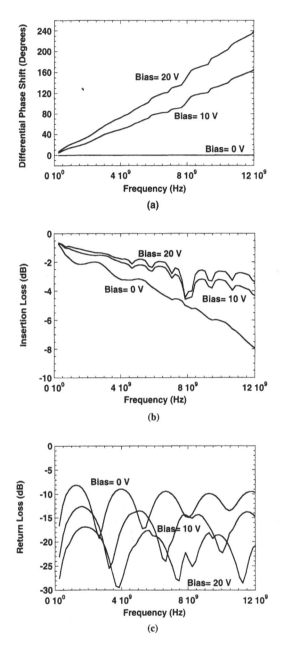

Figure 3: Measured performance of improved periodically loaded line phase shifter with BST capacitors- a) Differential phase shift b) Insertion loss c) Return loss

The measured performance of the first periodically loaded line phase shifter circuit did not meet the design expectations and it was determined that fabrication process related issues were responsible. Due to lift-off problems during top metal deposition, the BST capacitor yield on the first phase shifter process was only about 50% (capacitors failed open). As a result of the low device yield, the capacitive loading of the transmission line was much lower than designed resulting in reduced phase shift. The higher than expected losses were attributed to CPW losses in the high resistivity silicon substrate due the formation of a parasitic MIS (metal-insulator-semiconductor) capacitor.

These problems were addressed in subsequent fabrication runs. The thickness of the capacitor top electrode metal was reduced to 0.3 μm resulting in a substantially easier lift-off process and the BST capacitor yield improved to 100%. In order to reduce the additional CPW losses arising due to the formation of a MIS capacitor, the silicon nitride was etched away from the gap regions of the CPW lines as suggested in [6]. The phase shifters fabricated with the modified process showed improved performance as indicated in figure 3. The improved phase shifter circuit was capable of 200° of phase shift with 6.2 dB of insertion loss and return loss better than 10 dB at 10 GHz. Detailed microwave characterization of the BST capacitors was also performed and it showed that BST capacitor loss is the dominant loss mechanism in the phase shifters since the BST capacitor quality factor is only about 10 at the X-band. Further improvements in BST capacitor quality factors due to advances in BST film processing and growth should lead to phase shifters with even better insertion loss performance.

CONCLUSIONS

Periodically loaded line phase shifter circuits using voltage tunable $BaSrTiO_3$ (BST) parallel plate capacitors have been demonstrated at X-band. The best circuits demonstrated 200° of phase shift with an insertion loss of 6.2 dB at 10 GHz. Even better loss performance is expected with further improvements in BST capacitor quality factors.

ACKNOWLEDGEMENTS

The authors would like to thank Dr. O. Auciello, Dr. S. Streiffer and Dr. J. Im at Argonne National Labs for providing some of the ferroelectric thin films for this work. This work was supported by DARPA under the FAME program (Grant # 442530-23146).

REFERENCES

1. Vijay K. Varadan, K.A. Jose, Vasundara V. Varadan, R. Hughes, and James F. Kelly, "A Novel Microwave Planar Phase Shifter," *Microwave Journal*, pp. 244-54, April 1995.

2. Franco De Flaviis, N. G. Alexopoulos and Oscar M. Stafsudd, "Planar Microwave Integrated Phase Shifter Design with High Purity Ferroelectric Material," *IEEE Transactions on Microwave Theory and Techniques*, vol. 45, No. 6, pp. 963-9, June 1997.

3. F. W. Van Keuls, R. R. Romanovsky, D.Y. Bohman, M. D. Winters, F. A. Miranda, C. H. Mueller, R. E. Treece, T. V. Rivkin, and D. Galt, "($YBa_2Cu_3O_{7-\delta}$, Au)/$SrTiO_3$/$LaAlO_3$ Thin Film Conductor/Ferroelectric Coupled Microstripline Phase Shifters for Phased

Array Applications," *Applied Physics Letters*, vol. 71 (21), pp. 3075-7, 24 November 1997.

4. F. W. Van Keuls, R. R. Romanovsky, N. D. Varaljay, F. A. Miranda, C. L. Canedy, S. Aggarwal, T. Venkatesan, and R. Ramesh, "A Ku-Band Gold/$Ba_xSr_{1-x}TiO_3$/$LaAlO_3$ Conductor/Thin-Film Ferroelectric Microstrip Line Phase Shifter for Room Temperature Communications Applications," *Microwave and Optical Technology Letters*, vol. 20, No. 1, pp. 53-6, Wiley, Jan. 1999.

5. Amit S. Nagra and Robert A. York, "Distributed Analog Phase Shifters with Low Insertion Loss," *IEEE Transactions on Microwave Theory and Techniques,* vol. 47, No. 9, pp.1705-1711, Sept. 1999.

6. Y. Wu, H. S. Gamble, B. M. Armstrong, V. F. Fusco, and J. A. Carson Stewart," SiO_2 interface layer effects on microwave loss of high-resistivity cpw line," *IEEE Microwave and Guided Wave Letters,* vol. 9, No. 1, pp.10-12, Jan. 1999.

Electric-Field Tuning

TUNABLE DIELECTRIC THIN FILMS FOR HTS MICROWAVE APPLICATIONS

B. H. MOECKLY, Y. M. ZHANG
Conductus, Inc., 969 W. Maude Ave., Sunnyvale, CA 94086, moeckly@conductus.com

ABSTRACT

SrTiO$_3$ (STO) thin films are promising for a variety of applications requiring tunability. We describe the growth and characterization of STO thin films including their dielectric properties. We also present attempts at reducing the loss tangent of these films, and we discuss their integration with high-temperature superconductor (HTS) microwave filters for trimming purposes.

INTRODUCTION

There are many potential applications for tunable dielectric thin films in advanced wireless communications. Examples include tuning of HTS microwave filters (our immediate interest), input and output matching networks for tunable cryogenic low-noise amplifiers, automatic gain control for RF receivers, time delay devices (LC transmission lines), phase shifting antenna arrays, and voltage controlled oscillators in phase locked loops. STO thin films are ideal for cryogenic applications because their dielectric constant may be easily tuned at low temperatures. Furthermore, the compatibility of STO with YBa$_2$Cu$_3$O$_7$ (YBCO) is well known, and tunable dielectrics such as STO offer several advantages over semiconductor varactors. The superior performance provided by YBCO filters relies on their high quality factor (Q) values. So although the tunability of STO films is excellent, their relatively high loss tangent at present inhibits their widespread use. In addition, the dielectric properties of STO thin films are not well understood. We therefore desire a better understanding of the relationship between the microstructure of STO films and their dielectric properties with the goal of reducing the loss.

The first section of this paper describes the low-frequency dielectric properties of STO thin films we have grown by laser ablation. The second section describes attempts we have made to reduce the loss tangent carried out largely by a "shake-and-bake" approach. In the final section, we present results of our efforts to tune high-performance YBCO thin-film resonators with STO thin films while preserving the high Q values.

GROWTH AND PROPERTIES OF SrTiO$_3$ THIN FILMS

In this section we discuss the growth and characterization of STO thin films, including the effect of growth conditions on the low-frequency dielectric properties.

Dielectric properties

In order to routinely assess the dielectric properties of our films, we employ photolithographic liftoff processing to pattern coplanar interdigital capacitor structures on the films. These capacitors typically consist of 20 interdigitated fingers spaced 10 µm apart with a length of 1 mm. To extract the dielectric constant from the measured capacitance, we have used the analytic solution of Gevorgian et al.[1] which was derived using conformal mapping and partial capacitance techniques.

We have evaluated a variety of electrode materials, including in-situ lattice-matched oxides such as YBCO and SrRuO$_3$, but we have found that metals such as Ag or Au give similar results, and we routinely use them for electrodes. We measure the temperature dependence and electric field dependence of the dielectric constant (ε_r) and the loss tangent (tanδ) using an HP 4284A LCR meter. Note that the loss tangents we report include extrinsic losses such as might arise from the electrodes and the electrode/film interface. The thin-film sample is held in a cryostat fitted with coaxial cables, and the measurements are carefully calibrated. For routine evaluation, these measurements are made at low frequency, typically 10 or 100 kHz. Although this is far lower than the frequency at which we envision applications (~1 GHz), we believe these simpler measurements are indicative of the high-frequency behavior of STO, as has been reported.[2]

Mat. Res. Soc. Symp. Proc. Vol. 603 © 2000 Materials Research Society

Our measured temperature dependencies of ε_r and $\tan\delta$ are similar to those reported by other groups.[2-5] We observe an increase of both of these quantities with decreasing temperature until a temperature between 20 and 60 K is reached, at which point ε_r and $\tan\delta$ reach a maximum and then begin to decrease. The initial increase has been attributed to softening of the soft optical phonon mode,[6,7] and the subsequent decrease, seen only in thin films, may be due to the existence of a random internal electric field.[8] We occasionally observe another loss peak occurring at approximately 100 K; this peak is in addition to the ubiquitous lower-temperature loss peak. For our films with higher dielectric constant values, however, this higher-temperature loss peak is usually absent. These peaks in $\tan\delta(T)$ are variously ascribed to phase transitions[9] though there is strong evidence that they are due to thermally-activated defect modes.[10]

Effect of growth conditions

All films discussed in this paper were grown by laser ablation. The films were deposited in an oxygen ambient using either ceramic or single-crystal targets. A variety of substrates were employed and will be discussed in the next section. The excimer laser was operated at 248 nm with a typical repetition rate of 10 Hz and a fluence of 1 J/cm^2. We have observed a strong dependence of the film properties on substrate temperature and background oxygen pressure. As shown in Figure 1, the out-of-plane lattice constant of the films as measured by x-ray diffraction depends rather strongly on oxygen pressure during growth. The large scatter in the data is due to a variety of substrate temperatures (740 to 850 °C) and film thicknesses (0.1 to 4 μm). Nevertheless, there is a general tendency for the lattice constant to peak around 100 to 200 mTorr and to decrease for higher and lower oxygen pressures. We note that the trend we observe is similar to that reported by some other groups.[11] The reason for this dependence is not obvious, but is probably related to the complex dynamics of the ablated plume and the ensuing behavior of the adatoms upon arrival at the substrate. Note that most of our films do not possess the lattice constant of bulk single-crystal STO; this observation may be expected due to the stress imposed by the substrate. However, we also find that the films with maximum dielectric constants have a lattice constant well above that of single-crystal STO. Thus an increase in the thin-film dielectric constant of the films does not imply that they are becoming more "single-crystal-like," at least in terms of the c-axis lattice parameter, which does not appear to be a reliable indicator of the film properties.

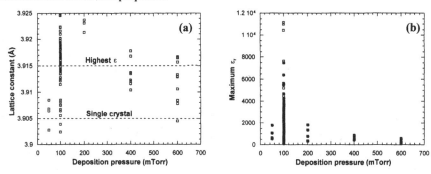

Fig. 1 Dependence of (a) out-of-plane lattice constant and (b) maximum dielectric constant on oxygen pressure during laser ablation of STO thin films. The data are measured at 10 kHz.

We have indeed found that the maximum values of dielectric constants occur for films grown around 100 mTorr, as shown in Figure 1b. Significantly lower values are achieved for higher background pressures. This trend is opposite to that reported by some other groups.[2] This observation likely emphasizes the system-dependent nature of the laser-ablation growth process.

We have also observed a strong effect of substrate temperature on the film properties. Though the lattice constant does not depend strongly on substrate temperature, the maximum dielectric constant does, as illustrated in Figure 2. Again, there is significant scatter in the data due to different deposition pressures, substrates, and film thicknesses, but the trend for the maximum ε_r to increase with temperature is clear. It is not obvious what the role of temperature is. The vast difference in kinetic energy of the deposited atoms due to this temperature difference could be responsible for better atomic ordering, though we have no direct evidence of such. We also do not know whether the differences we observe result from a change in growth mode of the STO, though that has been reported.[3] We do note, however, that our highest dielectric constants are higher than those reported in the literature, particularly for films deposited on non-lattice-matched substrates (i.e. non-STO as for all films discussed in this paper). These observations highlight the complex relation between film growth, microstructure, and dielectric properties. We speculate that the most difficult materials properties to quantify (i.e. defects) are likely responsible for these differences in bulk film properties.

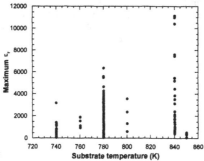

Fig. 2 Dependence of the maximum dielectric constant (at 10 kHz) on substrate temperature for laser-ablated STO films.

Substrates and Buffer Layers

We have evaluated the effect of a variety of substrate materials and buffer layers on the properties of the overlying STO thin films. We have investigated primarily the following substrate materials (with their respective lattice mismatches with single-crystal STO): $LaAlO_3$ (2.9%), MgO (7.9%), $NdGaO_3$ (1.9%), and Al_2O_3 (2.1%). We might anticipate that the growth mechanism of STO may be different depending on the lattice match with the substrate, e.g. so that in the case of MgO, the formation of dislocations at the interface may serve to relieve the stress so that the bulk of the film grows with less stress and fewer defects. In fact, we have found that we are able to grow the most satisfactory films on $LaAlO_3$ and $NdGaO_3$. It is much more difficult to deposit completely oriented films on MgO and Al_2O_3; care has to be taken not to nucleate (110) and (111) oriented grains. Furthermore, we always obtain lower values of ε_r and tuning ($\Delta\varepsilon_r/\varepsilon_r$) on MgO and Al_2O_3. At least for laser-ablated films, then, these substrates have thus far proven unsatisfactory. And though the overall goal is to reduce tanδ and not necessarily to obtain high ε_r values, we also find that the choice of substrate has no effect on the tanδ values for a given ε_r value.

We have additionally evaluated several buffer layers deposited between the substrate and STO film, which may also be expected to aid in stress relief or chemical/structural compatibility with the deposited film. We have deposited the following materials as buffer layers: $LaAlO_3$, $NdGaO_3$, $SrRuO_3$, and $SrAlNbO_3$. (It is well known that CeO_2 – widely used as a buffer layer for YBCO films – leads to poor STO growth.) We have deposited these buffer layers on each of the aforementioned substrates; note that in the case of $LaAlO_3$ and $NdGaO_3$, homoepitaxial growth is obtained, which has been reported to result in superior STO film growth.[12] Unfortunately, without discussing details, we have found that the addition of buffer layers usually results in poorer-quality films than sans buffer: we have never seen improvement upon adding a buffer layer, and usually the films were degraded. We have not investigated in detail why this is the case, i.e. whether the buffer layers are of poor quality (we expect not, based on previous investigations) or whether some chemical incompatibility exists.

Film Thickness

We have found a relatively strong dependence of STO film dielectric properties on film thickness, in accord with reports from other groups.[5] Figure 3 displays the thickness dependence of ε_r and $\tan\delta$ for many STO films grown under different conditions and on various substrates, hence the scatter. Note a strong tendency for ε_r to increase with increasing film thickness. Interestingly, we have found this effect to abate above ~2.5 μm. The cause of any critical thickness is unknown. Speculation is that as the films become thicker, the stress induced by the substrate is relieved, thereby leading to altered dielectric properties, perhaps due to reduced defect creation. Though possible, we have no direct evidence for this hypothesis. Preliminary TEM investigations of our films[13] have indicated nicely ordered growth from the substrate to the top of 2-μm-thick films, and they appear to be relatively free of extended defects throughout. It is clear, however, that the film properties are altered as the films become thicker. We again do not believe that the films are becoming more "single-crystal-like" as they become thicker, however. Note in Figure 3b that as ε_r increases with increasing thickness, so does $\tan\delta$. This is unlike single-crystal STO in which case ε_r is higher than thin films, but $\tan\delta$ is lower.

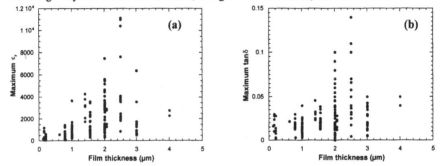

Fig. 3 Dependence on STO film thickness of (a) maximum dielectric constant and (b) maximum loss tangent. The measurement frequency is 10 kHz.

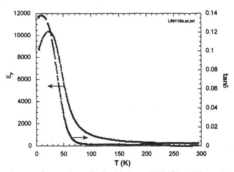

Fig. 4 Temperature dependence of ε_r and $\tan\delta$ of a high-ε_r STO film. The substrate is NdGaO$_3$, and the film is 2.5 μm thick. This sample was measured at f = 10 kHz.

High-ε_r Films

Through a combination of deposition conditions, substrate choice, and film thickness, we have been able to obtain heteroepitaxial growth of STO films which have very high values of dielectric constant, higher than reported in the literature. Figure 4 presents the temperature dependence of ε_r and tanδ for one such film. Our highest films display ε_r exceeding 11,000 at low temperature. Note, however, that the tanδ values of these films are also quite high, so as to probably render them unusable for our intended applications. Fortunately, we are also able to obtain films with significantly lower loss, as discussed in the next section.

Tuning and Loss

As discussed above, we observe a rather clear relationship between ε_r and tanδ. This is shown is Figure 5, which plots the maximum ε_r value vs. the maximum tanδ value for a variety of STO films of varying thickness and grown on various substrates. The tendency for tanδ to increase with ε_r is clear. This trend may imply that the same mechanism responsible for the high values of dielectric constant is also responsible for the loss. If so, this behavior is to be distinguished from single crystals, in which case the ε_r values are higher and the loss values lower than in thin films.

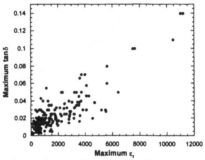

Fig. 5 Relationship between the maximum values of dielectric constant and loss tangent at 10 kHz for several STO thin films.

For our intended applications we are interested in the magnitude of the tuning, $\Delta\varepsilon_r/\varepsilon_r$, and not the absolute value of ε_r. Thus it is fortunate that a high value of dielectric constant is not required in order to obtain a high degree of tuning, as shown in Figure 6. Here $\Delta\varepsilon_r/\varepsilon_r \equiv [\varepsilon_r(0)-\varepsilon_r(E)]/\varepsilon_r(0)$. Since ε_r and tanδ are correlated, a high amount of tuning may still be obtained for lower-loss films.

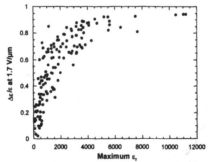

Fig. 6 Amount of tuning for several films at $E = 1.7$ V/μm vs. maximum dielectric constant.

What does determine the amount of tuning for a given STO film? Again, the precise mechanism is not known, but we have found a strong correlation of $\Delta\varepsilon_r/\varepsilon_r$ with the change of ε_r with temperature, as shown in Figure 7. This probably has to do with the frequency change of the soft optical phonon mode in STO, which softens with decreasing temperature and increasing electric field.

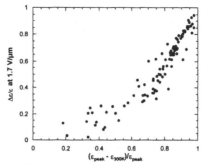

Fig. 7 Tuning of ε_r at $E = 1.7$ V/μm vs. change of ε_r with temperature.

We have found that the tunability of our STO films is excellent. As an example, the high-ε_r film shown in Figure 8a tunes by 95% (a change in ε_r of a factor of 18) with a modest electric field of less than 2 V/μm. (The field here is the average field in the film, given by $V/2.4d$, where V is the applied voltage and d is the spacing between the electrodes.[1]) Note in Figure 8b that this leads to a field-dependent figure of merit k of a few hundred at these voltages. Another example is shown in Figures 8c and 8d, in which case a lower-ε_r film still has excellent tunability and reasonable k-factor values. Particularly for high-ε_r films, we often observe hysteresis on the forward and reverse electric field sweeps, as shown in the figures.

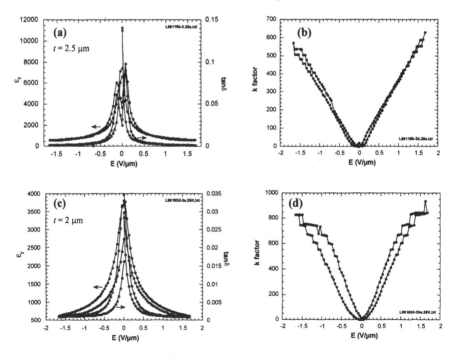

Fig. 8 Electric field dependence of (a) ε_r, tanδ, and (b) figure of merit k for a STO thin film at a temperature of 25 K. Data from another film at 25 K are shown in (c) and (d). The figure of merit is defined as $k \equiv [\varepsilon(0) - \varepsilon(E)]/[\varepsilon(0)\tan\delta(E)]$. The data were measured at 10 kHz.

There has been some debate on how to define k, so in Figure 9 we present the field dependence of the k-factor defined in two ways, relative to the tanδ value at zero field (the maximum value) or the tanδ value at the field E. Note that in both cases we are able to obtain k values over 100 at 70 K, which is the temperature of interest for applications. Note that this performance is actually better than single-crystal STO, which, despite its lower loss, displays virtually no tuning at this high temperature.

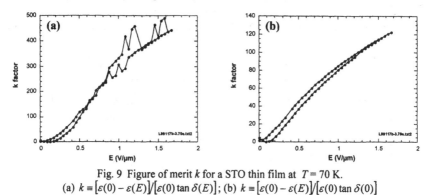

Fig. 9 Figure of merit k for a STO thin film at $T = 70$ K.
(a) $k \equiv [\varepsilon(0) - \varepsilon(E)]/[\varepsilon(0)\tan\delta(E)]$; (b) $k \equiv [\varepsilon(0) - \varepsilon(E)]/[\varepsilon(0)\tan\delta(0)]$

As we have discussed, there exists a tradeoff between tuning and loss. It is our intention to exploit somewhat this tradeoff in order to produce reasonable tunability with as low tanδ values as possible. Figure 10 shows $\varepsilon_r(T)$ and tan$\delta(T)$ for a film with a modest maximal ε_r, but a tanδ at 60 K which is just over 0.002. This value of tanδ is approaching an order of magnitude improvement in typical reported tanδ numbers, and is also approaching single crystal values measured by us in the same geometry (recall that extrinsic losses are included in our reported tanδ values).

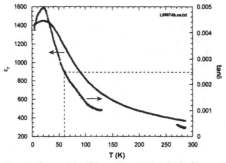

Fig. 10 Temperature dependence of ε_r and tanδ for a 3-µm-thick STO film on NGO measured at 10 kHz.

ATTEMPTS TO REDUCE THE LOSS

We would still like to produce tunable films with significantly lower values of tanδ than are now available. Loss values of ~10^{-4} would make these films more attractive for RF applications. In this section we describe several attempts to reduce the loss in our films.

Doping

We have altered the properties of our STO films by introducing dopants. In single crystal STO, it has been shown that point defects created by the introduction of dopants leads to an increase in the microwave loss.[16] However, the mechanism is different in STO films, and we may speculate that point defects, perhaps oxygen vacancies, play a large role in the nature of the dielectric properties. If so, attempts to reduce the number of oxygen vacancies may be reasonable. We have therefore tried doping in an attempt to perhaps change the valence of the ions in the lattice and thereby affect the number of oxygen vacancies. Furthermore, it has been reported that doping with Ca and W can reduce the loss in $BaSrTiO_3$ films,[14] and we decided to test this hypothesis with our STO films.

Table I We have deposited several doped STO films with the dopants listed below. The shot ratio gives the number of excimer laser pulses for each target. All films are 1 micron thick.

Dopant	Shot ratio (STO:dopant)	Atomic % (rough guess)	Peak tanδ	Peak ε_r	$\Delta\varepsilon_{max}/\varepsilon_{max}$(%) @1.7V/$\mu$m
Ce	20:1	5%	0.01	425	34
	10:1	10%	0.008	255	17
Ca	20:10	10–30%	0.011	446	16
	10:10	20–50%	0.01	323	7
Cr	20:2	1%	0.015 @ 60K	1150	46
	20:10	5%	0.01 @ 60K	435	10
Mo	200:1	0.5%	0.015 @ 60K	755	34
	100:1	1%	0.01 @ 60K	560	12
	40:1	2.5%	0.01 @ 60K	525	11
	20:1	5%	0.01 @ 60K	470	17
Mn	20:1	0.5%	0.024	1450	54
	20:5	2.5%	0.02	400	8
W	40:1	5%	12.5 @ 60K	-	-
	400:1	0.5%	0.25 @ 60K	334	13
	1000:1	0.2%	0.012	316	26
Co	20:5	2.5%	0.05 @ 60K	902	52
Y	20:1	3%	0.008	436	25
Ga	20:5	3%	0.001 @ 60K	167	0
Ir	20:5	2.5%	0.8 @ 60K	176	1

We accomplished the doping by alternately laser ablating from STO and from a second oxide target of the dopant element in question. In the past we have successfully used this approach to dope YBCO films; properties of films made by the two-target approach were shown to agree with those grown from a single doped-YBCO target. We have introduced a number of dopants, including Ca and Ce which presumably substitute for Sr, and Cr, Mo, Mn, and W which we expect to substitute for Ti. In addition, we have tried doping with Co, Y, Ga, and Ir. Table I lists the various dopants and a rough guess of their concentration. Only a small amount of doping on the Ti sites was necessary to dramatically alter the properties of the STO films. Doping the Sr sites produced less of an effect.

We have found that doping produces several interesting changes in the dielectric properties of the films. However, the main effects of doping STO in all cases were a reduction in ε_r (without a significant improvement in tanδ), very often increases in tanδ, and degraded tunability of ε_r and tanδ. In addition, some dopants readily produced films which were weakly conducting, and in many cases the temperature dependenceies of ε_r and tanδ were considerably altered. Despite the interesting behavior we have observed, the bottom line so far is that doping has not led to a reduction in tanδ and in fact has made things worse. We have not examined the microstructure of these doped films, so the implication is unclear, but it is possible that our pure STO films are already as defect-free as possible.

Superlattices

We have also made several attempts at reducing the loss by making superlattice thin film structures by forming repeated bilayers of STO and a lattice-matched dielectric with lower loss. We had hoped this would lead to a reduction in the overall tanδ of the films while preserving reasonable ε_r values and tunability. We chose SrAlTaO$_3$ and SrAlNbO$_3$ as interlayer materials. These dielectrics are very well lattice matched to STO and have low ε_r and tanδ values. An example is shown in Figure 11 of STO/SAT superlattice films. X-ray diffraction spectra of these films displayed very clear multiple satellite peaks, indicating excellent superlattice structure. Indeed, from these satellite peaks it was possible to calculate the thicknesses of the layers. In this example we formed bilayers of 15 nm of STO and 3, 10, and 20 nm of SAT; the total film thickness was 800 nm. As shown in the figure, the temperature dependence of ε_r and tanδ is dramatically altered from pure STO films. The peak in ε_r and tanδ shifts to higher temperatures as the SAT layer thickness is increased. We do not understand the origin of this behavior. Increasing amounts of SAT also leads to a reduction in the overall dielectric constant, as we may expect. Unfortunately, we also observe that for a given dielectric constant, the superlattices have higher loss than pure STO films. Again, our attempt to reduce the loss had the opposite effect.

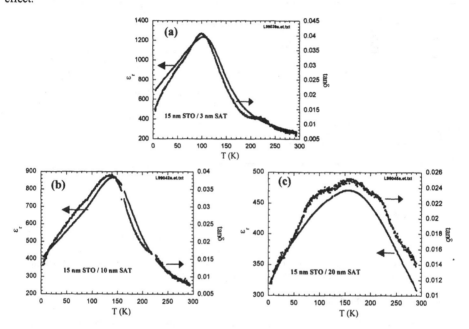

Fig. 11 Temperature dependence of dielectric properties for three STO/SAT superlattice thin films with the relative bilayer thicknesses noted.

Oxygen post-annealing

Finally, we have tried a number of oxygen annealing experiments in an attempt to affect the oxygen vacancy concentration or order in the STO films, thereby perhaps reducing the loss tangent. Table II lists some of the annealing experiments we have tried. Moderate temperature (200 to 400 °C) and high temperature (800 to 1000 °C) anneals have been carried out for periods of 1 hour to several hours. In most cases, anneals at 200 °C, even for many hours, have little or no effect on the dielectric properties. Oxygen anneals at 400 °C, on the other hand, often lead to a reduction in ε_r, $\Delta\varepsilon_r/\varepsilon_r$, and tan$\delta$. However, the trend of ε_r and tanδ displayed in Figure 5 remains, and a lower tanδ value is not obtained for a given ε_r. High-temperature anneals have the effect of almost always increasing ε_r and $\Delta\varepsilon_r/\varepsilon_r$, with the unfortunate effect of increasing tanδ

as well. This occurs for both "good" (high ε_r, high $\Delta\varepsilon_r/\varepsilon_r$) and "poor" (low ε_r, low $\Delta\varepsilon_r/\varepsilon_r$) films, as can be seen in the Table. Again, the attempt at reducing $\tan\delta$ has thus far been unsuccessful.

Table II The effect of *ex-situ* oxygen anneals for several STO thin films.

Sample	O$_2$ anneal	d-spacing (Å)	Peak tanδ	Peak ε	$\Delta\varepsilon/\varepsilon$ (%) @ 1.7 V/μm
L99047a		3.9142	0.032	3190	65.8
L99047a-2	400 °C 1 hr		0.032	2100	
L99047a-3	800 °C 1 hr	3.9136	0.025	3130	65.2
L99047a-4	900 °C 1 hr	3.9130	0.030	3810	71.4
L99047a-5	1000 °C 1 hr	3.9133	0.046	4270	74.5
L99047b		3.9164	0.033	3220	69.8
L99047b-2	400 °C 1 hr		0.023	1774	44.3
L99069a		3.9188	0.014	1887	
L99069a-2	800 °C 1 hr	3.9194	0.017	2215	61.6
L99069a-3	900 °C 1 hr	3.9192	0.025	2848	66.6
L99069a-4	1000 °C 1 hr	3.9182	0.032	2645	66.0
L99069a-5	1000 °C 1 hr	3.9183	0.039	3013	70.8
L99069b		3.9201	0.0054	950	29.5
L99069b-3	1000 °C 1 hr	3.9186	0.012	318	29.5
L99072a		3.9168	0.0105	327	45.9
L99072a-2	1000 °C 1 hr	3.9144	0.028	460	60.2
L99072b		3.9224	0.0030	148	14.2
L99072b-2	1000 °C 1 hr	3.9174	0.0138	631	73.2
L99073a		3.9164	0.0150	2795	65.2
L99073a-2	800 °C 6 hr	3.9168	0.0190	2795	65.2
L99073a-3	1000 °C 1 hr	3.9164	0.039	3398	71.4
L99074b		3.9152	0.005 .0025@60K	1523	37.6
L99074b-3	1000 °C 1 hr	3.9154	0.0115	1462	
L99075b		3.9172	0.007	1427	38.3
L99075b-2	1000 °C 1 hr	3.9158	0.0125	1430	37.7

RF APPLICATIONS OF SrTiO$_3$ THIN FILMS

While we continue our investigations into the properties of STO thin films and attempts to reduce their loss, it is imperative that we also evaluate their utility for the applications of interest. To that end, we have incorporated our present STO films into YBCO microwave resonators in order to demonstrate their feasibility for tuning applications.

Implementation Approaches

Presently available STO thin films are still too lossy to incorporate directly with high-Q YBCO thin film microwave filters; growing YBCO on top of a STO thin film or vice versa would reduce the total Q of the circuit to maybe a few hundred from several 10s of thousands, thereby defeating the YBCO performance advantage.[15] Therefore, we have taken the approach of decoupling the STO capacitor from the YBCO resonator. Although this approach will reduce the amount of available tuning, the tunability of the STO films is sufficiently great that we are able to trade off tuning for Q and can obtain performance which is sufficient for trimming applications and is on a par with, and perhaps better than, commercially-available varactors.

One approach we have taken is to make a STO thin film capacitor as described previously which is then flip-chip bonded onto a YBCO thin film resonator which has been designed so that the STO tuning capacitor is only weakly coupled to the resonator. Indium is used as the bonding material. This method is well suited to evaluation of the STO films, since the STO capacitors can be changed in a matter of minutes.

A second approach is a monolithic one in which STO is first grown onto a substrate in selected regions. This can be accomplished by using a shadow mask, or STO may be removed by etching. The YBCO thin film is then grown on the wafer and patterned in such a way as to simultaneously form STO capacitors and couple them into the resonator structure. At present,

54

this approach is a more complex process, but eventually will be superior for integration into a manufacturing process.

Performance

Figure 12 shows the performance we have achieved for the case in which a tunable STO capacitor is flip-chip bonded onto a YBCO resonator circuit chip. This resonator had an unloaded Q of 28,000 and a center frequency of about 841 MHz at 77 K. With applied dc bias, the resonant frequency is tuned to higher values due to the change in ε_r of the STO, and the Q also increases as expected because of the decrease in $\tan\delta$ of the STO film. For a bias of 40 V (E <2 V/μm), we are able to obtain a frequency shift of 0.64 MHz (0.075 %) while maintaining a Q of over 15,000. Despite the relatively little tuning, this performance is sufficient for trimming applications.

Figure 13 displays the performance of another YBCO resonator with greater tuning (Δf = 1.5 MHz) and smaller Q. In addijtional to the fundamental mode at 840 MHz (Figure 13a), we show the performance of the 3rd harmonic mode at 2.5 GHz. The relationship between the fundamental and the 3rd harmonic mode is given by $Q_3/Q_1 = (f_1/f_3)^2$, so because this resonator was not designed for 2.5 GHz, we expect the Q to be lower at this frequency. Note that we are able, however, to maintain a Q of over 3,000 while tuning by Δf = 5.3 MHz (0.2%).

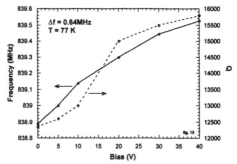

Fig. 12 Bias dependence of the resonant frequency and Q of a YBCO resonator coupled to a STO thin-film capacitor. A bias of 40 V corresponds to an electric field of 1.7 V/μm.

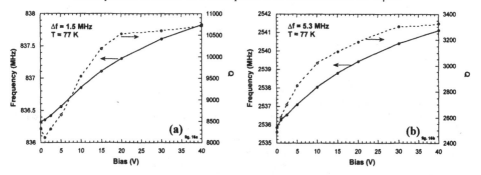

Fig. 13 Tuning of f_0 and Q for a YBCO resonator coupled to a STO thin-film capacitor. (a) The response of the fundamental mode. (b) The response of the 3rd harmonic mode.

CONCLUSIONS

We have discussed our efforts to optimize the dielectric properties of STO thin films, including the effects of growth conditions, doping, annealing, and multilayer structures. The films display excellent electric-field tunability (> factor of 10 in $E < 2$ V/μm), and we have grown heteroepitaxial STO films possessing very high dielectric constants (>10^4). We have found a significant and disappointing correlation between tuning and loss. Though the values of our loss tangents are on a par with the lowest reported for films with similar dielectric constants and tunabilities, the low-temperature loss tangent of STO films is still greater than single crystals. (Note that for the applications of interest, however, single crystals are probably not useful because their tuning is so small above ~60 K.)

Though some advantage may be gained by trading off tuning for loss, significant improvement of the loss of STO thin films (to <10^{-3}) is proving difficult. A detailed understanding of the microstructure of these films and its relation to dielectric properties is needed. We are thus beginning to use a variety of techniques to study the defects, stoichiometry, microstructure, and homogeneity of our films.

In order to measure the RF properties of the films and to evaluate their suitability for applications, we have incorporated tunable STO interdigital capacitors into YBCO lumped-element microwave resonators using approaches intended to minimize the loss. We have indeed produced STO films that are suitable for cryogenic microwave filter applications. At 840 MHz, the performance is similar to state-of-the-art GaAs varactors, while at 2 GHz and above, the performance is superior.

ACKNOWLEDGMENTS

This work was supported by DARPA, Contract No. N00014-98-C-0287. We are grateful for the technical assistance of S. Corrales.

REFERENCES

1. S. S. Gevorgian, T. Martinsson, P. L. J. Linnér, and E. L. Kollberg, IEEE Trans. Microwave Theory and Techniques **44**, 896 (1996).
NOTE: The solution in this paper contains errors in two of the three terms contributing to the capacitance. These errors have the effect of overestimating the dielectric constant, particularly for smaller numbers of interdigitated fingers. We have corrected these errors to the satisfaction of the authors.
2. M. J. Dalberth, R. E. Stauber, J. C. Price, C. T. Rogers, and D. Galt, Appl. Phys. Lett. **72**, 507 (1998).
3. M. Lippmaa, N. Nakagawa, M. Kawasaki, S. Ohashi, Y. Inaguma, M. Itoh, and H. Koinuma, Appl. Phys. Lett. **74**, 3543 (1999).
4. D. Galt, J. Price, J. A. Beall, and R. H. Ono, Appl. Phys. Lett. **63**, 3078 (1993).
5. H.-C. Li, W. Si, A. D. West, and X. X. Xi, Appl. Phys. Lett. **73**, 464 (1998).
6. R. A. Cowley, Phys. Rev. **134**, A981 (1964).
7. J. M. Worlock and P. A. Fleury, Phys. Rev. Lett. **19**, 1176 (1967).
8. O. G. Vendik, J. Appl. Phys. **82**, 4475 (1997).
9. R. Viana, P. Lunkenheimer, J. Hemberger, R. Böhmer, and A. Loidl, Phys. Rev. B **50**, 601 (1994).
10. M. Dalberth, Ph.D. thesis, University of Colorado (1999).
11. R. E. Treece, J. B. Thompson, C. H. Mueller, T. Rivkin, and M. W. Cromar, IEEE Trans. Appl. Supercon. **7**, 2363 (1997).
12. Q. X. Jia, A. T. Filndikoglu, D. Reagor, and P. Lu, Appl. Phys. Lett. **73**, 897 (1998).
13. D. Smith, N. Newman, and B. H. Moeckly, unpublished.
14. H. Chang, I. Takeuchi, and X.-D. Xiang, Appl. Phys. Lett. **74**, 1165 (1999).
15. A. T. Findikoglu, Q. X. Jia, I. H. Campbell, X. D. Wu, D. Reagor, C. B. Mombourquette, and D. McMurry, Appl. Phys. Lett. **66**, 3674 (1995).
16. G. Rupprecht and R. O. Bell, Phys. Rev. **125**, 1915 (1962).

RF-MAGNETRON SPUTTERED STRONTIUM TITANATE: STRUCTURE, PROCESSING AND PROPERTY RELATIONSHIPS

B.J. GIBBONS, Y. FAN, A.T. FINDIKOGLU, D.W. REAGOR, Q.X. JIA
Superconductivity Technology Center, Los Alamos National Laboratory, Los Alamos, NM
87545, gibbons@lanl.gov

ABSTRACT

The low frequency dielectric properties of epitaxial $SrTiO_3$ thin films deposited on $LaAlO_3$ are presented. The films were deposited using radio-frequency magnetron sputtering from stoichiometric targets in an Ar/O_2 atmosphere. For the first time, the effects of *in situ* ozone annealing during the early stages of deposition were explored. X-ray diffraction results indicated that the ozone treatment resulted in more symmetric and sharper diffraction peaks (2Θ FWHM decreased from 0.17° to 0.10°). In addition, the peaks for the ozone treated samples were shifted in 2Θ towards values approaching the bulk value. Rutherford backscattering measurements showed Sr/Ti ratios of 1:1 for these samples, indicating these peak shifts are not due to compositional variations. The dielectric constant of the ozone treated samples increased from 275 at room temperature to 1175 at 22 K (measured at 100 kHz). The effective loss tangent of the device remained between 1×10^{-4} and 1×10^{-3} down to 100 K, where it began to increase. The tunability was also measured. The ozone treated sample showed tunability of 46%, 43% and 38% at 22 K, 40 K and 60 K, respectively. Finally, similar measurements were completed at 1 MHz, indicating a minimal dependence of these properties on frequencies in this range.

INTRODUCTION

Electrically tunable microwave devices based on $SrTiO_3$ (STO) have been extensively investigated recently. In these designs, one takes advantage of the dc electric field tunability of the nonlinear dielectric $SrTiO_3$ film at cryogenic temperatures. For these devices to be practical, it is desirable to have the largest dielectric tunability in combination with the lowest dielectric loss achievable. These dielectric losses originate from various sources, including losses in the bulk of the STO film, losses at the interface between the substrate (typically $LaAlO_3$) and the STO film, and losses at the interface between the STO film and the electrode. Within the film, such defects as misorientation, grain boundaries, and oxygen vacancies can contribute to the loss. Ideally, elimination of as many of these factors as possible should enhance the dielectric properties of STO thin films.

Of the majority of the reported work on STO thin films for high frequency applications, pulsed laser deposition (PLD) is the most commonly used deposition technique [1-4]. Other methods, including chemical routes [5], metalorganic chemical vapor deposition (MOCVD) [6], molecular beam epitaxy (MBE) [7], and rf-sputtering (both on and off-axis) have been utilized [8-11], as well. In terms of PLD, it is well known that the process inherently results in the presence of macroparticles on the film surface. Such features may affect the dielectric properties of the final device. Thus, a technique which results in smoother surfaces/interfaces should beneficially influence the dielectric properties. Therefore, in this work rf-magnetron sputtering is used to deposit the STO-based heterostructures. In addition, in order to address the issue of oxygen non-stoichiometry in these dielectric films, ozone anneals at various points throughout the depositions are completed. Ozone has been used for MBE growth of oxide thin films, however it has never been implemented in conjunction with rf-sputtering. The effects on the

Mat. Res. Soc. Symp. Proc. Vol. 603 © 2000 Materials Research Society

resulting devices are exhibited through measurement and correlation of the structural and electrical properties.

EXPERIMENTAL PROCEDURE

Thin film heterostructures were prepared using standard 90°-off-axis rf-magnetron sputtering as developed by Eom *et al.* [12]. The high vacuum system used is equipped with four 2 inch diameter US Gun magnetron sputter sources. In addition, the system is equipped with a load lock/anneal chamber for sample loading/unloading and ozone treatment. Here, stoichiometric $SrTiO_3$ and $LaAlO_3$ (LAO) targets were used. The deposition temperature for the STO and the LAO was 750 °C, and the sputtering gas composition was a 3:1 ratio of Ar/O_2 at a total pressure of 90 - 95 mTorr. Sputtering was performed using an rf power of 125 W at a frequency of 13.56 MHz. Ozone was produced from a commercial ozone generator (Model GLS-1, PCI-Wedeco Environmental Technologies, West Caldwell, NJ). The typical output from this generator as specified by the manufacturer is 10% to 14% ozone, with the remainder O_2.

These conditions resulted in a deposition rate of approximately 1200 Å/hr, as determined by profilometry. For those films which were treated with ozone, the STO deposition was interrupted after 15 minutes ($\approx$ 300 Å) and the sample was removed to the load lock/anneal chamber. Subsequently, the O_3/O_2 mixture was flowed through the chamber at 50 – 70 Torr. This was done at various temperatures and for varying lengths of time. The optimal conditions were found to be at 500 °C for 10 minutes. After the anneal, the film was reintroduced into the deposition chamber (and the temperature returned to 750 °C) whereupon the deposition was continued. Upon the completion of the growth, the film was removed to the load lock/anneal chamber where the O_3/O_2 mixture was flowed over the sample while it was cooled to room temperature. The films used in this study were all 0.4 μm thick.

The film structural parameters were determined using conventional x-ray diffraction on a Siemens D5000 diffractometer. $\Theta/2\Theta$ scans, ω rocking curves and ϕ-scans were completed using this system. Out-of-plane lattice parameters were determined using a Nelson-Riley parameter analysis from the 00ℓ peaks of the STO. These values were confirmed using refinement procedures in the Winjade diffraction pattern analysis software (Materials Data Incorporated, Livermore, CA).

Low frequency dielectric properties were measured on sputter deposited Au coplanar capacitors with a 10 μm gap. The capacitors were fabricated using standard photolithography/liftoff procedures. An HP 4284A LCR meter was used for measurement. The small signal capacitance data reported here were obtained at 100 mV and 100 kHz (or 1 MHz). In order to estimate the dielectric constant of the STO films from these measurements, a conformal mapping analytical model was used [13].

RESULTS AND DISCUSSION

Figure 1 shows an x-ray pattern for an STO film deposited using the *in situ* ozone anneal process. For this particular sample the *c*-axis length was determined to be 3.912 Å. This was typical of the best films deposited using this process. It should be noted that the bulk *c*-axis lattice constant is 3.905 Å. Thus, the films described here do have an extended lattice (for this case, by 0.18%). One cause for this is the lattice mismatch that exists between the STO film and the LAO substrate. The LAO substrate has an in-plane pseudo-cubic lattice constant of 3.79 Å. Thus, there is a 3% lattice mismatch between the STO and the LAO. Therefore, the STO is likely in a state of in-plane compression, thus resulting (due to the Poisson effect) in an extended

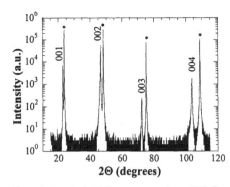

Figure 1: A typical Θ-2Θ x-ray pattern for an STO film deposited using the *in situ* ozone anneal process. The asterisks correspond to substrate peaks.

out-of-plane lattice constant. This same effect can be seen in films deposited by other techniques, as well [9, 11]. For comparison, films deposited using the same conditions aside from the ozone anneal typically showed *c*-axis lengths on the order of 3.92 Å and above. It is important to note that the objective of this study was not to specifically optimize the deposition conditions for STO films without using ozone, but to assess the effects of using an *in situ* ozone treatment on films deposited using standard conditions.

The effects of ozone on the growth of STO films is exhibited clearly in Fig. 2. Here, the 002 peak of various STO films are given (the 002 peak was chosen because it is the most intense of the 00ℓ peaks). Of particular concern is the shape of the peak resulting from different processing conditions. Using standard conditions (without ozone), it is clear that the peak has a significant distribution of lattice constants towards the higher end (> 3.93). That is, the peak is quite asymmetric. The effects of introducing an *in situ* ozone anneal to the deposition process are such that the resulting peak shape is more symmetric, sharper (full-width-at-half-maximum decreases from 0.17° to 0.1°) and shifted towards lattice constants more resembling bulk material.

In order to ensure that these differences in peak shape were not due to significant deviations from stoichiometry within the films, Rutherford backscattering measurements were performed on thin STO films deposited onto MgO substrates. MgO substrates were used in order to avoid interference in the spectra from the heavier La atoms in the LAO substrates. The results indicated that within the measurement limits of RBS, the films had a Sr/Ti ratio of 1. This is in contrast to the work of Wang *et al.* [11], who found a Sr deficiency of 20% for films deposited using rf-magnetron sputtering and at similar conditions.

Next, dielectric measurements at low frequencies were taken. Although the loss of STO films at microwave frequencies is likely different from that at low frequencies, these results can provide insight into the mechanisms of loss in these thin film

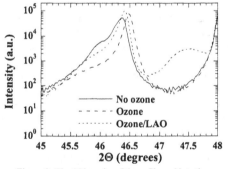

Figure 2: The 002 peaks of three films. Note the effects on the shape of the peak due to the ozone processing.

materials. Thus, leading to improvement of these properties at microwave frequencies. Fig. 3 shows the normalized capacitance and effective loss tangent (of the entire device) for three films as a function of temperature. The three films are: 1) deposited without ozone, 2) deposited with ozone and 3) deposited with a homoepitaxial LAO layer and with ozone. It has been shown that including the homoepitaxial LAO layer can result in lower dielectric loss while maintaining the same tunability for PLD films [14]. From this figure, it is clear that the sample processed using the *in situ* ozone anneal displays far superior properties. For the sample deposited without ozone, the dielectric constant increases from 65 to 160 between room temperature and 20 K. The

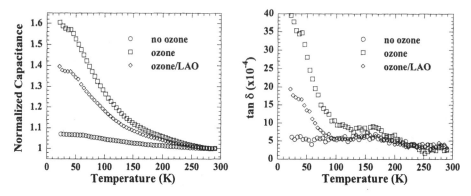

Figure 3: The capacitance and the loss tangent as a
function of temperature for three films.

dielectric constant of the ozone/LAO sample increases from 70 to 570 over the same range.
Finally, the dielectric constant of the ozone treated sample increases from 275 to 1175.

Fig. 3 also shows the effective loss tangent ($\varepsilon''/\varepsilon'$) of these samples as a function of
temperature. For all three samples, the tan δ stays between 1×10^{-4} and 1×10^{-3} from room
temperature to 100 K. These results compare very well to much of the data found in the
literature. Specifically, Li *et al.* show comparable tan δ levels in PLD deposited films [2].
However, those measurements were made through the thickness of the film (not a typical
microwave geometry). In addition, the bottom electrode layer ($SrRuO_3$) in those heterostructures
has a lattice constant similar to that of STO, thus reducing the resulting strain in the STO film.
Wang *et al.* show tan δ levels which appear to comparable down to 200 K for rf-magnetron
sputtered STO films of similar thickness [11]. Below 200 K, however, their data begin to
increase beyond the levels shown here. In addition, those films were deposited using a Sr
enriched target. Viana *et al.* completed a study on the dielectric properties of single crystal STO
[15]. There, they found the tan δ levels remain on the order of $3 - 8 \times 10^{-4}$ aside from two
relaxations at low temperatures where it rises to 10^{-3} (at 5.5 kHz). Thus, the results presented
here for the effective loss of the device are on the order of those in single crystal STO.

Fig. 4 shows tunability (normalized capacitance and loss tangent vs. applied voltage) data
for the same set of films. Here, voltages up to $\pm$ 40 V were applied across the 10 μm gap
between the Au pads. For the film deposited without ozone, there is very little tunability. The
sample with the homoepitaxial LAO layer shows intermediate tunability, with a factor of 2 lower
loss as compared to the ozone treated film. However (as with the capacitance vs. temperature
data), the dielectric constant of the film deposited using the *in situ* ozone treatment shows good
tunability. The relative dielectric constant changes from 1175 at 0 applied volts to 630 at $\pm$ 40
applied volts (46% change). The loss tangent behaves similarly. At relatively high voltages, it
appears to saturate for the two ozone treated samples.

For the ozone treated film, Fig. 5 shows the tunability of the capacitance and loss tangent
at 22 K, 40 K and 60 K. From these plots, it is clear that the sample still has significant
tunability at these higher temperatures (ε' changes by 43% at 40 K and 38% at 60 K). Again, the
loss tangent appears to saturate to the same level at all temperatures. Finally, in order to examine
the frequency dependence of these tunabilities, the same measurements were performed at
1 MHz. Fig. 6 is a plot of the capacitance as a function of bias for the ozone treated sample at
100 kHz and 1 MHz. From this plot, the lack of frequency dependence (in this frequency range)
is clear. The loss tangent shows a similar lack of frequency dependence.

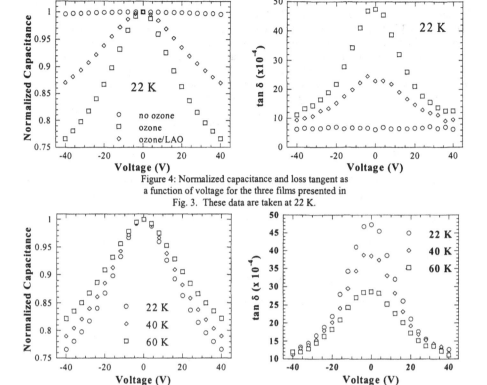

Figure 4: Normalized capacitance and loss tangent as
a function of voltage for the three films presented in
Fig. 3. These data are taken at 22 K.

Figure 5: Capacitance and loss tangent at three
temperatures for an ozone treated film.

Finally, it is important to emphasize that these tunabilites and effective loss values are obtained for films that are only 0.4 μm thick, and without any post-deposition high-temperature anneal process. Li *et al.* showed comparable loss levels for films 2.5 μm thick [4]. As the film thickness decreased, the loss levels increased to somewhat higher than those presented here. Other researchers have used high-temperature anneals to improve the tunability [11]. Here, these values were obtained on as-deposited films.

CONCLUSIONS

In this work, STO films were prepared via rf-magnetron sputtering. For the first time, ozone was incorporated into the deposition process in order to improve the issue of oxygen non-stoichiometry within these materials. X-ray diffraction data indicated a strong effect of using

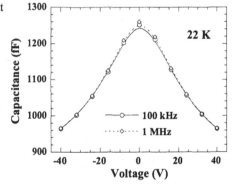

Figure 6: Capacitance as a function of voltage at 100 kHz and 1 MHz for the ozone treated sample. Note the lack of frequency dependence.

ozone on the structural properties of the films. Under the same conditions, using ozone resulted in films with c-axis lattice constants of 3.12 Å, whereas without ozone, the same lattice constant was typically 3.2 Å or greater. In addition, the appearance of the x-ray peaks was significantly improved. That is, the peaks were much more symmetric when using the ozone treatment, indicating a more even distribution of lattice constants. Dielectric measurements on these films were also completed. Capacitance data taken at 100 KHz indicated changes of up to 60% between room temperature and 22 K for the ozone treated samples. The effective dielectric loss of the device showed very promising behavior. Values measured were comparable (and sometimes better) than much of the data presented in the current literature for STO films deposited via all techniques. Tunability data were also presented.

REFERENCES

[1] A. T. Findikoglu, Q. X. Jia, I. H. Campbell, X. D. Wu, D. Reagor, C. B. Mombourquette and D. McMurry, Appl. Phys. Lett., **66**, p. 3674 (1995).

[2] H.-C. Li, W. Si, A. D. West and X. X. Xi, Appl. Phys. Lett., **73**, p. 190 (1998).

[3] H.-C. Li, W. Li, R.-L. Wang, Y. Xuan, B.-T. Liu and X. X. Xi, Materials Science and Engineering B, **56**, p. 218 (1998).

[4] H.-C. Li, W. Si, A. D. West and X. X. Xi, Appl. Phys. Lett., **73**, p. 464 (1998).

[5] U. Selvaraj, A. V. Prasadarao, S. Komarneni and R. Roy, Materials Letters, **23**, p. 123 (1995).

[6] K. Fröhlich, D. Machajdík, A. Rosová, I. Vávra, F. Weiss, B. Bochu and J. P. Senateur, Thin Solid Films, **260**, p. 187 (1995).

[7] D.-W. Kim, D.-H. Kim, B.-S. Kang, T. W. Noh, D. R. Lee and K.-B. Lee, Appl. Phys. Lett., **74**, p. 2176 (1999).

[8] H.-M. Christen, J. Mannhart and E. J. Williams, Physical Review B, **49**, p. 12095 (1994).

[9] A. B. Kozyrev, T. B. Samoilova, A. A. Golovkov, E. K. Hollmann, D. A. Kalinikos, D. Galt, C. H. Mueller, T. V. Rivkin and G. A. Koepf, Journal of Applied Physics, **84**, p. 3326 (1998).

[10] D. Fuchs, C. W. Schneider, R. Schneider and H. Rietschel, Journal of Applied Physics, **85**, p. 7362 (1999).

[11] X. Wang, U. Helmersson, L. D. Madsen, I. P. Ivanov, P. Münger, S. Rudner, B. Hjörvarsson and J.-E. Sundgren, Journal of Vacuum Science and Technology A, **17**, p. 564 (1999).

[12] C. B. Eom, J. Z. Sun, K. Yamamoto, A. F. Marshall, K. E. Luther, T. H. Geballe and S. S. Laderman, Appl. Phys. Lett., **55**, p. 595 (1989).

[13] S. Gevorgian, E. Carlsson, S. Rudner, L.-D. Wernlund, X. Wang and U. Helmersson, IEE Proc.-Microw. Antennas Propag., **143**, p. 397 (1996).

[14] Q. X. Jia, A. T. Findikoglu, D. Reagor and P. Lu, Appl. Phys. Lett., **73**, p. 897 (1998).

[15] R. Viana, P. Lunkenheimer, J. Hemberger, R. Böhmer and A. Loidl, Physical Review B, **50**, p. 601 (1994).

$Ba_{1-x}Sr_xTiO_3$ THIN FILM SPUTTER-GROWTH PROCESSES AND ELECTRICAL PROPERTY RELATIONSHIPS FOR HIGH FREQUENCY DEVICES

JAEMO IM [1], O. AUCIELLO[1], S.K. STREIFFER[1], P.K. BAUMANN[1], J.A. EASTMAN[1],
D.Y. KAUFMAN[2], A.R. KRAUSS[3], AND JIANXING LI[4]
[1] Materials Science Division, Argonne National Laboratory, Argonne, IL 60439
[2] Energy Technology Division, Argonne National Laboratory, Argonne IL 60439
[3] Materials Science and Chemistry Division, Argonne National Laboratory, Argonne, IL 60439
[4] Johnson-Matthey Electronics, Spokane, WA 99216

ABSTRACT

Precise control of $Ba_{1-x}Sr_xTiO_3$ (BST) film composition is critical for the production of high-quality BST thin films. Specifically, it is known that nonstoichiometry greatly affects the electrical properties of BST film capacitors. We are investigating the composition-microstructure-electrical property relationships of polycrystalline BST films produced by magnetron sputter-deposition using a single target with a Ba/Sr ratio of 50/50 and a (Ba+Sr)/Ti ratio of 1.0. It was determined that the (Ba+Sr)/Ti ratios of these BST films could be adjusted from 0.73 to 0.98 by changing the total ($Ar+O_2$) process pressure, while the O_2/Ar ratio did not strongly affect the metal ion composition. The crystalline quality as well as the measured dielectric constant, dielectric tunability, and electrical breakdown voltage of BST films have been found to be strongly dependent on the composition of the BST films, especially the (Ba+Sr)/Ti ratio. We discuss the impact of BST film composition control, through film deposition and process parameters, on the electrical properties of BST capacitors for high frequency devices.

INTRODUCTION

Recently $(Ba_xSr_{1-x})Ti_{1+y}O_{3+z}$ (BST) films are investigated as electric-field tunable elements for high frequency devices.[1,2] The high dielectric tunability (dependence of permittivity on electric field), high breakdown field, and relatively low loss tangent of BST at microwave frequencies[3] make it attractive for application in high frequency devices such as varactors, frequency triplers, tunable phase shifters, etc.[1,2]

Radio frequency (RF) magnetron sputter deposition of BST thin films, using a multi-component BST oxide target, is a suitable processing route for the production of high-quality BST thin films. However, it is important to understand the effect of film processing parameters on the composition and microstructure of the BST films in order to achieve optimum properties. It is typically found that sputtering from a single multicomponent stoichiometric oxide target produces a non-stoichiometric oxide thin film. For the BST system, BST targets with various (Ba+Sr)/Ti ratios[4] and Ba/Sr ratios[5] were utilized to produce compositionally-adjusted BST films by physical vapor deposition. In this paper, we report that the (Ba+Sr)/Ti ratios of BST films can be adjusted simply by using a single stoichiometric BST target in conjunction with accurate control of the total process gas ($Ar+O_2$) pressure.

EXPERIMENT

BST thin films were deposited using an Ar-O_2 gas mixture in a RF magnetron sputtering system equipped with a 3" diameter target. The substrate and target (a sintered stoichiometric $Ba_{0.5}Sr_{0.5}TiO_3$ disc provided by Johnson Matthey) were positioned parallel to each other in an on-axis configuration with 10 cm separation. Prior to the initial deposition, the BST target was pre-sputtered for more than 20 hours to stabilize the surface composition. All BST films were deposited at 650°C on Pt(120nm)/SiO_2(300nm)/Si substrates. The thickness of all BST films was approximately 80nm. Capacitors for electrical testing were produced by depositing 100nm Pt top electrodes at 350°C through a shadow mask with circular openings of 100 µm diameter. After top electrode deposition, the samples were annealed in a quartz tube furnace at 550°C for 30 minutes in air in order to improve the top Pt electrode/BST interface.

Rutherford backscattering spectrometry (RBS), x-ray diffraction (XRD), field emission scanning electron microscopy (FESEM), and scanning probe microscopy (SPM) were used to investigate the thickness, composition, crystallographic orientation, microstructure, and roughness of the BST films, respectively. Dielectric properties (relative permittivity ε and dielectric loss $\tan\delta$ as a function of applied electric field) of the Pt/BST/Pt capacitors were measured at 10 kHz and 0.1 Vrms oscillation level using a Hewlett-Packard 4192A impedance analyzer.

RESULTS AND DISCUSSION

Fig. 1 (a) shows the variation of the (Ba+Sr)/Ti and Ba/Sr ratios as a function of the total (Ar+O_2) pressure used to deposit various BST films. The O_2/Ar ratio was fixed at 1/5. The ratio of (Ba+Sr)/Ti changed from 0.73 to 0.98 for a variation in the total gas pressure from 22 to 58 mTorr. In contrast to the (Ba+Sr)/Ti ratio, the Ba/Sr ratio demonstrated a much weaker trend with pressure. The Ba/Sr ratio was 0.73 for BST films deposited in 22 mTorr total process pressure and 0.84 for films grown in 30 mTorr total process pressure; this ratio remained constant for films deposited at pressures higher than 30 mTorr. As expected, the growth rate of the magnetron sputtered BST strongly depends on the total process pressure (Fig. 1 (b)). In separate experiments, the O_2/Ar ratio was varied from 1:1 to 1:5, but no changes were observed in the metals composition of the BST films. These results are consistent with a mechanism based on the pressure dependencies of the individual fluxes of Ba, Sr, and Ti impinging on the substrate (i.e., the different constituents have different, pressure dependent angular scattering distributions in the gas phase, due to their different masses).[6]

BST films with (Ba+Sr)/Ti ratios of 0.9 or higher exhibited a polycrystalline structure characterized by the appearance of (100), (110), and (111) peaks in the XRD spectra (see Fig. 2). Note that the BST (111) peak for these films is not well resolved from the very intense Pt (111), and is therefore not shown in Fig. 2. The BST peaks disappeared as the (Ba+Sr)/Ti ratio is reduced below 0.85 (Fig. 2). The disappearance of the BST- XRD peaks, coupled with the fact that relatively high permittivities are still found for these samples, indicates that the high-Ti BST films are nanocrystalline.

Significant differences were also found in the surface morphology of BST films with various (Ba+Sr)/Ti ratios. As shown in Fig. 3, the root mean square (RMS) roughness measured by SPM over a scan area of 14 µm x 14 µm was approximately 30 Å for BST films with (Ba+Sr)/Ti ratios of 0.9 or higher, for which BST-related XRD peaks are clearly visible. The RMS roughness was reduced to approximately 20 Å for BST films with (Ba+Sr)/Ti ratios less than 0.85, correlating with the structural changes indicated by the XRD spectra (Fig. 2 and Fig. 3).

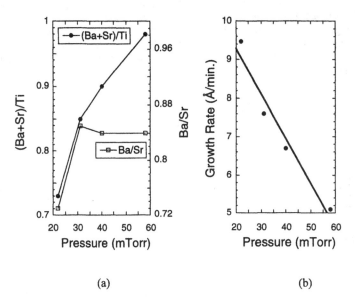

(a)   (b)

Figure. 1. (a) Variation of (Ba+Sr)/Ti and Ba/Sr ratios for BST films deposited at 650°C from a stoichiometric target with a Ba/Sr ratio of 50/50; and (b) growth rate as a function of the total pressure (Ar+O_2) in the magnetron deposition chamber.

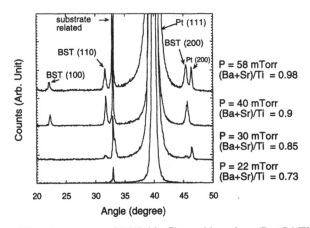

Figure. 2. X-ray diffraction spectra of BST thin films with various (Ba+Sr)/Ti ratio as a function of the total pressure (Ar+O_2) in the magnetron deposition chamber.

BST films with (Ba+Sr)/Ti ratios > 0.9 exhibited a dense and granular microstructure (Fig. 4(a)). In contrast, BST films with (Ba+Sr)/Ti ratios < 0.85 exhibited a featureless microstructure given the resolution of the FESEM techniques (Fig. 4 (b)). These surface morphology changes of BST films also support that the high-Ti BST films ((Ba+Sr)/Ti ratios < 0.85) are nanocrystalline. Further work, including TEM studies, is underway to elucidate the nature of these structural changes.

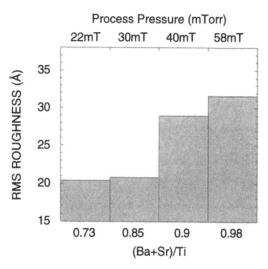

Figure 3. RMS roughness for BST films with various (Ba+Sr)/Ti ratio as a function of the total pressure (Ar+O$_2$) in the magnetron deposition chamber.

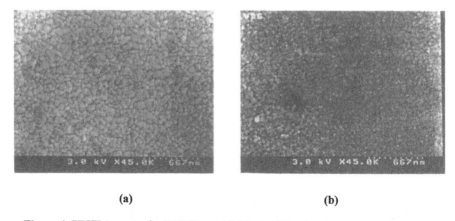

(a)   (b)

Figure 4. FESEM spectra for BST films with (Ba+Sr)/Ti ratio of (a) 0.9 and (b) 0.85

The dielectric constant, and dielectric loss at zero bias also exhibited a clear dependence on the total process pressure or (Ba+Sr)/Ti ratio[7]. As shown in Fig. 5 (a), the near-stoichiometric BST film ((Ba+Sr)/Ti = 0.98) exhibited the highest dielectric constant as well as the largest tunability (defined as $(\varepsilon_{max}-\varepsilon_{min}/\varepsilon_{max})$, where ε_{max} and ε_{min} are the maximum and minimum measured permittivity) of approximately 74%. Tunability tracks the general trends in dielectric constant, as shown in Fig. 5 (a). A significant reduction in permittivity and dielectric loss at zero bias was observed for films with (Ba+Sr)/Ti ratio < 0.85 (Fig. 5). The lowest dielectric loss of 0.0047 at zero bias was found for the sample with a (Ba+Sr)/Ti ratio of 0.73. Given the observed correlations between tunability and loss, a figure of merit K is frequently used, and is defined as:

$$K = \text{tunability}/\tan \delta \qquad (1)$$

As shown in Fig. 5 (b), the highest K value was found for the sample with a (Ba+Sr)/Ti ratio of 0.73. This demonstrates that the decrease in zero-bias permittivity for the highly nonstoichiometric samples is more than compensated for by their lower losses and higher breakdown fields, such that they may offer superior performance for some applications.

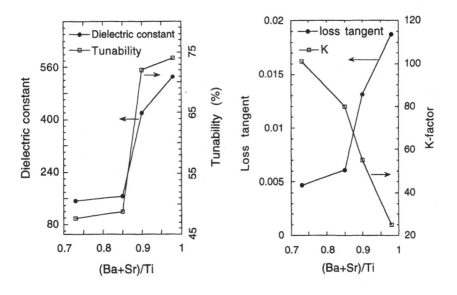

(a) Dielectric Constant and Tunability   (b) Loss tangent and K-factor

Figure 5. (a) Dielectric constant, tunability and (b) loss tangent and K-factor measured for BST capacitors with various (Ba+Sr)/Ti ratios, deposited at various total pressures (Ar+O_2) in the magnetron deposition chamber.

CONCLUSIONS

In conclusion, we have shown that BST films with high tunabilities, low losses, and high-dielectric breakdown fields can be grown using magnetron sputter deposition with judiciously chosen process parameters. Specifically, control is demonstrated of the composition (i.e., the (Ba+Sr)/Ti ratio) of BST films grown from a single stoichiometric target by use of tailored target-substrate geometry and deposition pressure. Figures of merit, defined as the ratio of tunability to dielectric loss, of approximately 100 have been obtained under optimized deposition conditions, with among the lowest losses at zero bias (0.0047) reported for physical vapor deposited BST films.

ACKNOWLEDGMENTS

This work was supported by the U.S. Department of Energy, BES-Material Sciences and the Office of Transportation Technology, under Contract W-31-109-ENG-38, and by the DARPA-FAME Program under contract AO G016. The authors acknowledge Dr. Pete Baldo for RBS measurements. We are also grateful to Dr. J. Lee (Johnson Matthey) for providing BST targets.

REFERENCES

1. J.M. Ponds, S.W. Kirchoefer, W. Chang, J.S. Horwitz, and D.B. Chrisey, Integr. Ferroelectr., **22**, p. 317 (1998).
2. F.A. Miranda, F.W. Van Keuls, R.R. Romanofsky, and G. Subramanyam, Integr. Ferroelectr. **22**, p. 269 (1998).
3. J.D. Baniecki, R.B. Laibowitz, T.W. Shaw, P.R. Duncombe, D.A. Neumayer, D.E. Kotecki, H. Shen, Q.Y. Ma, Appl. Phys. Lett. **72**, p. 498 (1998).
4. S. Yamamichi, H. Yabuta, T. Sakuma, and Y. Miyasaka, Appl. Phys. Lett. **64**, p. 1644 (1994).
5. B.A. Baumert, T.L. Tsai, L.H. Chang, T.R. Remmel, M.L. Kottke, P.L. Fejes, W. Chen, E.P. Ehlert, D.F. Sullivan, C.J. Tracy, and B.M. Melnick, Mater. Res. Soc. Symp. Proc. **493**, p. 21 (1998).
6. G.K. Wehner, J. Vac. Sci. Technol., **A 1**, 487 (1983).
7. S.K. Streiffer. C. Basceri, C.B. Parker, S.E. Lash, and A.I. Kingon, J. Appl. Phys. **86**, p. 4565 (1999).

Magnetic-Field Tuning

FERRITES FOR TUNABLE RF AND MICROWAVE DEVICES

GERALD F. DIONNE
Lincoln Laboratory, Massachusetts Institute of Technology, 244 Wood St., Lexington, MA 02420, dionne@ll.mit.edu

ABSTRACT

Microwave systems for communications and radar require control of propagation of the rf signal. Devices that accomplish this function include phase shifters, isolators and circulators, and tunable filters. In many instances, these devices are magnetic and are based on the variable permeability of electrically insulating ferrimagnetic oxides (ferrites). Recent advances in microwave ferrite devices have featured superconductor circuitry that promises to virtually eliminate insertion losses due to rf surface resistance. Lower conduction losses allow the use of small lightweight microstrip configurations in place of traditional bulky waveguide structures. For operation at cryogenic temperatures ferrimagnetic spinels and garnets will require chemical alteration to realize the full potential of these devices. Challenges include the reduction of magnetocrystalline anisotropy to optimize switching energies and speeds, and the elimination of fast-relaxing impurities in the magnetic garnets that can increase magnetic losses and degrade resonator Q factors at low temperatures.

INTRODUCTION

Ferrite devices that operate with magnetically controlled propagation velocity to alter the signal wavelength and phase have been widely used in microwave systems for more than fifty years. Adjustable phase shifters are the principal beam-steering components of agile-beam phased-array radars. Ferrite circulators and isolators are employed in radars to protect transmitters from unwanted reflected signals, and circulators are necessary to direct the return signals into independent receiver channels. Traditionally, these devices have been in waveguide configurations, which are typically large, heavy and expensive. In recent years, ferrite components in planar microstrip geometries have been advanced through the adoption of superconductor circuits [1]. In these devices, the losses due to surface resistance of the circuit are reduced dramatically. Other benefits can be realized from improved ferrite properties because the devices must be operated in a temperature-controlled cryogenic environment. One new concept that the use of superconductors has spawned is the application of conventional ferrites to an innovative tunable filter in which the Q and tuning speeds can be raised greatly as a result of lower circuit losses and switching energies [2].

One of the main features of these ferrite device concepts is operation of the ferrite in a partially magnetized state, in which the magnetic permeability is controlled directly by the state of magnetization. As a consequence, the hysteresis loop of the ferrite is of fundamental importance. The second and equally important aspect of ferrites is their magnetic and dielectric losses that must be minimized to reap the full benefits of the virtually lossless superconductors.

In this paper, relations for the rf permeability created by a linearly polarized signal interacting with a ferrite will first be reviewed, with emphasis on the application to a high-Q tunable microstrip resonator. The phenomenon of ferrimagnetic resonance (FMR) as influenced

Mat. Res. Soc. Symp. Proc. Vol. 603 © 2000 Materials Research Society

by materials properties will then be discussed. Finally, basic magnetochemistry of spinels and garnet ferrites will be described with emphasis on composition design tradeoffs expected in the adaptation of ferrite technology to cryogenic environments.

RF PERMEABILITY AND RESONATOR THEORY

From the theory of gyromagnetic effects in ferrites, the real and imaginary parts of the complex effective permeability $\mu = \mu' - j\mu''$ for an rf wave of linear polarization interacting with a ferrite in a state of magnetization $4\pi M$ are given by [3]

$$\mu' = 1 + \left\{ \frac{(\gamma 4\pi M)v_0\left(v_0{}^2 - v^2 + (\Delta v)^2\right)}{\left(v_0{}^2 - v^2 - (\Delta v)^2\right)^2 + 4(v_0\Delta v)^2} \right\} \quad , \tag{1}$$

and

$$\mu'' = \left\{ \frac{(\gamma 4\pi M)\Delta v\left(v_0{}^2 + v^2 + (\Delta v)^2\right)}{\left(v_0{}^2 - v^2 - (\Delta v)^2\right)^2 + 4(v_0\Delta v)^2} \right\} \quad , \tag{2}$$

where the gyromagnetic constant $\gamma = 2.8$ GHz/kOe, v is the signal frequency, v_0 is the frequency of the Lorentzian-shaped ferrimagnetic resonance line of intrinsic half-width Δv from spin-lattice relaxation damping, which is typically small enough to be neglected in comparison with v. Following the perturbation analysis of Lax and Button [3], the relations appropriate for a planar resonator in a circular toroidal configuration with closed magnetic circuit sketched in Fig. 1 when operated in the regime where $v_0 \ll v$ (in the upper wing of the resonance line) are [4],

$$\mu' \approx 1 - \xi\left(\frac{\gamma 4\pi M}{v}\right)\left\{\frac{\gamma\left[H + N_y(4\pi M)\right]}{v}\right\} \quad , \tag{3}$$

and

$$\tan\delta_\mu = \frac{\mu''}{\mu'} \approx \xi\left(\frac{\gamma 4\pi M}{v}\right)\left(\frac{\Delta v}{v}\right) \quad , \tag{4}$$

where the H and $4\pi M$ vectors are directed along the axis of propagation z, $0 \leq N_y \leq 1$ is the y-axis demagnetizing factor, and $\xi \leq 1$ is a microwave filling factor [5]. In this configuration the magnetic field component of the rf signal is along the x axis.

From the basic theory of electromagnetic waves, the rf propagation velocity $v = c/(\mu'\varepsilon')^{1/2}$, where c is the velocity of light and ε' is the real part of the electric permittivity (or the dielectric constant) and is usually constant with frequency. For the half-wavelength resonator structure of length L in Fig. 1, the resonator fundamental frequency v_r would be $v/2L$. An approximate relation for the frequency tuning ratio derived from Eq. (3) was reported previously [4]:

$$\frac{\nu_r}{\nu_{r0}} = \frac{\nu_{r0} + \Delta\nu_r}{\nu_{r0}} \approx 1 + \frac{\xi}{2}\left\{\frac{(\gamma 4\pi M)\gamma\left[H + N_y(4\pi M)\right]}{\nu_{r0}^2}\right\} \quad , \tag{5}$$

where ν_{r0} ($>> \nu_0$) is the resonator frequency at $4\pi M = 0$. For high-Q filter applications, the magnetic loss contribution to the Q factor is approximately the inverse $\tan\delta_\mu$ of Eq. (3). In Fig. 2, tuning performance with existing ferrite is contrasted with that projected for improved ferrite [6]. Data were from a niobium superconductor resonator on an iron garnet substrate at 4 K.

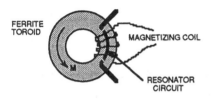

FERRITE TOROID — MAGNETIZING COIL — RESONATOR CIRCUIT

Fig. 1. Circular toroidal device configuration with magnetizing coil encircling the resonator.

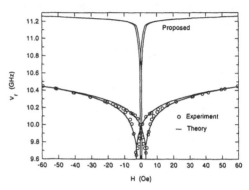

Fig. 2. Comparison of theory and experiment for ferrite with $4\pi M_s = 1800$ G and $H_c = 3$ Oe. Upper curves indicate proposed tunability improvement with ferrite $4\pi M_s = 2400$ G and $H_c = 0.5$ Oe.

By combining Eqs. (4) and (5), a tunable filter figure of merit K can be defined in terms of the product of $\Delta\nu_r/\nu_{r0}$ and the Q factor. For a closed magnetic circuit at low field, $H \approx 0$ with the result

$$K = 2\left(\frac{\Delta\nu_r}{\nu_{r0}}\right)\frac{1}{\tan\delta_\mu} = N_y\left(\frac{\gamma 4\pi M_s}{\nu_r}\right)\left(\frac{\nu_r}{\Delta\nu}\right) \quad . \tag{6}$$

For feasible values of $\nu_r/\Delta\nu > 10^4$, $N_y \approx 1$, and $\gamma 4\pi M_s/\nu_r \approx 0.7$ (which is usually considered as an upper limit to avoid losses that occur because of inter-domain fields that exist in the partially magnetized state), $K \sim 10^4$ could be a practical limit for a filter with frequency controlled by the state of magnetization.

FERRITE PROPERTIES

For optimum performance of a linear-polarization ferrite resonator that is tunable in the partially magnetized state, the objective would be to employ the ferrite with the largest practical $4\pi M$ and the narrowest $\Delta\nu$. These parameters determine the static performance. When the tuning frequency must be changed, the hysteresis properties of the ferrite determine the energy and switching speed of the device. Design of the ideal ferrite for a particular application can be a challenging exercise owing to the complex interrelations and tradeoffs that must be considered. In addition to magnetization and ferrimagnetic resonance linewidth, other important properties of these oxides include magnetoelastic (anisotropy and magnetostriction) and dielectric properties, all of which are sensitive to temperature. In some cases the reduction in temperature to the cryogenic range can raise additional challenges [7,8].

Magnetization

A review of the temperature influence on the magnetization of ferrimagnetic oxides begins with the theory of molecular fields [9], refined to account for the local weakening of magnetic superexchange to cause sublattice canting [10]. The magnetization M may be deduced from the resultant of the individual opposing sublattice magnetic moments per mole, i.e., $|\mathcal{M}_i - \mathcal{M}_j|$. For sublattice i,

$$\mathcal{M}_i(T) = \mathcal{M}_i(0)\mathcal{B}_{S_i}(x_i) \qquad , \qquad (7)$$

where $\mathcal{B}_{S_i}$ is the Brillouin function, $x_i = (g\mu_B S_i / kT)(N_{ii}\mathcal{M}_i + N_{ij}\mathcal{M}_j)$, g is the spectroscopic constant (≈ 2), μ_B is the Bohr magneton, k is the Boltzmann constant, T is the temperature, S_i is the spin, and N_{ii}, N_{jj} are the intra- and N_{ij} the inter-sublattice molecular-field coefficients. Solutions are carried out by simultaneous iteration over the various sublattices.

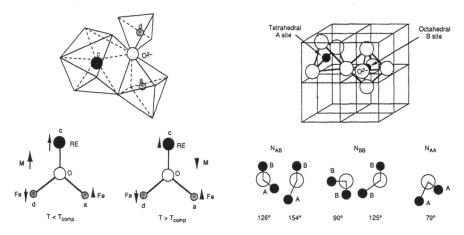

Fig. 3. Garnet crystal sites with bond arrangements. c-Sites can be occupied by rare-earth ions.

Fig. 4. Spinel crystal structure with bond angle diagrams.

For microwave applications, frequencies below 10 GHz favor lower magnetization magnetic garnets based on $\{Y_{1-kc}c_{kc}\}_3[Fe_{1-ka}a_{ka}]_2(Fe_{1-kd}d_{kd})_3O_{12}$ (YIG). The brackets in the formula designate the dodecahedral, octahedral and tetrahedral sites according to the generic formula $\{c\}_3[a]_2(d)_3O_{12}$. The three crystallographic sites of the garnet lattice are related as sketched in Fig. 3. Relations for the molecular-field coefficients as functions of dilution are [10]:

74

$$N_{dd} = -30.4\left(1 - 0.87k_a\right)$$

$$N_{aa} = -65.0\left(1 - 1.26k_d\right) \qquad \text{moles/cm}^3 \qquad , \qquad (8)$$

$$N_{ad} = +97.0\left(1 - 0.25k_a - 0.38k_d\right)$$

where k_a and k_d represent the dilution fractions of the a and d sublattices.

For the higher magnetization ferrites required at millimeter wavelengths, e.g., 35 GHz, lithium spinel ferrite $Fe[Li_{0.5}Fe_{1.5}]O_4$ with dilution of the weaker sublattice by Zn^{2+} ions is an appropriate choice. The site designations are indicated by the generic formula $A[B]_2O_4$, with A for the tetrahedral and B for the octahedral sites. Figure 4 illustrates part of a spinel unit cell in which, unlike the garnet case, the cubic site symmetry axes are coincident with the lattice axes. The corresponding relations for the N_{ij} coefficients are [11]:

$$N_{AA} = -150\left[1 - \frac{4}{3}k_B\right] = -150(1 + z/3)$$

$$N_{BB} = -60\left[1 - \frac{1}{3}k_A\right] = -60(1 - z/3) \qquad \text{moles/cm}^3 \qquad , \qquad (9)$$

$$N_{AB} = +273\left[1 - \frac{1}{5}k_A\right] = +273(1 - z/5)$$

which apply to the formula $Fe^{3+}_{1-z}Zn^{2+}_z[Li^{1+}_{0.5-z/2}Fe^{3+}_{1.5+z/2}]O_4$ up to $z = 0.4$. Solutions for the family of rare-earth iron garnets, as well as nickel and manganese spinel ferrites, are also available in the literature [12,13].

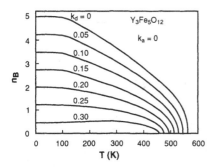

Fig. 5. Computed thermomagnetization curves of YIG with fractional tetrahedral site dilution $k_d \leq 1$.

The design of an iron garnet for low temperatures involves a broader range of magnetization values. With the exception of magnetic rare-earths such as Gd^{3+} replacing Y^{3+} in the c-sublattice, every composition increases in M_s as T decreases. Families of computed thermomagnetization curves for tetrahedral and

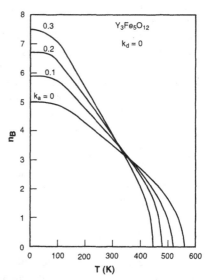

Fig. 6. Computed thermomagnetization curves of YIG with fractional octahedral site dilution $k_a \leq 1$.

octahedral sublattice dilution are reproduced in Figs. 5 and 6, respectively [10]. In the LiZn ferrite case, n_B increases sharply because the A-sublattice moment reduces while the B sublattice moment rises from the additional Fe^{3+} content needed to restore the charge balance upset by the removal of Li^{1+}. Computed curves for this family are plotted in Fig. 7 [11].

In the very low temperature regime dilution effects on the Fe^{3+} sublattices can be estimated in some cases without temperature corrections to the individual sublattice magnetizations, i.e., $\mathscr{B}_{Si}(x_i) \sim 1$. At temperatures below the onset of thermally induced spin canting, the net magnetic moment of a three-sublattice garnet expressed in Bohr magnetons per formula unit is

$$n_B = \left| n_{Bc}(1-k_c) - n_{Bd}(1-k_d)C_d + n_{Ba}(1-k_a)C_a \right| \qquad , \qquad (10)$$

where C_d, $C_a < 1$ represent zero-temperature canting factors that were discussed by several authors [14-17] for the octahedral and tetrahedral sublattices that become significant at large values of the respective dilution fractions k_d and k_a. With the c sublattice occupied by diamagnetic ions and the d and a sites occupied by Fe^{3+} ions with $n_B = 5$, Eq. (10) is reduced to

$$n_B = \left| 15(1-k_d)C_d - 10(1-k_a)C_a \right| \qquad . \qquad (11)$$

For the spinel case $A[B]_2O_4$, the corresponding relation is

$$n_B = \left| 10(1-k_B)C_B - 5(1-k_A)C_A \right| \qquad , \qquad (12)$$

where the C_A, $C_B < 1$ are the corresponding canting factors [14,15].

In Tables I and II summaries of magnetic dilutants and their effects on magnetic properties are listed for the YIG and lithium ferrite families, respectively. Since Curie temperatures are of less concern in cryogenic environments, tailoring of the magnetization in the garnets simplifies to substituting Ga^{3+} or Al^{3+} that favor tetrahedral sites to lower n_B and In^{3+} or Sc^{3+} to raise n_B. In spinels, Al^{3+}, Ga^{3+} (which favor octahedral sites in this latttice) or the three-ion $Li^{1+}+2Ti^{4+}$ combination replacing one tetrahedral and two octahedral Fe^{3+} ions may be used to reduce n_B. The substitution of Zn^{2+} into the tetrahedral sublattice (up to 30%) will increase n_B.

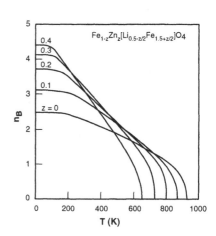

Fig. 7. Computed thermomagnetization curves of Li spinel ferrite with concentrations z of Zn^{2+} in the tetrahedral A sublattice.

Magnetoelastic Properties

Besides the magnetization, the ability to magnetize a magnetic material is strongly dependent on magnetic field strength required to rotate the magnetization vector into alignment with the field. It is the property that defines the easy and hard directions of magnetization.

The first-order magnetocrystalline anisotropy characterized by the parameter K_1 derives its influence from excited orbital terms of the Fe^{3+} ion [18]. These effects are dependent on the cubic crystalline electric field and therefore differ between octahedral and tetrahedral sites. A rule that seems to apply well in most cases is that the octahedral sites produce magnetic anisotropy that favors the $\langle 111 \rangle$ axes (easy directions). Work reported on spinels and garnets indicate that the tetrahedral site contributions are of opposite sign but sufficiently smaller that they can be generally ignored [19,20]. For parameter design, an effective strategy may be implemented by matching selected dilutants to the appropriate sublattices.

The anisotropy constant controls a number of ferrite properties, including the domain wall energy. In most cases, the temperature sensitivity of K_1 is substantially greater than that of M_s, as shown in Fig. 8 for YIG [21,22]. Of equal importance to that of K_1 or M_s is the ratio $|K_1|/M_s$, which is used as a design parameter of hysteresis loops [23]. As discussed above, $|K_1|$ is decreased by octahedral B sublattice dilution which can be accomplished by the diamagnetic substitutions listed in Tables I and II. The effects of dilution on $|K_1|/M_s$, however, differ between garnets and spinels. If the B sublattice of spinels

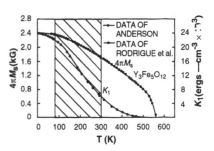

Fig. 8. Variation of $4\pi M_s$ and K_1 of YIG as a function of T. $4\pi M_s$ data of Anderson [21] and K_1 data of Rodrigue et al. [22].

is diluted, $|K_1|/M_s$ should increase only slightly at first, but would grow to a peak value as M_s passes through zero at the compensation point where the two opposing sublattices cancel each other. However, if the octahedral sublattice of the garnet is diluted, $|K_1|/M_s$ will decrease sharply. Dilutions of the corresponding tetrahedral sublattices have their main influence on M_s and therefore also produce opposite results. Substitutions into A sites of spinels, for example, will decrease $|K_1|/M_s$ because of the rise in M_s, but an equivalent change in the d sublattice of garnets will cause $|K_1|/M_s$ to increase.

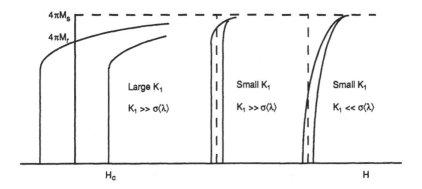

Fig. 9. Hysteresis loop models indicating the influence of K_1 and $\langle \lambda \rangle$.

Table I. Initial effects of selected dilutant ions in garnet ferrites $\{c^{3+}\}_3[a^{3+}]_2(d^{3+})_3O_{12}$.

Ions	Site	M_s	$|K_1|$	$|K_1|/M_s$	$|\langle\lambda\rangle/K_1|$	Footnote
Al^{3+}, Ga^{3+}	d(a)	⇓	—	⇑	—	1
$2Ca^{2+}+V^{5+}$	2c + d	⇓	—	⇑	—	2
$Ca^{2+}+Si^{4+}$	c + d	⇓	—	⇑	—	2
$Ca^{2+}+Ge^{5+}$	c + d	⇓	—	⇑	—	2
In^{3+}, Sc^{3+}	a	⇑	⇓	⇓⇓⇓	⇑	2
$Ca^{2+}+Zr^{4+}$	c + a	⇑	⇓	⇓⇓	⇑	2
$Ca^{2+}+Sn^{4+}$	c + a	⇑	⇓	⇓⇓⇓	⇑	2
Bi^{3+}, Pb^{2+}	c	—	—	—	—	3

1. Al^{3+} and Ga^{3+} occupy both octahedral and tetrahedral sites, but initially have strong preference for the tetrahedral sites in the garnets.
2. Sr^{2+} can be used instead of Ca^{2+} in c sites, as well as La^{3+} instead of Y^{3+} for lattice parameter adjustments.
3. Bi^{3+} and Pb^{2+} are important for magnetooptical applications but can increase microwave losses.

Table II. Initial effects of selected dilutant ions in spinel ferrite $A^{3+}(B^{3+})_1(B^{2+})_1O_4$.

Ions	Site	M_s	$|K_1|$	$|K_1|/M_s$	$|\langle\lambda\rangle/K_1|$	Footnote
Al^{3+}, Ga^{3+}	B(A)	⇓	⇓	—	—	1
Mg^{2+}	B	⇓	⇓	—	—	2
$Li^{1+}+2Ti^{4+}$	A + 2B	⇓	⇓	—	—	3
Zn^{2+}	A	⇑	—	⇓	—	

1. Al^{3+} and Ga^{3+} occupy both octahedral and tetrahedral sites, but opposite to the garnets, they initially have strong preference for the octahedral sites in the inverted spinels.
2. Mg^{2+} is the divalent-ion basis for the magnesium ferrite family.
3. $Li^{1+}+2Ti^{4+}$ combination is used in Li ferrite. The added Li^{1+} replaces Fe^{3+} in the tetrahedral sites, but the net result is a lowering of M_s by the $2Ti^{4+}$ ions diluting the octahedral sublattice.

The magnetostrictive expansion or contraction of the lattice that occurs during the magnetization process is the reaction of the lattice to rotation of the magnetization vector from easy to hard directions. Where magnetostrictive stress effects are large enough to influence anisotropy, and to alter the shape of the hysteresis loop, the important parameter is the ratio of the average magnetostriction constant $\langle\lambda\rangle$ to the anisotropy constant $|\langle\lambda\rangle/K_1|$. Consequently, magnetostriction tends to track with the amount of Fe^{3+} in the octahedral sites.

Hysteresis and Switching Properties

Coercive fields of the hysteresis loops of ceramic ferrites increase monotonically with $|K_1|$ because of the higher domain wall energies and their stronger pinning at pores and grain boundaries. For this reason, larger $|K_1|$ values that occur at lower temperatures can be detrimental where rapid switching of the magnetic state is required. With reference to the hysteresis loop models shown in Fig. 9, the relation between switching energy E_{sw} and coercive field H_c is essentially the area of the hysteresis loop,

$$E_{sw} \cong 2H_c\left(4\pi M_r\right) \qquad , \qquad (12)$$

where $4\pi M_r$ is the remanent magnetization (at $H = 0$). The other variable of importance is the switching time t_{sw} for an applied magnetic field H,

$$t_{sw} \cong \frac{\beta d}{H - H_c} \qquad , \qquad (13)$$

where d is the distance through which a domain wall must traverse, e.g., the thickness of a toroid wall, and $\beta \sim (K_1 + \sigma\langle\lambda\rangle)^{1/2}$ is the domain wall energy under stress of magnitude σ [24]. As a result, the minimization of $|K_1|$ and the stress energy $|\sigma\langle\lambda\rangle|$ would be a major objective in optimizing the switching performance.

At cryogenic temperatures, the coercive field increases dramatically, as shown in the example for YIG in Fig. 10. Where superconductor circuitry is involved in the device design, the necessity of reducing $|K_1|$ is magnified accordingly.

Magnetic Loss

For microwave applications, there are two spin-lattice relaxation mechanisms that contribute to the magnetic loss [25]. In the ferrites commonly employed for control devices, the microwave magnetic energy resides in the spin system of the

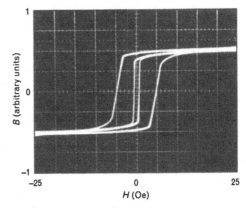

Fig. 10. Magnetic flux density B versus H loops of polycrystalline YIG at 300 K and 77 K, showing the increase in coercive field H_c at low temperatures.

79

collective exchange-coupled Fe^{3+} ions. Because Fe^{3+} has a ground state with zero orbital angular momentum, spin-lattice interaction occurs through the excited state interactions mentioned earlier. As the phonon density decreases at lower temperatures, the microwave energy remains in the spin system for increasingly longer times. The relation for the temperature dependence of the intrinsic linewidth is $\Delta H \sim T^n$, where n is an integer that is determined by whether the interaction between spins and the lattice is through a Raman (n = 7 or 9) or the direct process (n = 1). It is the decrease in the density of phonons at lower temperatures that lengthens the relaxation time and reduces ΔH.

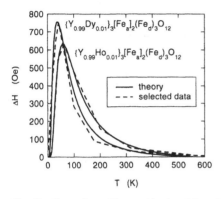

Fig. 11. Comparison of theory with selected data of FMR linewidth as a function of temperature for $(Y_{0.99}Dy_{0.01})Fe_5O_{12}$ and $(Y_{0.99}Ho_{0.01})Fe_5O_{12}$. Data of Seiden [27].

Fig. 12. Theoretical estimates of FMR linewidth versus temperature for three orders of Dy^{3+} impurity levels in YIG.

The second source of magnetic loss comes from fast-relaxing rare-earth ions that are common impurities in the starting chemicals Y_2O_3 used in the preparation of magnetic garnets. Rare-earth ions such as Dy^{3+} or Ho^{3+} have strong spin-orbit coupling that leads to fast spin-lattice relaxation controlled by an Orbach process with an exponential temperature dependence [26]. For the conventional ferrimagnetic resonance fast-relaxation case, Seiden's data and analysis [27] led to the conclusion that the rare-earth contribution to ΔH_i reaches a peak value at a temperature well below 300 K. In Fig. 11, theory and experiment are compared for a typical case, (Y,Dy)IG and (Y,Ho)IG, where it is seen that ΔH reaches a peak value at T ~ 77 K that is more than an order of magnitude greater than the value at 300 K [28]. For most applications, particularly those at lower temperatures, fast-relaxing ions must be removed from the ferrite to obtain the highest resonator Q values. Theoretical estimates of the reduction of ΔH of YIG by purifying Y_2O_3 are shown in Fig. 12 [28].

Dielectric Loss

For all microwave applications, a major concern is power loss through dielectric absorption mechanisms. These effects are reflected by the imaginary part of the complex permittivity $\varepsilon = \varepsilon' - j\varepsilon''$, which is usually expressed as the dielectric loss tangent $\tan\delta_\varepsilon = \varepsilon''/\varepsilon'$. In ferrites where thermally activated electron hopping by the reaction $Fe^{2+} \leftrightarrow Fe^{3+} + e^-$ causes a conductivity σ, the dielectric loss can be expressed as the temperature dependent function

$$\tan\delta_\varepsilon = \frac{\sigma}{2\pi\nu\varepsilon'} = \frac{ne^2D}{2\pi\nu\varepsilon'kT} \exp\left(-\frac{E_{hop}}{kT}\right) \tag{14}$$

where n is the density of carriers, D is the diffusion constant, E_{hop} is the activation energy, and ν is the frequency of the microwave signal. Typically $E_{hop} \sim 0.1 - 0.5$ eV, and the electron hopping activity should decrease by several orders of magnitude as T is lowered from 300 K to 77 K. Suppression of hopping conduction should reduce dielectric loss tangents particularly at lower frequencies, as indicated by Eq. (14).

FERRITE DESIGN ISSUES

Investigations of the low temperature properties of ferrites have been limited to specific generic compounds. The early studies were carried out to determine the physical mechanisms responsible for the behavior of magnetization, magnetocrystalline anisotropy, magnetostriction, and spin-lattice relaxation. Renewed interest in the values of these basic parameters in compositions for practical applications at reduced temperatures has arisen because of the introduction of phase shifters and circulators that utilize the low-resistance capabilities of high-T_C superconductor circuits. If the conduction losses can be virtually eliminated by this new technology, propagation losses originating in the ferrite and the loss from hysteresis loop switching will become increasingly more important design issues for the stringent thermal dissipation budgets of cryogenically refrigerated enclosures.

At frequencies in the standard microwave bands (1 to 12 GHz), compositions of the YIG family can be tailored for the application once the desired parameters are known. As listed in Table I, Al^{3+} or Ga^{3+} substitutions into the d sites (or three-ion $2Ca^{2+}+V^{5+}$ combinations into the c and d sites respectively) would lower the magnetization into the range of interest. Dilutions levels are estimated in Fig. 5. The attendant increase in $|K_1|/M_s$, however, would raise the coercive fields and switching energies. To compensate for the increase in $|K_1|/M_s$, In^{3+} may be substituted into a sites; alternatively, $Ca^{2+}+Zr^{4+}$ combinations into respective c and a sites can be used to dilute the a sublattice moment. Because this action would raise M_s, thereby offsetting the effect of Al^{3+}, iterative adjustments between a and d sublattice dilution could be necessary before the desired parameters are obtained. Where a narrow ΔH is critically important, such as in high-Q filter applications, Ca^{2+} in place of Y^{3+} may also be preferred in YIG-based ferrites because of traces of fast-relaxing rare-earth impurities that occur in the Y_2O_3, as mentioned earlier.

At millimeter-wave frequencies of 35 GHz and above, spinel ferrites offer the advantage of substantially higher magnetization, particularly with Zn^{2+} substitutions that are indictated by Fig. 7 and listed in Table II. The corresponding increase in $|K_1|/M_s$, however, may be more difficult to offset by chemical substitution because dilution of the B sublattice would be required.

To establish a catalog of microwave ferrites for cryogenic applications, renewed programs of fundamental studies of low-temperature properties should be undertaken. During the course of these investigations, further insights are likely to be gained into the physics of fundamental processes such as domain wall energies and movement, and stress sensitivity of hysteresis loops. This work should be focused on establishing the chemical designs for the range of properties necessary to solve the complex tradeoff problems involved in device performance optimization.

ACKNOWLEDGMENT

The author is grateful Dr. D.E. Oates for the opportunity to collaborate on the device aspects of this program, and to G.L. Fitch for computer programming of the molecular field theoretical model. This work was sponsored by Defense Advanced Research Projects Agency.

REFERENCES

1. G.F. Dionne, D.E. Oates, D.H. Temme, and J.A. Weiss, *IEEE Trans. Microwave Theory Tech.* **44**, 1361 (1996).
2. D.E. Oates and G.F. Dionne, *1997 IEEE MTT-S Digest*, p. 303.
3. B. Lax and K.J. Button, *Microwave Ferrites and Ferrimagnetics* (McGraw-Hill Book Co, New York, 1962), Chapter 4.
4. G.F. Dionne and D.E. Oates, *IEEE Trans. Magn.* **33**, 3421 (1997).
5. R.A. Pucell and D.J. Masse, *IEEE Trans. Microwave Theory Tech.* **20**, 304 (1972).
6. G.F. Dionne and D.E. Oates, *J. Appl. Phys.* **85**, 4856 (1999).
7. G.F. Dionne, *J. Phys. IV France* 7, C1-437 (1997).
8. G.F. Dionne, *J. Appl. Phys.* **81**, 5064 (1997).
9. L. Néel, *Ann. Phys. (Paris)* **3**, 137 (1948).
10. G.F. Dionne, *J. Appl. Phys.* **41**, 4874 (1970).
11. G.F. Dionne, *J. Appl. Phys.* **45**, 3621 (1974).
12. G.F. Dionne and P.F. Tumelty, *J. Appl. Phys.* **50**, 8257 (1979).
13. G.F. Dionne, *J. Appl. Phys.* **63**, 3777 (1988).
14. M.A. Gilleo, *J. Phys. Chem. Solids* **13**, 33 (1960).
15. C. Borghese, *J. Phys. Chem. Solids* **28**, 2225 (1967).
16. I. Nowik, *J. Appl. Phys.* **40**, 5184 (1969).
17. A. Rosencwaig, *Can. J. Phys.* **48**, 2868 (1970).
18. J.H. Van Vleck and W.G. Penney, *Philos. Mag.* **17**, 961 (1934).
19. K. Yosida and M. Tachiki, *Prog. Theor. Phys.* **17**, 331 (1957).
20. G.F. Dionne, *J. Appl. Phys.* **40**, 1839 (1969).
21. E.E. Anderson, *Phys. Rev.* **134**, A1581 (1964).
22. G.P. Rodrigue, H. Meyer, and R.V. Jones, *J. Appl. Phys.* **31**, 376S (1960).
23. D.J. Craik and R.S. Tebble, *Ferromagnetism and Ferromagnetic Domains*, (John Wiley, New York, 1965), Chapter 2.
24. N. Menyuk and J.B. Goodenough, *J. Appl. Phys.* **26**, 8 (1955).
25. K.J. Standley and R.A. Vaughn, *Electron Spin Relaxation Phenomena in Solids*, (Plenum Press, New York, 1969), p. 5.
26. R. Orbach, *Proc. R. Soc.* **A264**, 458 (1961).
27. P.E. Seiden, *Phys. Rev.* **133**, A728 (1964).
28. G.F. Dionne and G.L. Fitch, paper AT-01 at 1999 Magnetism & Magn. Matls. Conf., *J. Appl. Phys.* **86** (2000), in press.

MATERIALS ISSUES IN THE APPLICATION OF FERROMAGNETIC FILMS IN TUNABLE HIGH Q VHF/UHF DEVICES

R.M. WALSER, A. P. VALANJU, and W. WIN
*ECE Department, University of Texas, Austin, TX 78712, walser@mail.utexas.edu

ABSTRACT

Thin ferromagnetic films have narrow ferromagnetic resonance bandwidths, and large RF permeabilities that are attractive for tunable high Q device applications. They can also be processed at low temperatures and integrated with most RF and microwave materials. At VHF/UHF frequencies, where discrete devices can be realized, magnetic loss tangents (Q^{-1}) less than 10^{-3} can be obtained, if the effects of spin resonance and eddy current losses are reduced. The material figure of merit requires a tradeoff of the permeability, and Q. The highest FOM is obtained by maximizing the magnetization, anisotropy, resistivity, and minimizing the magnetic loss. This paper discusses the interplay of these properties, and the fill factor and magnetic bias required for tunable inductors with Q>1000 in the ~50-500 MHz range. The most promising candidates for these applications are shown to be reactively sputtered nanocrystalline films. The large resistivity, magnetization, and thickness of these films can be used to attain large film permeances, and increased shape anisotropy fields from patterned film objects. The latter eliminates the need for using a large bias magnet to achieve Q's ~1000 at VHF/UHF operating frequencies.

INTRODUCTION

Many current applications in telecommunications and radar require tunable circuits with Q >1000 that are difficult to achieve using existing components and materials. At higher microwave frequencies (≥ 1 GHz), this Q cannot be achieved without using superconducting transmission lines to reduce the conductor losses of distributed devices. In the VHF/UHF range, the same materials can be used in discrete, resonant LC circuits to increase the ratio of the component reactance to the conductor resistance beyond that realizable in a distributed circuit. A high circuit Q can then be achieved without the use of superconductors, if the Q of the L, C, and tunable material property (μ or ε), are all large enough.

Due to their large magnetization, small crystal anisotropy, and two-dimensional form, thin ferromagnetic films have a robust, planar, uniaxial, single-domain state that can be induced at low deposition temperatures. The processing of this state is compatible with other planar processed materials and devices. Metallic ferromagnetic films have a large magnetic exchange, itinerant magnetic moments, and low crystal anisotropy that reduces the spin resonance damping in polycrystalline films well below that observed in polycrystalline ferrites. These properties suggest that ferromagnetic films could be of value in the design of tunable, high Q devices, if they can be made thin enough to reduce the effect of eddy currents, yet thick enough to provide adequate permeances in single, or laminated films.

The following sections develop a model for the Q, tunability, permeance, and bias and tuning fields of single-domain magnetic films. The results of the modeling are used to guide an assessment of the properties of optimum films, and to select promising candidate ferromagnetic films for tunable high Q device applications. In a later section, the results are compared with measurements aimed at verifying many of the important model predictions.

83

VHF/UHF PROPERTIES OF FERROMAGNETIC FILMS

Ferromagnetic spin resonance.

The magnetization of a ferromagnetic film with a large saturation value ($4\pi M_s \sim 10^4$ Gauss) has a nearly planar single-domain configuration resulting from a large perpendicular demagnetization and small crystalline anisotropy. The small-signal permeability of the film, derived from the equation of motion of the vector magnetization **M** in response to an applied microwave field **h**, is given by

$$\mu_{sr} = \mu_{sr}' - j\mu_{sr}'' \approx 1 + \frac{\gamma^2(4\pi M_s)(H_O + 4\pi M_s)}{f_{res}^2 - f^2 + j\alpha\gamma f \cdot (4\pi M_s)} \qquad (1)$$

where μ_{sr} is the spin resonance contribution to the relative permeability, measured when **h** is along the in-plane, hard magnetic axis perpendicular to **M**. The ferromagnetic resonance occurs at a frequency ,

$$f_{res}^2 \approx \gamma^2(H_O + H_k + H_{dh})(H_O + 4\pi M_s) = \gamma^2 H_T(H_O + 4\pi M_s) \qquad (2)$$

where $H_T = H_o + H_k + H_{dh}$, H_k and H_o are the internal uniaxial and applied fields parallel to **M**, f and f_{res} are the signal and ferromagnetic resonance frequencies in MHz, the gyromagnetic constant $\gamma = 2.8$ MHz/Oe, and α is the dimensionless Landau-Lifshitz damping constant. All fields are in Oersteds. H_k is induced in the film during deposition [1], and H_{dh} is the demagnetization field along the hard magnetic axis given by

$$H_{dh} = D_h(4\pi M_s) \qquad (3)$$

where D_h is the shape demagnetizing factor along the hard axis of the film object. Equations (1) and (2) assume that the demagnetization perpendicular to the film is $\approx 4\pi M_s$.

To illustrate, the natural ferromagnetic resonance ($H_o=0$) of an infinite thin film ($D_h=0$) is computed from (1), and shown in Figure (1a) for a 0.1 μm thick, amorphous film CoZrNb (CZN) alloy film. The magnetic constants used in the calculation were derived from the measured in-plane hysteresis loops shown in Figure (1b), and by adjusting $\alpha=0.012$ to produce the best match between theory and experiment.

Figure (1a) shows that for a spin resonance response, a high magnetic Q is obtained at frequencies well below the FMR frequency. Assuming that $f \ll f_{res}$ and $\mu_{sr} \gg 1$, (1) and (2) yield:

$$\mu_{sr}' \approx \frac{4\pi M_s}{H_T} \qquad (4)$$

$$\mu_{sr}'' \approx \frac{\alpha \cdot f(4\pi M_s)}{\gamma \cdot H_T^2} \qquad (5)$$

$$Q_{sr} = \frac{\mu_{sr}'}{\mu_{sr}''} = \frac{\gamma \cdot H_T}{\alpha \cdot f} \qquad (6)$$

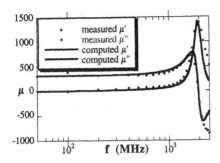

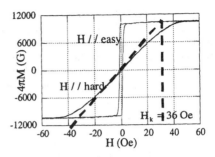

Figure (1a). Natural FMR of 0.1 μm thick CZN film. An α=0.012 produced the best agreement between theory and measurement.

Figure (1b). Easy and hard axis hysteresis loops of CZN film with anisotropy determined by extrapolating the hard axis loop from H≈0.

Variations in the μ_{sr} and Q_{sr} of a CZN film with H_T computed using (4) and (6) are shown in Figure (2) below. This data shows that at frequencies of 100 MHz and 500 MHz, a Q_{sr} > 1000 can be obtained for a CZN film with α=0.01, for $H_T \approx 300$ Oe and 1800 Oe respectively.

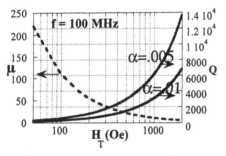

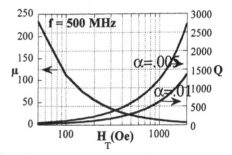

Figure (2a). Variation of μ_{sr} and Q_{sr} of CZN film with H_T as computed from (4) and (6) at 100 MHz.

Figure (2b). Variation of μ_{sr} and Q_{sr} of CZN film with H_T as computed from (4) and (6) at 500 MHz.

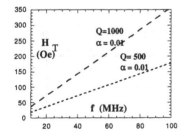

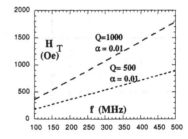

Figure (3a). The total field H_T needed for Q=1000 (500) for CZN films in VHF range.

Figure (3b). The total field H_T needed for Q=1000 (500) for CZN films in UHF range.

Equation (4) shows that with increasing H_T, Q_{sr} increases, while (5) shows that μ_{sr} decreases. Equation (6) can be used to calculate, for a given material, the total (internal +external) field H_T required to increase Q to a desired level at a particular operating frequency. The results [Figures (3a) and (3b)] for the CZN film with the natural FMR of Figure (1a), show that a magnetic Q >1000 can be realized throughout the VHF range with bias fields ranging from ~50 to 350 Oe, and throughout the UHF range with bias fields ranging from ~400 to 1800 Oe.

Eddy Current Loss

The maximum thickness of a ferromagnetic film that can be used in a device is limited by eddy currents. To include eddy currents, the μ of a ferromagnetic film with thickness d is written [2],

$$\mu = \mu' - j\mu'' = \mu_{sr} \left[\frac{\sinh k + \sin k}{k(\cosh k + \cos k)} - j \frac{\sinh k - \sin k}{k(\cosh k + \cos k)} \right] \qquad (7)$$

where the skin depth δ is given by $\delta = (\rho/\pi f \mu_{sr} \mu_0)^{1/2}$. In practical units (d in μm, f in MHz, and ρ in $\mu\Omega\cdot$cm), $k = 2\pi d(10^3 f \mu_{sr}/\rho)^{1/2}$. Expanding the terms in (7) as power series in k, and retaining the lowest order terms, the first order approximate μ of a high Q film is,

$$\mu' \approx \mu_{sr}' \quad \text{and} \quad \mu'' \approx \mu_{sr}'' + \frac{(2\pi d)^2 10^3 f}{6\rho} (\mu_{sr}')^2 \qquad (8)$$

which gives:

$$\frac{1}{Q} = \frac{\mu''}{\mu'} = \frac{1}{Q_{sr}} + \frac{1}{Q_{eddy}} \qquad (9)$$

where

$$Q_{sr} = \frac{\mu_{sr}'}{\mu_{sr}''} \quad \text{and} \quad Q_{eddy} = \frac{6\rho}{(2\pi d)^2 10^3 f \ \mu_{sr}'} \qquad . \qquad (10)$$

Thus, the total film Q will be less than either Q_{sr} and Q_{eddy}.

The variation in Q with film thickness was computed from (7) for the CZN film and plotted in Figures (4a) and (4b). The value of Q_{sr} is found by extrapolating the curves to zero thickness. The results show that a film Q > 1000 can be obtained in these amorphous films for thicknesses of ~1 μm.

Magnetic Bias

The fields required for high Q are 10-100 times larger than the typical values of H_k reported for ferromagnetic films. To eliminate the necessity of using a large permanent magnet to provide these fields, the film is patterned into objects small enough to provide the bias through an internal shape demagnetizing field. For a film object with d<<h<<e (d = thickness, h = hard axis length, and e = easy axis length),

$$H_{dh} \approx \frac{d}{h} 4\pi M_s \qquad (11)$$

so that a field of $H_{dh} \approx 500$ Oe could be obtained in a 1-μm thick CZN film with h=20 μm. The results of film patterning experiments are discussed later.

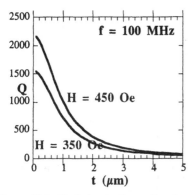

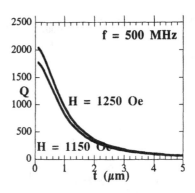

Figure (4a). Variation in Q as a function of the thickness of a CZN film at 100 MHz.

Figure (4b). Variation in Q as a function of the thickness of a CZN film at 500 MHz..

Tunability

For $f \ll f_{res}$, where a high Q can be obtained, the figure of merit (FOM) of a magnetic film tuned by a field variation $\pm\Delta H/2$ about H_{dh}, is defined as the product of $<Q>$, the average Q over the bias range, and the fractional change in μ' ($\Delta\mu'/\mu'$) as follows:

$$FOM = \left| \frac{\Delta\mu'}{\mu'} \cdot \langle Q \rangle \right|_{H_T} \tag{12}$$

Equations (4) and (5) can be used to calculate the figure of merit (FOM) of tunable devices fabricated with ferromagnetic thin films. For a small tuning field ΔH, (4) and (8) yield

$$\frac{\Delta\mu'}{\mu'} = \frac{\Delta H}{H_{dh}} \qquad H_0 = \pm\frac{\Delta H}{2}, \qquad H_T \approx H_0 + H_{dh} \tag{13}$$

where ΔH is a quasi-static tuning field, and we have assumed that the static bias field is derived from the internal field H_{dh}. Equation (12) shows that for a $<Q>=1000$, a FOM $=200$ can be achieved if μ' can be tuned by 20%.

The variation in μ was calculated from (4) for a CZN film and plotted in Figures (5a) and (5b). At 100 MHz, a tunability of 20% was obtained for $H_{dh}=300$ and $\Delta H\approx60$, in agreement with (13). A similar close agreement was obtained at 500 MHz, where a tunability of 5% was obtained for $H_{dh}=1.5$ kOe and $\Delta H\approx80$ Oe.

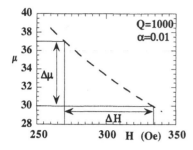

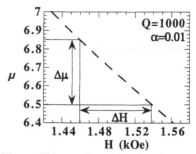

Figure (5a). A Q=1000 and $\Delta\mu/\mu\sim$20% is calculated for a CZN film biased to H_T =300 ± 30 Oe at f=100 MHz.

Figure (5b). A Q=1000 and $\Delta\mu/\mu\sim$5% is calculated for a CZN film biased to H_T =1500 ± 40 Oe at f=500 MHz.

DEVICE FILL FACTOR

The magnetic fill factor of an inductive device is maximized with the film permeance Λ (Λ =μd). If the desired device tunability can not be achieved with a single film with maximum Λ, film laminates must be used and the maximum fill factor obtained by maximizing both Λ, and the volume fraction V_f of laminate in the core.

Maximum Film Permeance

A maximum Λ corresponds to a maximum μd product. By combining (4), (8) and (11), it can be shown that

$$\Lambda_{max} = h_{max} \qquad (14)$$

i.e. the maximum permeance corresponds to the maximum hard-axis dimension used to obtain the film demagnetizing field. The maximum value of h can be determined as follows. For a given film, the maximum film thickness d_{max} is obtained by requiring

$$Q_{eddy} = \frac{6\rho}{(2\pi d)^2 10^3 f \cdot \mu_{sr}'} = \frac{3\rho}{(2\pi^2 d)10^3 f \cdot h} \geq Q \qquad (15)$$

and the minimum film thickness d_{min} is obtained by requiring

$$Q_{sr} = \frac{\gamma \cdot H_{dh}}{\alpha \cdot f} = \frac{\gamma \cdot d \cdot 4\pi M_s}{\alpha \cdot f \cdot h} \geq Q. \qquad (16)$$

Since $d_{max} \sim 1/f$, and $d_{min} \sim f$, there is a frequency f_{max} where they are equal, and above which the desired Q cannot be obtained for any film thickness. Equating (15) and (16) yields

$$f_{max} = \frac{1}{\pi h Q}\left(\frac{3\rho\gamma\left(4\pi M_s\right)10^{-3}}{2\alpha}\right)^{1/2}. \qquad (17)$$

From (14), Λ_{max} is obtained by maximizing h, which from (17), is obtained at the smallest value of f_{max}, corresponding to setting the signal frequency $f=f_{max}$. Figure (6) graphically summarizes the results of the optimization process.

For a CZN film with Q=1000 at 100 MHz, a value of h_{max}=23 μm is found from (17). The corresponding film thickness, found by substituting h_{max} into either (15) or (16), is d≈1.0μm. The corresponding bias field obtained using (11) is H_{dh}=(d/h)(4πM$_s$)=435 Oe, which agrees with the bias field independently determined from the results of the general calculation from (7) shown in Figure (4a).

The results of this analysis can be used to determine the optimum properties of films that maximize the permeance, and hence the fill factor. Solving for h_{max} from (17), it follows that

$$\Lambda_{max} = h_{max} = \frac{1}{\pi f Q}\left(\frac{3\rho\gamma(4\pi M_s)10^{-3}}{2\alpha}\right)^{1/2} \qquad (18)$$

which, for a given f and Q, is maximized by maximizing ρ and 4πM$_s$, and minimizing α.

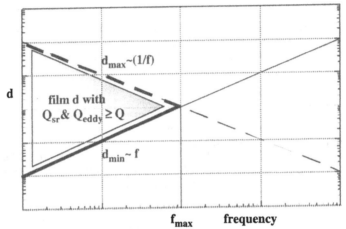

Figure (6). The range of film thickness for which Q_{eddy} & Q_{sr} > Q is bounded by d<d_{max}, d>d_{min}, and f<f_{max}. The maximum permeance is obtained when the signal frequency f =f_{max}.

Table I lists the properties of some candidate ferromagnetic films and, for comparison, the properties of some commonly used ferrites. The chief advantage of the ferrites is their large resistivity. Their main shortcoming is the difficulty of processing high quality films at low temperatures with a single domain state. The latter is used to engineer a wide range of internal bias fields in ferromagnetic films. The chief advantages of thin ferromagnetic films have been discussed previously. The table shows that, in comparison with CZN, reactively sputtered nanocrystalline films have larger values of 4πM$_s$ and ρ, and are attractive candidates for tunable devices.

Maximizing the product μρ in ferromagnetic films for high frequency sensor applications is the goal of considerable current research [2-5]. In the previous research, nanocrystalline films with large values of μ (>100) and ρ (>1300 μΩ·cm) were obtained by sputtering alloy targets of Fe and/or Co with Cr, Hf, or Ta in reactive oxygen/nitrogen gas mixtures. An initial study of the optimization of the properties of reactively sputtered CoCrTa:O/N alloy films for tunable device applications is reported in a companion paper [6].

T₿ble I. Comparison of properties of ferromagnetic films and ferrites.

Composition	Micro-structure	$4\pi Ms$ (kG)	α	ρ ($\mu\Omega\cdot cm$)	d_{max} (μm)	Process
YIG	polycrystal	2	<0.01	>10^{14}	large	(a.b.c)
NiZn	polycrystal	<5.2	~0.05	>10^{14}	large	(a,b,c)
Co_2Z	polycrystal	3	~0.05	>10^{14}	large	(a,c)
$Co_{50}Fe_{50}$	polycrystal	24	-	10	>10 μm	(c,d)
$Co_{80}Zr_{10}Nb_{10}$	amorphous	11	0.015	150	~1 μm	(e)
$Fe_{87.2}Cr_{4.6}Ta_{0.26}N_{7.4}$	nanocrystal	19	0.02	120	>3μm	(f)
$Co_{44}Fe_{19}Hf_{15}O_{22}$	nanocrystal	18	-	1300	-	(f)

(a) high T sintering (b) laser ablation (c) chemical (d) electroplating (e) RF sputtering (f) reactive sputtering

Volume Fraction

The device permeance is increased by laminating films with a thickness corresponding to the maximum permeance consistent with the desired Q. The maximum film thickness d_{max} is found by substituting h_{max} into either (15) or (16). The maximum magnetic volume fraction V_f

$$V_f = A_f \cdot T_f \tag{19}$$

is found by maximizing the area fraction A_f of the patterned film objects, and the thickness fraction $T_f = d_f/(d_f + d_s)$ where d_f and d_s are the thickness of film and substrate respectively.

Generally, $T_f \ll A_f$, and with d_f bounded by d_{max}, V_f is maximized by reducing the substrate thickness. For example, a $V_f \approx 14\%$ was achieved by laminating 1.9 μm thick amorphous CoFe alloy films deposited on 12 μm thick mylar substrates. The μ spectrum of a 10 layer composite is shown in Figure (7) for two values of magnetic bias.

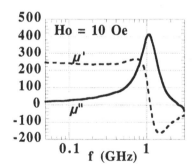

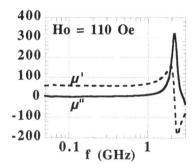

Figure (7a) μ spectrum of 10-layer composite of 1.9-μm CoFe amorphous film on a 12 μm mylar substrate. (Bias H=10 Oe)

Figure (7b) μ spectrum of 10-layer composite of 1.9-μm CoFe amorphous film on a 12 μm mylar substrate. (Bias H=110 Oe)

EXPERIMENTAL RESULTS

The measurement of a magnetic Q ~1000 requires the use of a resonant technique. Since cavity resonators are too large for use in the VHF range, discrete resonant LC circuits were used. The thin film material under test was inserted into the core of a small solenoid which was resonated by tuning it with a high Q capacitor. The inductive impedance was found by curve fitting the impedance response of the circuit modeled as a series RLC circuit. Accurate measurement of the resistive part of the circuit impedance was carried out at circuit resonance i.e. when the reactive impedance is zero.

Measurements at each field were made relative to those made with the test sample saturated at a high field where $\mu_{eff}'=1$ and $\mu_{eff}''=0$. The impedance at the resonance frequency f_{sat} of the saturated sample yields the circuit resistance R_{sat}. For smaller bias fields, where $\mu_{eff}'>1$ and $\mu_{eff}''>0$, an impedance $R=R_{sat}+R(H)=R_{sat}+\omega L_{sat}\mu_{eff}''$ is measured at the circuit resonance frequency f ($\sim f_{sat}$). For both measurements, the inductances L_{sat}, and $L=L_{sat}+L(H)=L_{sat}+\mu_{eff}'L_{sat}$ are found by curve fitting the impedance responses. The magnetic Q of the material is then obtained as:

$$Q = \frac{\mu'}{\mu''} = \frac{\mu'_{eff}}{\mu''_{eff}} = \frac{L(H)/L_{sat}}{R(H)/\omega L_{sat}} = \frac{\omega L(H)}{R(H)} \qquad (20)$$

The Q and tunable inductance L(H) of a 10-layer CZN laminate measured at f=25 MHz and 80 MHz, are shown in Figures (8a) and (8b). Since L(H) ~ μ_{eff}, the tunability $\Delta\mu_{eff}/\mu_{eff}$ = $\Delta L(H)/L(H)$ over a field tuning range ΔH. Figures (9) shows that the measured data were in good agreement with (13). These results show that a magnetic Q in excess of 1000, and a 30% tunability for moderate values of ΔH can be obtained at VHF frequencies with CZN films with an easy axis bias H_o.

A series of patterned CZN and CoFe films were fabricated to test the biasing of these samples using the hard-axis demagnetization field H_{dh}. The patterned CZN films had easy axes of e=100, 300, 750 and 1000 μm. The hard axes were fixed at h=22 μm. An anisotropy field of $H_{dh}\approx400$ Oe was determined from the hysteresis loops [Figures (10a) and (10b)] of the CZN sample with e=300 μm. The H_{dh} of all four samples were identical to within 4% as would be

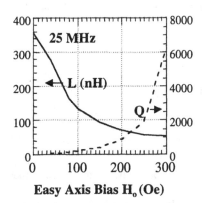

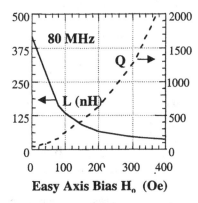

Easy Axis Bias H_o (Oe)   Easy Axis Bias H_o (Oe)

Figure (8a). Tunable inductance (solid) and Q (dotted) of 10 layer CZN laminate at 25 MHz.

Figure (8b). Tunable inductance (solid) and Q (dotted) of 10 layer CZN laminate at 80 MHz.

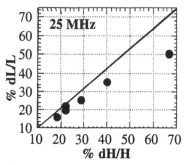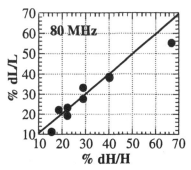

Figure (9). Correlation of % tunability (100xΔL/L) and % change in tuning field (100xΔH/H) for a 10 layer CZN laminate at 25 MHz (left) and 80 MHz (right). The measured points are in close proximity to the straight line with slope=1, indicating good agreement with theory.

expected when e>>h. The measured value of H_{dh} agreed with the theoretical approximation H_{dh} $\approx$(d/h)(4πMs)=(0.9 μm/22 μm)(10,480 G)=428 Oe to within the ~10% error in the approximation. Similarly, the calculated easy axis demagnetization field $H_{de}\approx$(d/e)(4πMs) agreed with measurement [Figure(10a)] to within the experimental error.

The agreement of the measured and computed anisotropy implies that the small signal response of the patterned films is equivalent to that of a single planar domain. This would require that the magnetized state of each film object be a single domain oriented along its easy axis. On the other hand, the easy axis loops of the patterned film has zero remanent magnetization [Figure (10a)]. A multi-domain state consisting of alternating stripe domains is compatible with both observations [7] if the easy axis demagnetization of the domains, and the coupling of spins through the Bloch domain walls are assumed to be negligible. Under these conditions, the small signal spin resonance of the multi-domain and single-domain states are identical [8].

Similar results were obtained with the patterned, 1.55 μm thick electroplated CoFe film shown in Figure (11). This film had a 4πM$_s$=15.6 kGauss and an expected $H_{dh}\approx$1.1 kOe which agreed well with the ~1 kOe measured value. The anisotropy of the patterned film was increased by ~45 times and the FMR frequency (11.6 GHz) was calculated to increase by a factor ~7.

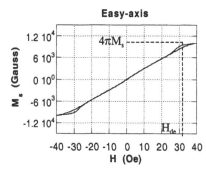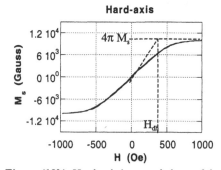

Figure (10a). Easy axis hysteresis loop of the patterned CZN film. $H_{de}\approx$32 Oe.

Figure (10b). Hard axis hysteresis loop of the patterned CZN film. $H_{dh}\approx$400 Oe.

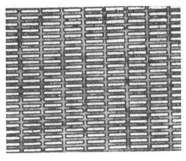

Figure (11). Microphotograph of electroplated CoFe film patterned with e=100μm x h=22μm film objects. A hard axis demagnetizing field of ~1000 Oe was measured for the 1.55 μm thick film.

bars = 22 μm x 100 μm

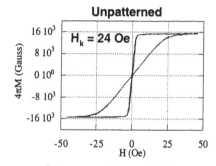

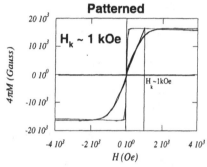

Figure (12). Easy and hard axis hysteresis loops of an unpatterned (left) and patterned (right) 1.55 μm thick CoFe film. The measured hard axis anisotropy of 1 kOe of the patterned film agreed well with the computed value. The patterned objects are shown in Fig. 11.

CONCLUSIONS

At frequencies well below FMR, the Q of a ferromagnetic film can be written as $1/Q = 1/Q_{sr} + 1/Q_{eddy}$, where Q_{sr} and Q_{eddy} are the spin-resonance and eddy current Q. To obtain a high Q, the resonance and eddy current losses must be reduced so that both Q_{sr} and Q_{eddy} are larger than the desired film Q. Q_{sr} is increased by shifting the FMR frequency far above the signal frequency, and Q_{eddy} is increased by reducing the film thickness well below the skin depth at the signal frequency. The bias fields required to reduce the resonance loss are 10-100 times the intrinsic anisotropy field and must be derived from the shape anisotropy of patterned film objects. The use of shape demagnetization establishes a lower bound on the film thickness, while an upper bound is established by the requirement for a low eddy current loss. The permeance of the film object is maximized by setting the signal frequency equal to the maximum frequency for which the desired Q can be attained. At this frequency, the maximum and minimum thicknesses are equal, and the maximum film permeance is identical to the maximum hard-axis dimension of the film object permitted by these constraints.

An assessment of films, based on properties that maximize the permeance, indicates that nanocrystalline films obtained by reactively sputtering CoFe/Ta,Hf,or Cr alloys in O/N gas mixtures, are the most attractive candidates for high Q applications. Model calculations, made using measured magnetic properties, show that ferromagnetic films with Q>1000, and a tunability ranging from 30% to 5%, can be obtained at VHF/UHF (~50 MHz to 500 MHz) frequencies using bias fields ranging from ~100 to 1500 Oe. Experimentally, it was shown that the required shape anisotropies can be derived from microlithographically patterned film objects.

Discrete resonant circuit measurements of the Q and tunability were made on CoZrNb alloy films and laminates and found to agree with the model calculations. This research has shown that ferromagnetic films have great promise for applications in high-Q tunable devices operating at VHF/UHF, and possibly higher, frequencies.

ACKNOWLEDGMENT

The authors gratefully acknowledge the support received from the DARPA-Frequency Agile Materials for Electronics (FAME) Program under the direction of Dr. Stuart Wolf.

REFERENCES

[1] R. F. Soohoo, *Magnetic Thin Films*, Harper and Roe (1965), Section 7.2.

[2] R. M. Bozorth, *Ferromagnetism*, D. Von Nostrand Co., 4th Printing, (1978) pp. 772-773.

[3] A. Makino and Y. Hayakawa, Matls. Sci. Eng. A, **A181/A182**, 1020, (1994); S. Ohnuma, H. Fujimori, S. Mitani, and T. Matsumoto, J. Appl. Phys. **79**, 5130, (1996).

[4] E. van de Reit and R. Roozeboom, J. Appl. Phys. **81**, 350, (1997).

[5] S. Jin, W. Zhu, R.B. van Dover, T.H. Tiefel, V. Korenivski, and L.H. Chen, Appl. Phys. Lett., **70**, 3162 (1997).

[6] W. Win, et al., "Nanocrystalline thin films for tunable high QVHF/UHF devices", these proceedings.

[7] J. Smit and H. P. J. Wijn, *Ferrites*, John Wiley and Sons (1959), pp. 82-84.

[8] D.Polder and J. Smit, Rev. Mod. Phys. **25**, 89 (1953).

IN-PLANE UNIAXIAL MAGNETIC ANISOTROPY OF
COBALT DOPED Y$_3$Fe$_5$O$_{12}$ EPITAXIAL FILMS

DARREN DALE, G. HU, VINCENT BALBARIN, Y. SUZUKI*
Department of Materials Science and Engineering, Cornell University, Ithaca, NY 14853
*suzuki@ccmr.cornell.edu

ABSTRACT

In an effort to develop a magnetic biasing layer for potential applications in integrated devices, we have grown thin films of Y$_3$Fe$_5$O$_{12}$ with increased uniaxial anisotropy by doping with varying Co^{2+} concentration. To compensate for the charge differential between Co^{2+} and Fe^{3+}, Ge^{4+} and Ce^{4+} are substituted for Fe^{3+} and Y^{3+}, respectively. These garnet films, prepared using pulsed laser deposition on (110) oriented Gd$_3$Ga$_5$O$_{12}$ substrates, exhibit excellent crystallinity as determined from X-ray diffraction and Rutherford backscattering spectroscopy. The addition of Co^{2+} in Y$_3$Fe$_5$O$_{12}$ films enhances the in-plane uniaxial anisotropy over an order of magnitude, depending on composition.

INTRODUCTION

In recent years, there has been interest in growing magnetic oxide thin films for use in high frequency inductance devices. Yttrium iron garnet, Y$_3$Fe$_5$O$_{12}$ (YIG), has been studied for its interesting magneto-optical properties and its potential use in optoelectronics applications. What is interesting for our purposes, is that YIG has been shown to increase the inductance in a device significantly [1]. The device performance at high frequencies, however, is severely hindered by unwanted power dissipation due to domain wall motion in YIG. An alternative device design is comprised of exchange-coupled magnetic film bilayers [2]. One layer is a magnetically soft layer with high permeability. The second layer is a biasing layer with an in-plane uniaxial anisotropy. Previous investigations have explored exchange coupling in garnet and spinel thin film structures. Strong exchange coupling has been observed for isostructural couples with comparable lattice parameters, but coupling is not observed for heterostructural couples, regardless of similarities in lattice parameter [3]. In order to investigate this two layer device design, we have recently developed a magnetically hard doped YIG thin film material to use as an exchange-bias layer for magnetically soft YIG films.

It has been found in bulk that Co^{3+} is not as effective as Co^{2+} in enhancing the uniaxial magnetic anisotropy in tetrahedral, octahedral or dodecahedral sites (4), and this is found to be the case in our thin films as well. Previous work in bulk has shown that the incorporation of Co^{2+} on the preferred octahedral site of the garnet structure will increase magnetic anisotropy [4,5]. Previous bulk work has been in the low Ge^{4+}/Co^{2+} limit (up to 0.8%) where single ion theory of magnetic anisotropy is valid [5,6].

We present here our work on the fabrication and characterization of Co^{2+} doped YIG films with uniaxial anisotropy in the plane of the film. In order to dope the octahedral site over a significant range of Co^{2+} concentrations, we have charge compensated Co^{2+} ions with either Ce^{4+} or Ge^{4+} ions, extending the limits of our solid solution. Previous work with polycrystalline YIG films prepared by MOCVD, with Ce^{4+} and Ge^{4+} charge compensated Co^{2+}, have shown that with increasing dopant levels the Curie temperature (T$_c$) and saturation magnetization (M$_s$) decrease, while the coercive field (H_c) increases [7,8]. Our work is focused on enhancing the in-plane magnetic anisotropy of YIG while not reducing the M_s significantly, in light of the exchange biasing role that the material should perform.

EXPERIMENT

Synthesis

We have chosen as our substrate (110) oriented $Gd_3Ga_5O_{12}$ (GGG) for several reasons. The substrate and film both have the garnet structure. The lattice parameter of GGG, 12.376Å is comparable to the 12.380Å lattice parameter of undoped YIG. Most work with YIG on GGG for magneto-optical storage applications have been with (111) oriented substrates, but this orientation counters our objective, as growth induced anisotropy results in an easy axis along the film normal. The (110) film orientation is ideal for our study, as we can study the magnetization along the (100), (110), and (111) directions, all in the plane of the film.

The films are deposited using the pulsed-laser deposition technique. Targets are prepared from stoichiometric amounts of Y_2O_3, Fe_2O_3, $Co(NO_3)_2*6H_2O$, GeO_2, and CeO_2. The mixed powders are heated in air for 12 hours at 600°C, ground and pressed into cylindrical pellets 1" in diameter. The pellets are placed on powders of their own composition and fired at 1050°C in air for 60 hours. Single phase polycrystalline targets are successfully prepared for $Y_3Fe_{5-2x}Co_xGe_xO_{12}$ (x=0, 0.25, 0.5) and $Y_{3-x}Ce_xFe_{5-x}Co_xO_{12}$ (x=0.25, 0.5, 0.75).

The films are deposited at 600°C in 10mTorr O_2. We use a KrF excimer laser with λ=248nm. The laser energy is 150mJ and the frequency is 10Hz. Under these conditions we have a deposition rate of approximately 30μ/min. The samples are cooled in 10^{-6}mTorr O_2 for 2 hours. Film thickness ranges from 750 to 2000Å, well below the approximate 1μm estimate where plastic deformation begins to occur [9]. Three film compositions are observed to be single phase ferrimagnetic as grown, they are $Y_{2.75}Fe_{4.5}Co_{0.25}Ge_{0.25}O_{12}$, $Y_{2.5}Fe_4Co_{0.5}Ge_{0.5}O_{12}$ and $Y_{2.75}Ce_{0.25}Fe_{4.75}Co_{0.25}O_{12}$. All film compositions, except $Y_3Fe_{3.5}Co_{0.75}Ge_{0.75}O_{12}$, are ferrimagnetic at room temperature after annealing in air for 2 hours.

Characterization

High resolution X-ray diffraction indicates that film reflections are found to overlap the substrate. Within the resolution of the diffractometer, the location of the actual film peak is indistinguishable from the substrate peak for thin films of all dopant concentrations. In-plane film orientation is determined by phi-scans of the (400) reflections. The results of the phi-scans do not indicate any evidence of in-plane misorientation of the film lattice and substrate lattice (figure 1).

Since we were unable to resolve a film peak using X-ray characterization, we have quantified the crystallinity of our films with Rutherford backscattering spectroscopy (RBS) channeling experiments. Channeling experiments using He^{4+} at 3 MeV result in a figure of merit of crystallinity is $\chi_{min} \sim 10$ %, suggesting excellent crystallinity. Figure 2 shows an example of a nominally $Y_{2.5}Ce_{0.5}Fe_{4.5}Co_{0.5}O_{12}$ film. Film thickness and composition is also determined by RBS experiments with films prepared on Si wafers placed next to the single crystal GGG substrates. Simulations of the RBS spectra indicate that to within 0.5%, the composition of the film is the same as the composition of the bulk target. Atomic force microscopy (AFM) provides a measure of the surface topography of our films. Ex-situ films below the critical thickness limit typically exhibit an rms roughness of 0.78 nm (as compared to the unit cell parameter of 1.2nm), so it is very difficult to distinguish any surface structure. Post-annealed films are significantly rougher with rms roughness of 1.9nm.

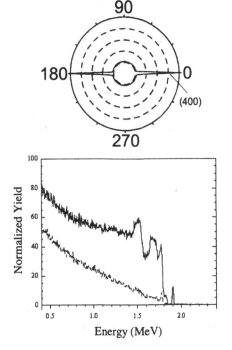

Fig. 1: Phi-scan of $Y_3Fe_4Co_{0.5}Ge_{0.5}O_{12}$ indicates good in-plane alignment of doped YIG film and GGG substrate crystal axes.

Figure 2: Rutherford channeling analysis of a $(Y_{2.5}Ce_{0.5})(Fe_{4.5}Co_{0.5})O_{12}$ film reveals a figure of merit of crystallinity of $\chi_{min}\sim$ 10 %, suggesting excellent crystallinity. The peaks at high channel numbers are due to the Au/Pd overlayer we sputtered on our samples to prevent charging of the insulating films.

Comparison of magnetization data with observed bulk values provides us with evidence as to the ordering of magnetic moments within the garnet crystal structure. Magnetic characterization is made with a Lakeshore vibrating sample magnetometer (VSM). Magnetization vs. applied field measurements made along the film plane normal indicate that the film normal is a hard axis. There is no evidence of growth induced anisotropies perpendicular to the film as observed in (111) single crystalline YIG films. Figure 3(a) shows the out of plane magnetization loop of a (110) $Y_3Fe_{4.5}Co_{0.25}Ge_{0.25}O_{12}$ film saturates at fields corresponding to the demagnetization field $4\pi M_s \sim 1400$ Oe.

As stated above, ex-situ films of compositions $Y_{2.75}Ce_{0.25}Fe_{4.75}Co_{0.25}O_{12}$, $Y_3Fe_{4.5}Co_{0.25}Ge_{0.25}O_{12}$, and $YFe_4Co_{0.5}Ge_{0.5}O_{12}$, exhibit ferrimagnetic behavior. Only the $Y_3Fe_{3.5}Co_{0.75}Ge_{0.75}O_{12}$ films never show any ferrimagnetic character at room temperature, even after post-deposition annealing. This is not too surprising since there is a substantial amount of Ge^{4+} that is substituting for Fe^{3+}, which could interrupt the order and magnetic exchange interactions. Dhara et al. have shown that the Curie temperature is indeed depressed substantially with increased Ge^{4+} doping. The Ge^{4+} ion substitutes for the Fe^{3+} with $5\mu_B$ while the Ce^{4+} substitutes for Y^{3+} ions which contributes no electron magnetic moment. Therefore the magnetic moment of the germanium doped samples is reduced by cobalt as well as germanium substitution

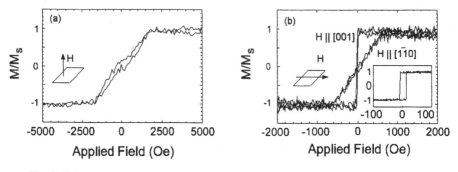

Fig. 3: $Y_3Fe_{4.5}Co_{0.25}Ge_{0.25}O_{12}$ MH loops (a) perpendicular to the plane and (b) in-plane easy and hard directions. The inset of 3(b) is an enlarged view of the hysteresis in the easy direction.

of Fe^{3+}, while the moment of cerium doped samples is reduced by cobalt only. This has been verified in the saturation magnetization values of the various compositions.

Magnetic anisotropy of undoped YIG in the (110) plane exhibits twofold symmetry. If the magnetic anisotropy were due to crystal structure, fourfold symmetry should exist in the plane of the film instead of the twofold symmetry that we observe. We have grown and characterized films of undoped YIG on (110) GGG substrates in order verify that the twofold symmetry of the anisotropy is due to the effect of the slight lattice mismatch between the film and the substrate. The substrate lattice parameter is smaller than that of the pure YIG film, albeit by 0.03%, and the compressive stress appears to dominate the anisotropy, shifting the axis of easy magnetization from [111] (observed in bulk [4,5]) to [110]. The twofold symmetry we observe in the (110) films is evidence of the dominant role of stress anisotropy. We estimate a stress anisotropy term of $K_{stress} = 3\lambda\sigma/2$, where λ is the magnetostriction coefficient and $\sigma = Y\epsilon$ is the stress [10]. For YIG, Young's modulus Y is 2×10^{12} dynes/cm², the strain ϵ is 0.03%, and $\lambda \sim 4 \times 10^{-6}$. Therefore the undoped control films, which exhibit an in-plane uniaxial anisotropies of K= 3.8×10^3 ergs/cm³, are consistent with the estimated K_{stress} value.

The doping of cobalt ions, which have anisotropic crystalline fields, enhances spin orbit coupling and hence the magnetostrictive coefficients and magnetic anisotropy. Figure 3(b) shows magnetization loops along the easy and hard directions in a (110) $Y_3Fe_{4.5}Co_{0.25}Ge_{0.25}O_{12}$ film. The axis of easy magnetization in Co doped films prior to annealing is found to be [001]. Fratello et al. have observed that cobalt doping of the garnets in liquid phase epitaxy films is limited to ~ 5 cation percent of the iron concentration by the precipitation of $CoFe_2O_4$ [11]. Since we use a nonequilibrium deposition technique, it is not surprising that we can dope YIG further without precipitating $CoFe_2O_4$.

We have imposed the divalency of Co by charge compensating with 4+ ions, so the contribution to anisotropy could be due to Co^{2+} ions in tetrahedral, octahedral or dodecahedral sites. Although we have no direct measure of the amount of Co on each site, it has been shown that the Co^{2+} prefers the octahedral site over the tetrahedral site [4]. For comparison, we have grown YIG films with similar amounts of cobalt doping but no charge compensation. These films which contain Co^{3+} ions show no measurable change in anisotropy from the undoped YIG.

The saturation magnetization of the doped YIG films decreases with increasing cobalt concentration, as seen in Figure 4. Since the net moment in YIG is due to the antiferromagnetic interaction among the cations on the 16 octahedral and 24 tetrahedral sites per unit cell and since the Y^{3+} contributes no magnetic moment, the reduction of the moment suggests the presence of tetrahedral Co^{2+} and/or Fe^{2+}.

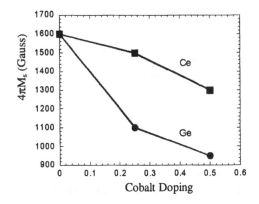

Figure 4: M_s variation with increasing cobalt doping. Samples charge compensated with nonmagnetic Ge^{4+} decrease more dramatically due to substitution on Fe^{3+} sites. Ce^{4+} substitutes for nonmagnetic Y^{3+}.

Measurements of doped YIG films show that the easy direction shifts from [100] to [110] when we anneal the films. Figure 5 shows for annealed films with various degrees of Ge/Co and Ce/Co doping, we observe significant contribution of the Co^{2+} to the magnetic anisotropy. We have defined the "anisotropy field" as the field corresponding to $Ms/\left(\dfrac{dM}{dH}\right)_{H=0}$. This anisotropy field H_K is in turn related to a uniaxial anisotropy constant by $K_u = M_s H_K/2$. The in-plane uniaxial anisotropy remains twofold symmetric and is similar in magnitude to the *ex-situ* films. The shift in the easy axis cannot be explained in terms of microstructure but may be attributed to the rearrangement of the Co^{2+} ions since bulk work has shown that Co^{2+} ions in the octahedral site and tetrahedral sites exhibit crystal anisotropies of opposite sign [4]. This hypothesis is further supported by magnetization results in undoped YIG films, which indicate no shift in the easy direction.

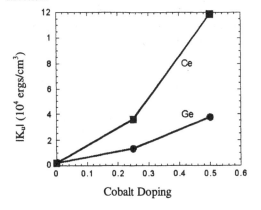

Figure 5: Magnetic anisotropy of Cobalt doped YIG films as a function of Ge/Co and Ce/Co doping.

In future work we will replace Ce^{4+} and Ge^{4+} with cations of varying ionic radii and electronic structure, in order to investigate their effect on magnetization, coercive field, and anisotropy. We expect that preferential substitution on the octahedral site with a nonmagnetic cation may prevent decreases, or perhaps even cause an increase in magnetization.

CONCLUSIONS

We have grown highly crystalline YIG films on [110] GGG using pulsed laser deposition. We have demonstrated that the uniaxial anisotropy of [110] oriented YIG films can be enhanced substantially by doping with divalent cobalt. Such magnetically harder garnets are promising as an exchange biasing layer of high permeability YIG for high frequency inductor applications.

ACKNOWLEDGEMENTS

We would like to thank Maura Weathers and Peter Revesz for their technical assistance. This work was supported in part by an ONR Young Investigator Award (Y.S.) and an REU grant (D.D.) from the Cornell Center for Materials Research (NSF-MRSEC). Structural characterization was carried out at the central facilities of the Cornell Center for Materials Research.

REFERENCES

1. K.I. Arai, M. Yamaguchi, H. Ohzeki and M. Matsumoto, IEEE Trans. Magn. **27** 5337 (1991).
2. E.M. Gyorgy, J. M. Phillips, Y. Suzuki and R.B. van Dover, US Patent #5665465.
3. Y. Suzuki, R. B. van Dover, E. M. Gyorgy, Julia M. Phillips and R.J. Felder, Phys. Rev. **B53** 14016 (1996).
4. M.D. Sturge, E.M. Gyorgy, R.C. LeCraw and J.P. Remeika, Phys. Rev. **180** 413 (1969).
5. P. Hansen, W. Tolksdorf and R. Krishnan, Phys. Rev. **B16** 3973 (1977).
6. W.P. Wolf , Phys. Rev. **108** 1152 (1957).
7. S. Dhara, A.C. Rastogi and B.K. Das, J. Appl. Phys. **79** 953 (1996).
8. M. Gomi, K. Kishimoto and M. Abe, J. Magn. Soc. Jpn. **11** 309 (1987).
9. J.W. Matthews, A.E. Blakeslee, S. Mader, Thin Solid Films, **33** 253 (1976).
10. Y. Suzuki, H.Y. Hwang, S-W. Cheong, R.B. van Dover, Appl. Phys. Lett. **71** 140 (1997).
11. V.J. Fratello, E.M. Gyorgy and R.B. van Dover, unpublished.

SUBSTITUTED BARIUM HEXAFERRITE FILMS
FOR PLANAR INTEGRATED PHASE SHIFTERS

S.A. OLIVER *, S.D. YOON **, I. KOZULIN **, P. SHI **, X. ZUO **, C. VITTORIA **
* Center for Electromagnetic Research, Northeastern University, Boston MA 02115,
saoliver@neu.edu.
** Department of Electrical and Computer Engineering, Northeastern University, Boston MA
02115.

ABSTRACT

Thick films of scandium-substituted $BaSc_xFe_{12-x}O_{19}$ (x=0, 0.4, 0.6) were deposited by pulsed laser ablation deposition onto A-plane ($11\underline{2}0$) sapphire (Al_2O_3) substrates to yield highly oriented films having the crystallographic c-axis ([0001]) in the film plane. Selected films were characterized by x-ray diffraction, electron microscopy, magnetometry and torque magnetometry measurements. These structural and magnetic measurements show that dense films have excellent in-plane uniaxial anisotropies and good square loop behaviors. Such films will be useful for planar latched phase shifters that can be integrated into microwave integrated circuits.

INTRODUCTION

Reciprocal and nonreciprocal phase shifter devices that use ferrite materials have been well developed for waveguide systems operating at radio or microwave frequencies. The extension of ferrites for use in planar microwave devices such as stripline or microstripline-based phase shifters has also been explored, although devices based on these geometries have not yet seen widespread deployment.[1] All of these devices obtain their phase shift through the employment of the enhanced effective permeability that occurs in a partially or fully magnetized ferrite material, such that the propagation constant of the transmitted signal is greatly increased relative to the value in the same material when demagnetized. Typically these ferrites are magnetized through the application of an external magnetic field generated by coils or latching wires. However, for frequency applications above 12 GHz this magnetic biasing technique starts to become impractical because of the magnetic field strengths needed to bias standard ferrite materials to the operational frequency.

Hexaferrite materials are currently being developed to overcome this requirement of needing large magnetic fields for the operation of ferrite phase shifters at higher microwave frequencies. Here, the primary advantage of the hexaferrite is their large uniaxial magnetocrystalline anisotropy that gives rise to an substantial internal uniaxial anisotropy field (H_A).[2][3] This magnetically self-biasing anisotropy field acts to replace the external magnetic field typically used to bias ferrites, so that, in principle, hexaferrite phase shifters can be designed to operate at frequencies above 50 GHz while using only a small external magnetic field to overcome the hexaferrite coercive field and latch the film magnetization at remanence.

To be useful for planar phase shifters, hexaferrite films must have thicknesses above 25 micrometers, and must be highly oriented with the crystallographic c-axis lying within the film plane. To date, only pulsed laser deposition techniques have proven capable of producing highly oriented thick hexaferrite films at reasonably high growth rates.[4] Here, we report on the deposition of highly oriented scandium-substituted barium hexaferrite ($BaSc_xFe_{12-x}O_{19}$)

films on sapphire substrates, where the substitution of scandium cations for iron in the hexagonal close-packed structure acts to systematically reduce the film uniaxial anisotropy field. [5][6]

EXPERIMENTAL

Hexaferrite films were deposited onto polished A-plane (11$\underline{2}$0) sapphire (Al$_2$O$_3$) substrates by pulsed laser deposition at temperatures of from 840^{0}C to 925^{0}C. In order to obtain films having different magnetic properties, films were deposited using several pressed and sintered BaSc$_x$Fe$_{12-x}$O$_{19}$ targets where x= 0, 0.4 and 0.6. The densities of the pure BaFe$_{12}$O$_{19}$ targets were approximately 5 g/cm^3, approaching the bulk value, while the densities of the pressed scandium-substituted targets were close to 4.2 g/cm^3. All films reported on here were deposited in an oxygen background gas at a pressure of either 20 mTorr or 50 mTorr. The laser beam was focused to a spot to obtain a laser fluence on the target of 4 - 5 J/cm^2. The deposition time was varied between films, depending upon the net desired film thickness, with the thicknesses of the resulting films ranging from 2 micrometers to 14 micrometers.

The crystallographic structure of selected films was characterized by x-ray diffraction using a Rigaku 300 diffractometer, while the surface morphology was examined by a JEOL 6100 scanning electron microscope. The magnetic parameters and orientation of the film easy axis were measured on a vibrating sample magnetometer and a torque magnetometer, where measurements were performed both in the film plane and out of the film plane.

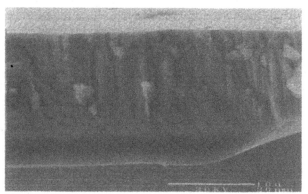

Figure 1 Cross-section micrograph of a 14 μm thick BaFe$_{12}$O$_{19}$ film on a fractured sapphire substrate. The c-axis lies in the plane of the page.

RESULTS

In general, the surface morphology and crystallographic orientation of hexaferrite films deposited on sapphire can vary dramatically between films depending upon the growth rate, deposition temperature, and target material. It has now been reasonably well established that barium hexaferrite films deposited onto c-plane (0001) sapphire must be grown at temperatures of 900^0C or above to obtain highly oriented films.[4][7] Although this result also holds true for pure barium hexaferrite films having thicknesses below about 5 micrometers deposited on $(11\underline{2}0)$ Al_2O_3, it was found that thicker films deposited at 840^0C have a denser, more homogenous texture, and much better crystallographic orientation than films the films deposited at 900^0C. This is demonstrated in Figure 1, which shows a 14 micrometer thick $BaFe_{12}O_{19}$ film on sapphire. Because of the substantial mismatch in coefficients of thermal expansion between barium hexaferrite and sapphire, these thick films have large tensile stresses, such that fracture or delamination are often observed. For the film shown in Figure 1 fracture has occurred within the sapphire substrate, and this film section has delaminated. In particular, the face exposed in Figure 1 has fractured roughly parallel to the film [0001] crystallographic axis.

The high degree of crystallographic orientation in the denser films is shown by the x-ray diffraction pattern shown in Figure 2, which was taken on a 8 micrometer thick $BaFe_{12}O_{19}$ film. Here, excellent orientation is revealed due to the presence of the dominant $(20\underline{2}0)$, $(30\underline{3}0)$ and $(40\underline{4}0)$ peaks, indicating that the film [0001] axis lies within the film plane.[8]

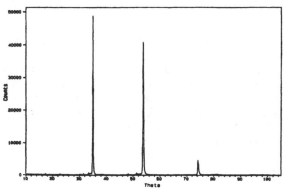

Figure 2 X-ray diffraction pattern for a $BaFe_{12}O_{19}$ film.

The magnetic properties of these highly oriented films were evaluated through torque magnetometry and magnetometry measurements, taken both in and normal to the film plane. Figure 3 shows the results of a torque measurement taken on a 2 micrometer thick film of $BaSc_{0.4}Fe_{11.6}O_{19}$ where a 10 kOe magnetic field (H) was rotated in the film plane. Both clockwise (solid line) and counterclockwise (dashed lines) film rotations are shown, and the two-fold symmetry in the torque measurements distinctly shows that the uniaxial easy axis (crystallographic [0001] axis) lies in the film plane at an angle of approximately 60^0 (and 240^0). Meanwhile, the overlap of the data for the two rotation directions indicates that the value of H is approaching the value of the uniaxial anisotropy (H_A) for this film composition.

The hysteresis loop behavior for the $BaSc_{0.4}Fe_{11.6}O_{19}$ film of Figure 3 is shown in Figure 4, where measurements were taken both along the in-plane easy axis and hard axis. Excellent square-loop behavior is shown for the easy axis, with a remanence value approaching 95% of the full saturation magnetization, as is desired for a latching phase shifter. Here the apparent decrease in moment at higher field values arises due to an over-correction

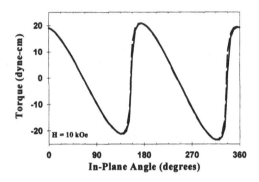

Figure 3. In-plane torque results for a $BaSc_{0.4}Fe_{11.6}O_{19}$ film.

for the sapphire diamagnetic contribution. The coercive field (H_c) corresponding to the easy axis is 500 Oe. Meanwhile, the substantial field needed to magnetize the film along the hard axis is also apparent, as is the saturation that occurs at approximately 11.5 kOe, a value that roughly corresponds to H_A. The saturation magnetization value ($4\pi M_s$) for this film was 3.7 kG, which is lower than that expected for bulk materials having this composition.[5]

When the film easy axis lies in the film plane, the film normal should also be a hard axis. Such hard axis behavior is shown in Figure 5, which shows the magnetization behavior along the film normal of the same $BaSc_{0.4}Fe_{11.6}O_{19}$ film as Figures 3 and 4. In this orientation, the magnetization cannot be saturated with a 13 kOe field because of the shape demagnetizing field that acts to decrease the value of H within the film by approximately $4\pi M_s$.

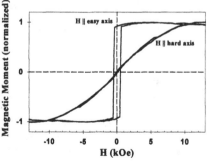

Figure 4. In plane hysteresis loops along the hard axis and easy axis of a $BaSc_{0.4}Fe_{11.6}O_{19}$ film.

104

CONCLUSIONS

Scandium-substituted barium hexaferrite films having excellent structural and magnetic properties have been grown by pulsed laser deposition onto oriented sapphire substrates. Results were shown for one film composition, $BaSc_{0.4}Fe_{11.6}O_{19}$, which showed that this film had the excellent in-plane uniaxial anisotropy orientation and good square loop magnetic properties needed for future planar integrated latching phase shifters. However, thicker as-produced films show significant tensile stresses, and additional effort is needed to obtain the 25 micrometer or thicker films required for low loss devices.

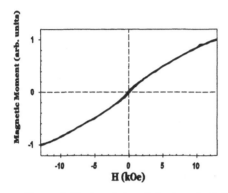

Figure 5. Hysteresis loop behavior of a $BaSc_{0.4}Fe_{11.6}O_{19}$ film with H along the film normal.

ACKNOWLEDGMENTS

This research was supported by the Office of Naval Research and the Defense Advanced Research Projects Agency under the 1996 Multidisciplinary University Research Initiative.

REFERENCES

1. For a review of ferrite based phase shifters, see for example, S.K. Koul and B. Bhat, *Microwave and Millimeter Wave Phase Shifters, vol. 1*, Artech House, Boston, 1991, pp. 315-384, and references therein.
2. H.L. Glass, Proc. IEEE **76**, p. 151-158 (1988).
3. G.F. Dionne and J.F. Fitzgerald, J.Appl.Phys. **70**, p. 6140-6142 (1991).
4. P.C. Dorsey, D.B. Chrisey, J.S. Horwitz, P. Lubitz and R.C.Y. Auyeung, IEEE Trans. Magnetics **30**, p. 4512-4517 (1994).
5. P. Röschmann, M. Lemke, W. Tolksdorf and F. Welz, Mat. Res. Bull. **19**, p. 385-392 (1984).
6. L.M. Silber and W.D. Wilber, IEEE Trans. Magn. **22**, p. 984 (1986).
7. S.R. Shinde, S.E. Lofland, C.S. Ganpule, S.B. Ogale, S.M. Bhagat, T. Venkatesan and R. Ramesh, J. Appl. Phys. **85**, p. 7459-7465 (1999).
8. International Center for Diffraction Data (ICDD) card numbers 39-1433 and 43-0002.

FERROMAGNETIC RESONANCE
IN SINGLE CRYSTAL BISMUTH IRON GARNET FILMS

V. DENYSENKOV [1], A. JALALI-ROUDSAR [1], N. ADACHI [1,2],
S. KHARTSEV [1], A. GRISHIN [1], T. OKUDA [2]

[1] Department of Condensed Matter Physics, Royal Institute of Technology, S-100 44 Stockholm, Sweden

[2] Nagoya Institute of Technology, Nagoya 466, Japan.

ABSTRACT

Among other magneto-optic materials bismuth substituted yttrium iron garnet $(Bi_xY_{1-x})_3Fe_5O_{12}$ has the highest Faraday rotation effect in visible region. Completely substituted bismuth iron garnet $Bi_3Fe_5O_{12}$ (BIG) films have been grown by pulsed laser deposition technique onto (111) $(NdGd)_3(ScGa)_5O_{12}$ single crystal with lattice constant of 12.623 Å. X-ray diffraction proves epitaxial film quality. VSM measurements yield the saturation magnetization $4\pi M_s = 1100$ G and coercive field is about 50 Oe. Ferromagnetic resonance (FMR) method reveals perpendicular magnetic anisotropy in fabricated films. Angular measurements of FMR give the constants of uniaxial and cubic anisotropy in BIG film: $K_u = 5.9 \times 10^4$ erg/cm^3 and $K_1 = -6.95 \times 10^3$ erg/cm^3. The Faraday rotation has been found to reach - 7.8 deg/μm at 630 nm.

INTRODUCTION

The increasing demands for performance of signal processing technique in optical communications stimulate the investigation of new materials for fiber optic applications. Single crystals of rare earth iron garnets have remarkable transparency in the near infrared and exhibit effects of nonreciprocity for microwaves and optics: large rotation of polarization vector associated with the magnetization parallel to the light propagation direction, known as Faraday effect.

Completely substituted bismuth iron garnet (BIG) films have attracted much attention for magneto-optic applications because they have giant Faraday rotation effect reaching –8.4 deg/μm at 633 nm wavelength [1] and low losses in the infrared and visible ranges used for optical communications. Commercial magneto-optical devices are all based on bulk crystals or thick film technology. Completely Bi-substituted iron garnet bulk crystal is a nonequilibrium phase at high temperatures, thus only thin film of this material can be obtained clamping $Bi_3Fe_5O_{12}$ on the appropriate single crystal substrate. Although thin BIG films exhibit high magneto-optic performance and waveguide technology shows great promise, nevertheless sufficient cost/performance advantages have been yet achieved.

Using BIG one can perform simple thin film optical isolators since giant Faraday rotation makes possible to use orthogonal geometry and overcome difficulties regarding in-film plane light propagation. Just 6 μm thick BIG film is required to provide 45° of rotation necessary to isolate tunable laser from back-scattered light. Besides this, such device has minimal optical losses compared with in-plane light propagation case. Another possible application of such film is magneto-optical Bragg cell modulators for tunable RF spectrum analyzers [2]. Such modulators can be tuned in wide frequency range by varying bias magnetic field.

We report on pulsed laser deposited BIG (111) films and their magnetic, high frequency and magneto-optic properties. Highly crystalline film structure has been proved by x-ray diffraction (XRD), vibrating sample magnetometer (VSM) has been used to measure saturated magnetization while Ferromagnetic Resonance (FMR) spectrometry gives complete information on uniaxial and cubic magnetic anisotropy in fabricated films.

EXPERIMENT

BIG films have been grown by pulsed laser deposition (PLD) technique using KrF excimer laser (LPX300, Lambda Physik, $\lambda = 248$ nm) with energy of 750 mJ per pulse and repetition rate up to 50 Hz.

The precursor ceramic target used for laser ablation is a pellet, which consists of powders with the composition $3Bi_2O_3 + 5Fe_2O_3$ and was made by cold pressing at 600 kg/cm^2 and sintering at temperature of 800 °C for 4 hours. The density of the target is 6.09 G/cm^3. The target was spinning during deposition to prevent its fast erosion and consumption. Also, it has been resurfaced by grinding before each deposition.

The (111) oriented $(NdGd)_3(Sc,Ga)_5O_{12}$ (NGSGG) 5×5 mm^2 single crystal has been used as a substrate to grow BIG films. It has a light violet color and exhibits paramagnetic properties at room temperature. NGSGG lattice constant was found to be 12.623 Å. This value is very close to BIG lattice constant thus NGSGG substrate is believed to be the most appropriate to fabricate high performance BIG films.

Films were synthesized in the presence of oxygen. Growing conditions are summarized below:

Target	Precursor $3Bi_2O_3 + 5Fe_2O_3$ ceramics
Background pressure, Torr	1×10^{-7}
Working pressure of oxygen, mTorr	10
Substrate temperature, °C	550
Deposition time, hours	2
Laser pulses repetition rate, Hz	40
Density of laser energy, J/cm^2	2-3

The crystalline structure of the grown films has been investigated by x-ray diffraction using *Siemens D-5000* powder diffractometer. Bragg-Brentano θ-2θ scan, rocking curves (ω-scans) and off-normal planes φ-scans for the film side of the sample and for the substrate side have been recorded. To measure magnetic moment of BIG/NGSGG structure and calculate saturation magnetization $4\pi M_s$, the *M-H* hysteresis loops have been traced in in-plane magnetic field by *Oxford Instruments* vibrating sample magnetometer. Film volume was calculated after film thickness measurement. Thickness was measured by the atomic force microscope.

The FMR spectrometer used for this work consists of 12 inch electromagnet from the *Varian Inc.* which provides DC magnetic field up to 2.4 T with uniformity better than 10^{-5}.

The intensity of the magnetic field was measured by the Hall effect probe placed between the magnet poles. Microwave part of the spectrometer consists of the X-band waveguide in the magnet gap with the Gunn diode inside. Changing the Gunn diode we have been able to vary frequency from 8 to 40 GHz. It operates in a marginal oscillator regime. To calibrate spectrometer and determine operating frequency the paramagnetic salt $MnSO_4$ has been used before and after each measurement. FMR results presented in this paper have been obtained at 9.159 GHz operating frequency. Magnetic field was modulated while the signal across the Gunn diode was measured by *Stanford Research Systems* lock-in amplifier. Magnetic field sweep has been automatically controlled by PC/Keithley-2002/2400 platform. Special routine under *LabView 5.0* has been used to sweep magnetic field H, read and record H, and derivative of absorption $d\chi''/dH$ data. Two-circle cradle with a holder for ferromagnetic film provides angular FMR measurements varying both polar angle θ and asimuthal angle φ.

To find constants of uniaxial and cubic magnetic anisotropy K_u and K_l, we have modified the method described by H. Makino and Y. Hidaka [3]. Firstly, the orientation of $[1\,1\,2]$ crystalline axis in BIG film has been found by XRD. Then the angular dependence of resonance magnetic field H_{res} on azimuthal angle φ has been measured. The φ angle is counted from $[112]$ axis to the direction $[001]$. During these measurements the polar angle θ (between the magnetic field and normal to the film plane $[111]$) has been fixed at $50°$. This orientation corresponds to the maximum variation of resonance magnetic field versus azimuthal angle. Then the film has been specially oriented to align magnetic field in $(1\,10)$ crystallographic plane and this azimuthal angle has been fixed to measure the resonance magnetic field H_{res} dependence on polar angle θ. Full set of experimental data has been fitted to theoretical angular H_{res} vs. θ dependence to compute both K_u and K_l constants.

RESULTS

XRD θ-2θ scan of BIG film grown onto $(NdGd)_3(ScGa)_5O_{12}$ single crystal (see Fig. 1) shows the film is single $Bi_3Fe_5O_{12}$ phase despite the use of precursor oxide target. Only

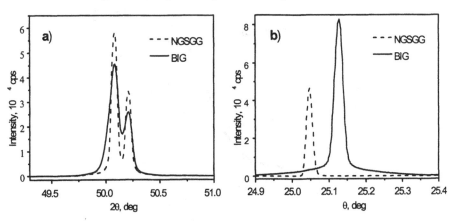

Fig. 1. XRD pattern in CuK_α radiation for (444) Bragg reflections from 0.84 μm thick $Bi_3Fe_5O_{12}$ film deposited onto $(NdGd)_3(ScGa)_5O_{12}$ substrate: **a)** θ-2θ scan from BIG film and substrate; **b)** (444) rocking curves for BIG film and NGSGG substrate.

(444) and multiple integer Bragg reflections from BIG and NGSGG have been seen in Bragg-Brentano geometry. The lattice constant of (111) oriented BIG film was found to be 12.623 Å. Rocking curves of the (444) reflections from film and substrate confirm the mismatch between film and substrate lattice is very small (less than 0.09%). Fig. 2 shows the φ-scan from off-normal (664) plane measured at oblique geometry. 120° symmetry proves the cubic crystalline structure and epitaxial quality of (111) oriented BIG film.

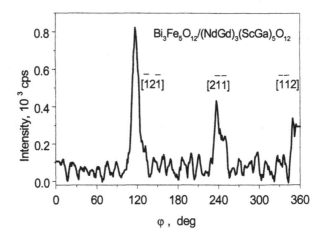

Fig. 2. XRD φ-scan of off-normal BIG (664) plane.

The VSM measurements have been performed to determine saturated magnetization $4\pi M_s$ value in BIG film. Fig. 3 shows *M-H* hysteresis loop for BIG film measured in magnetic field parallel to the film surface. $4\pi M_s$ was found to be equal to 1100 G. The saturated

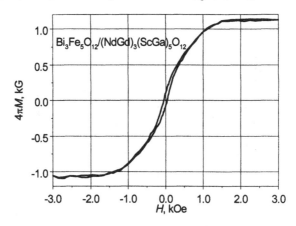

Fig.3. *M-H* hysteresis loop of BIG (111) film.

magnetization for BIG film grown by reactive ion beam sputtering (RIBS) technique reported earlier was 1500 G [1]. Among ferromagnetic films having equal Faraday rotation the films with lower saturated magnetization are much more preferable since they have advantage requesting lower external magnetic field biasing for device applications.

In Fig. 4 and Fig. 5 the results of FMR measurements are presented. Figure 4 shows 120° periodic azimuthal dependence of resonance magnetic field H_{res} and linewidth ΔH measured in magnetic field tilted from the film normal at the polar angle $\theta = 50°$.

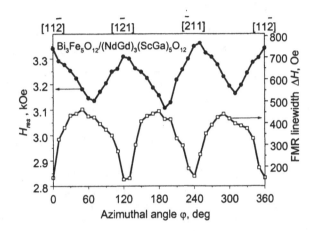

Fig. 4. Azimuthal dependence of FMR field (solid dots) and FMR linewidth (open squares) for BIG film at polar angle $\theta = 50°$.

120° symmetry is caused by magnetocrystalline anisotropy in (111) BIG plane. Usually to obtain so clear periodic dependence film sample must be etched to get a circular shape. In our case 120° periodicity has been observed for square shaped sample without etching. It is evident shape anisotropy is negligible in this film. The reason could be that sufficient positive growth induced anisotropy overcomes shape anisotropy which tends to keep the magnetization in film plane. Measured $H_{res}(\varphi)$ dependence and comparison with XRD show that higher resonance field and narrower FMR linewidth correspond to the main crystalline directions [112], [121], and [211], which are easily recognized in FMR experiments. The narrow linewidth appears at the same azimuth angles as XRD reflections in φ-scan measurements. The polar angle θ dependence of H_{res} when magnetic field aligned in (110) crystallographic plane (azimuth angle $\varphi = 0$) is shown in Fig. 5. Lower value of the resonance magnetic field corresponds to the perpendicular orientation of magnetic field to the film plane. Thus the easy magnetization axis is perpendicular to the film and coincides with the [111] crystallographic direction.

To determine the anisotropy constants, K_l and K_u, macroscopic expression for the energy of uniaxial and cubic magnetic anisotropy and magnetostatic energy have been used to derive equilibrium orientation of the magnetization and FMR field in external magnetic field parallel to (110) plane. Computer routine has been developed using *Mathematica*, which gave the best fit of experimental data (solid circles in Fig. 5) to the theoretical curve when the fitting parameters $K_u = 5.9 \times 10^4$ erg/cm^3 and $K_l = -6.95 \times 10^3$ erg/cm^3 have been chosen.

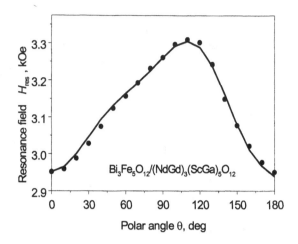

Fig. 5. FMR field versus polar angle θ in magnetic field aligned in ($\bar{1}$10) plane.

The positive value of the effective constant of uniaxial anisotropy $K_u^* = K_u - 2\pi M_s^2 = 1.09 \cdot 10^4$ erg/cm³ suggests easy axis is perpendicular to the film plane. Obtained value of cubic magnetic anisotropy K_I is smaller than the extrapolated value for bismuth substituted rare earth iron garnet films grown by liquid phase epitaxy (LPE) method [4] and experimental value for BIG film grown by RF sputtering [5].

CONCLUSIONS

Completely substituted (111) oriented bismuth iron garnet films of epitaxial quality have been grown onto (111)-oriented $(NdGd)_3(ScGa)_5O_{12}$ single crystal by KrF pulsed laser deposition technique. Films exhibit perpendicular magnetic anisotropy, saturated magnetization $4\pi M_s = 1100$ G, constants of uniaxial and cubic magnetic anisotropy $K_u = 5.9 \times 10^4$ erg/cm³ and $K_I = -6.95 \times 10^3$ erg/cm³.

REFERENCES

1. T. Okuda, T. Katayama, H. Kobayashi, N. Kobayashi, K. Satoh, and H. Yamamoto, J. Appl. Phys. **67**, 4944-4946 (1990).
2. C. S. Tsai, IEEE Trans. Magn. **32**, 4118 (1996).
3. H. Makino and Y. Hidaka, Mat. Res. Bull.**16**, 957-966 (1981).
4. P. Hansen, K. Witter, and W. Tolksdorf, J. Appl. Phys. **55**, 1052 (1984)
5. A. Thavendrarajah, M. Pardavi-Horvath, P.E. Wigen, and M. Gomi, IEEE Trans. on Magn., **25**, 4015-4017 (1989).

MAGNETICALLY TUNED SUPERCONDUCTING FILTERS*

D. E. OATES, A. C. ANDERSON, G. F. DIONNE
MIT Lincoln Laboratory, 244 Wood St., Lexington MA 02140-9108 oates@ll.mit.edu

ABSTRACT

We have demonstrated magnetically tunable superconducting filters consisting of microwave circuits coupled to ferrite substrates in monolithic structures using both niobium at 4 K and YBCO at 77 K. A three-pole 1% bandwidth filter with 10-GHz center frequency, 1-dB insertion loss, and greater than 10% tunability has been demonstrated. Operation of the ferrite in the partially magnetized state prevents degradation of the surface resistance of the superconductor. This technology has the potential for compact, light weight, multipole filters with very low insertion loss, <0.5 dB, and very rapid tunability, <1 μs, due to the low inductance of the magnetic circuits. We also discuss the use of lightly coupled ferroelectric films to introduce trimming of the individual resonators in the magnetically tuned filters. The materials issues that limit performance and means to improve present performance are discussed.

INTRODUCTION

Tunable filters are an important element of the microwave front end of frequency-hopped wireless communications and radar systems. Many such systems could benefit from a bandpass filter placed in front of the low-noise amplifier (LNA) to reduce the possibility of strong interfering signals reaching the LNA, where they can generate in-band spurious signals and distortion that would prevent the reception of desired signals. Insertion loss in this filter is an important issue because any losses directly degrade the system noise figure and dynamic range. Sharp cutoff of the filter passband is also desirable to eliminate signals nearby in frequency.

In the last few years, high-performance, very low loss, fixed-frequency filters utilizing high-transition-temperature superconductor (HTS) thin films have been developed [1]. These filters, often in combination with cooled LNAs, can improve the performance of many microwave systems. Filters with similar characteristics, and with the added advantage of fast tunability, would combine the improved sensitivity and selectivity with the agility required in many microwave systems.

We have previously shown [2,3] that low loss and tunability could be realized in superconductor/ferrite devices. The tunability results from changing the effective permeability of the ferrite material that is varied by changing the magnetization, rather than the traditional approach of changing the ferrimagnetic resonance, which requires large applied magnetic fields of several thousand Oe. Magnetic tuning can be accomplished without adversely affecting the low loss of the superconductor by using the partially magnetized state. The magnetic fields are low, yet interaction of the microwave fields with the ferrite can be strong. The applied fields need only be of order of the coercive field, which can be as low as a few Oe and can be obtained with simple coils of a few turns. The devices also have low inductance allowing fast switching of the filter frequency.

We have previously reported [2] results from measurements on tunable resonators showing that high Qs and adequate tunability are achievable using this novel approach. In evaluating the results of the resonator measurements we use a figure of merit K that has been defined [4] as

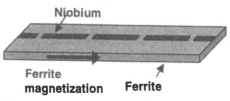

Niobium

Ferrite
magnetization **Ferrite**

Fig. 1. Schematic view of tunable filter, showing niobium resonators on the ferrite substrate. The coil for providing the magnetization is not shown.

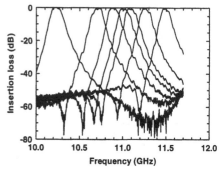

Fig. 2. Measured insertion loss for the filter shown in Fig. 1 for various values of applied magnetic field ranging from 0 to 300 G.

$$K = 2Q\frac{\Delta f}{f} \qquad (1)$$

where Q is the unloaded Q including all magnetic and dielectric losses of the tunable resonator, Δf is the tunable range and f is the center frequency of the resonator. The quantity is an indication of the material merit in making high-performance tunable devices. Measurements on niobium-on-ferrite resonators yielded K factors of 300. Using a hybrid geometry of YBCO on LaAlO$_3$ clamped to a ferrite substrate, we obtained K factors of 170 [2]. These figures are to be compared with a value of approximately 50 for the K measured for ferroelectric tuning with SrTiO$_3$ thin films [5]. A model for the tunability of the resonators has been presented previously [6].

RESULTS OF TUNABLE FILTERS

Niobium on Ferrite

For the first demonstrations, niobium superconductor was chosen because of the simplicity of fabrication. Niobium can be deposited directly on the ferrite, while the high-T$_c$ materials require epitaxial growth on lattice-matched substrates, ruling out direct deposition on ferrite. We fabricated a filter using linear, end-coupled, $\lambda/2$, niobium resonators on polycrystalline YIG. A niobium ground plane was used. The filter was operated at 4 K. Fig. 1 shows schematically the layout of the filter.

Fig. 2 shows the results of the measurements of the filter insertion loss (S_{21}) vs frequency for a number of applied external magnetic fields. The magnetic field ranges from zero (unmagnetized substrate) for the lowest frequency passband up to the maximum shown which is 300 G for the highest frequency passband. The tunable range shown is approximately 1.3 GHz, giving a tunability of 13%. Above 300 G the shape of the filter response begins to degrade, the insertion loss increases, and the shape becomes more rounded. The degradation of the passband results from decreases in the resonator Q and changes in the resonator coupling as the center frequency moves. The insertion loss shown is about 0.3 dB.

The tuning speed of a filter is an important consideration. Modern frequency-hopped radar and communication systems often change the operating frequency at a rapid rate, often in times less than 1 ms. The tunable filter technology described here is capable of rapid tuning because of the compact geometry and because the tuning of the filters requires only moderate

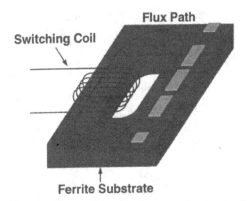

Switching Coil

Flux Path

Ferrite Substrate

Fig. 3. Filter using closed magnetic circuit for tuning. This device was used for the tuning speed measurements.

(~ 300 G) values of magnetic field, and the inherent switching time of the ferrite is of the order of 1 μs. Thus rapid low-energy tuning is possible. This is to be contrasted with conventional magnetically tuned filters that operate by changing the frequency of the ferromagnetic resonance and thus require tuning fields of several kG. To evaluate the tuning speed of the superconductor ferrite filters, we have fabricated a filter on a toriodally shaped substrate so that a closed magnetic circuit can be used to tune the filter, as shown in Fig. 3.

The measurements of the switching speed are shown in the oscilloscope photograph in Fig. 4. The measurements were carried out by applying a CW tone to the filter. We then observed the rise and decay of the filter output when a pulse of current is applied to the tuning coil. The pulse shifts the filter passband such that the CW input is in the passband of the filter when the pulse is on and out of the passband when the pulse is off. The top trace in the figure is the envelope of the filter output as a function of time. The bottom trace is the current in the switching coil. The horizontal scale of the photo is 5 μs/div. From the figure it can be seen that the filter response follows that of the current pulse. In this experiment the speed is limited by the rise and fall time of the current pulse used for the switching, which was determined by the pulse generator used. From this result the switching time of the filter can be estimated to be 5 μs or less. Further experiments are underway to improve the current pulse shape to better determine the true switching speed of the filter.

YBCO on Ferrite

As is well known, it is not possible to deposit YBCO or other high-T_c materials directly on ferrite substrates, because to obtain low microwave losses, the film must be grown epitaxially on a lattice-matched substrate. However, biaxially oriented buffer layers can be deposited on ceramic and polycrystalline substrates by the Ion-Beam-Assisted-Deposition (IBAD) process [7]. Epitaxial YBCO with low microwave losses can then be deposited on the IBAD layer by any convenient process such as laser ablation or sputtering [8].

Fig. 5 shows schematically the filter fabricated from IBAD YBCO on the polycrystalline ferrite YIG with a YSZ buffer layer as described in [8]. This is the same design and layout used for the niobium filter. The differences in dielectric constant were expected to be small enough that no adjustments of the design were needed. Fig. 6 shows the measurement of the filter insertion loss vs. frequency at 77 K for approximately the same values of the externally applied

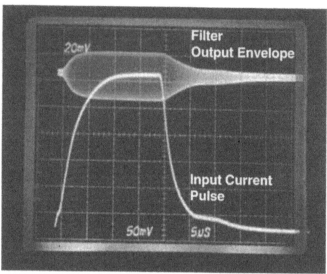

Fig. 4. Results of the switching speed measurements. The upper trace is the envelope of the microwave signal at the output of the filter. The lower trace is the current pulse applied to the coil of Fig. 3. The horizontal scale is 5 μs/div.

magnetic field as shown in Fig. 2. The tunability of 9.5% is consistent with the filter on YIG with niobium resonators when account is made for the lower saturation magnetization of the YIG at 77 K compared with 4 K [6]. The input power was –10 dBm. The insertion loss in the unmagnetized state is about 2 dB which is larger than expected from the surface resistance measured previously in the unpatterned IBAD YBCO [8].

Additional measurements of the temperature dependence of the Q of $\lambda/2$-resonators of IBAD YBCO have confirmed that the source of the excess loss is the ferrite. We have demonstrated that the losses in the ferrite increase as the temperature is lowered and reach a maximum between 70 and 20 K. The temperature dependence of the ferrite losses has been measured previously [9] and found to peak at low temperatures and depend strongly on the impurity level. We are continuing investigations of ferrites with higher purity starting materials in order to reduce the low-temperature losses. Modeling indicates that ferrite compositions optimized for cryogenic temperatures are expected to provide Qs of 10^5 [9].

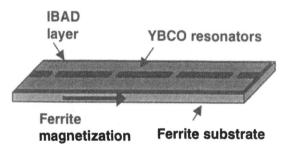

Fig. 5. Schematic view of filter fabricated from IBAD YBCO. The IBAD buffer layer is YSZ.

116

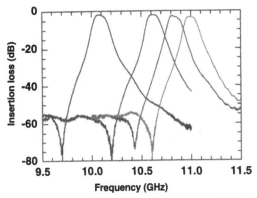

Fig. 6. Insertion loss and tunability of IBAD deposited YBCO on a polycrystalline YIG ferrite substrate. The temperature was 77 K and the input power was −10 dBm.

FERROELECTRIC TRIMMING

The filters discussed in the previous paragraphs are tuned by changing in unison all of the resonant frequencies of the resonators that comprise the filter. As seen in Fig. 2 and Fig. 6, this is sufficient to maintain the passband shape for the 3-pole 1%-bandwidth filters that have been demonstrated to date. For filters with more poles, however, or for filters with narrower bandwidths, the filter response will not maintain a constant shape over the whole tunable band. To do so it will be necessary to trim the resonant frequencies of the individual resonators. Trimming means that the frequencies of the individual resonators are changed by 0.2% or less. The proposed structure to accomplish the tunable filter with trimming is shown in Fig. 7. The majority of the tuning is accomplished using the magnetization of the ferrite exactly as discussed in the examples shown above. The trimming is accomplished by ferroelectric material such as strontium titanate (STO) deposited over a gap in the YBCO resonators, and variable capacitive coupling provides the tuning. The STO is located at a minimum of the electric field in the resonator so that the coupling to the resonator is weak. Because the coupling is weak it will not load the resonator or spoil the resonator Q. The figure of merit required for 0.2% trimming with resonator $Q = 10,000$ is 40, which is well within the range of values reported in the literature for STO. Preliminary attempts to deposit and measure the STO figure of merit have yielded values of 20 which was not limited by the STO itself but by the test fixture. Further optimization will reach the target value for the figure of merit.

CONCLUSIONS

We have shown the feasibility of implementing tunable superconducting filters using ferrites as the tuning element. We have demonstrated compact, low-loss filters with a tunability of greater than 10% at 10 GHz. The filters can be rapidly retuned with a tuning time of approximately 5 μs. We have also demonstrated integration with YBCO and operation at 77 K. We have considered the possibility of adding ferroelectric trimming elements so that specified filter response can be maintained over the entire tunable band. The insertion loss of the YBCO filters reported here is limited by the Q of the commercially available ferrites that we have been

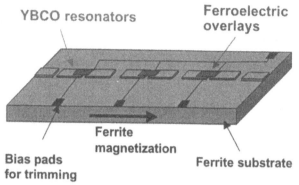

YBCO resonators

Ferroelectric overlays

Ferrite magnetization

Bias pads for trimming

Ferrite substrate

Fig. 7. Tunable ferrite/superconductor filter with ferroelectric trimming of the individual resonators.

using. Based upon our measurements and previous reports in the literature [9], ferrite Q values of greater than 10^5 should be possible at low temperatures. To achieve these Qs, high-purity ferrites free of rare-earth ions will be required. This effort is underway. We believe therefore that a figure of merit of several thousand will be possible.

ACKNOWLEDGMENTS

This work was supported by DARPA. The IBAD buffer layers and the YBCO were deposited at Los Alamos National Laboratory by Paul Arendt and Quanxi Jia. We thank R. P Konieczka for packaging, R. Boisvert for help with the measurements and G. Fitch for computer programming.

REFERENCES

[1] M. J. Lancaster, *Passive Microwave Device Applications of High Temperature Superconductors*, (Cambridge University Press Cambridge, 1997).

[2] D. E. Oates and G. F. Dionne, 1997 MTTS Digest. 303 (1997).

[3] D. E. Oates and G. F. Dionne, *IEEE Trans. Appl. Supercond,* 9, 4170 (1999).

[4] O. G. Vendik, L. T. Ter-Martirosyan, A. I. Dedyk, S. F. Karmanenko, and R. A. Chakalov, *Ferroelectrics* 144, 33 (1993).

[5] F. A. Miranda, C. H. Mueller, G. A. Koepf, and R. M. Yandrofski, *Supercond. Sci. Technol.* 8, 755 (1995).

[6] G. F. Dionne and D. E. Oates, *IEEE Trans. Magn.* 33, 3421 (1997).

[7] A. T. Findikoglu, S. R. Foltyn, P. N. Arendt, J. R. Groves, Q. X. Jia, E. J. Peterson, X. D. Wu, and D. W. Reagor, *Appl. Phys. Lett.* 69, 1626 (1996).

[8] Q. X. Jia, A. T. Findikoglu, P. Arendt, S. R. Foltyn, J. M. Roper, J. R. Groves, J. Y. Coulter, Y. Q. Li, and G. F. Dionne, *Appl. Phys. Lett.* 72, 1763 (1998).

[9] R. C. LeCraw and E. G. Spencer, *J. Phys. Soc. Japan* 17, 401 (1962)

NARROW-LINEWIDTH YTTRIUM IRON GARNET FILMS
FOR HETEROGENEOUS INTEGRATION

M. Levy,* R.M. Osgood, Jr.,* F. J. Rachford,** A. Kumar,*** H. Bakhru***
*Applied Physics Department, Columbia University, New York, NY 10027
** Naval Research Laboratory, Washington, DC 20375
***Physics Department, SUNY at Albany, Albany, NY 12222

ABSTRACT

We report on the fabrication and characterization of single-crystal films of yttrium iron garnet (YIG) for heterogeneous integration onto growth-incompatible substrates. The process entails the implantation of energetic helium ions into the material, resulting in the formation of a fast etching sacrificial layer. Separation from the growth substrate is followed by an annealing step, yielding a ferromagnetic resonance (FMR) linewidth of 0.70 Oe from a virgin sample of linewidth 0.55 Oe. Magnetization measurements show a distinct softening of the in-plane response to an applied field.

INTRODUCTION

Narrow-linewidth magnetic garnets are of interest in microwave systems because of their use as tunable low-loss filters. Of particular importance is the development of low-cost integrated technologies for microwave resonators and phased-array systems. Recently epitaxial liftoff techniques, entailing the separation of liquid-phase-epitaxy grown single-crystal ferrite films, have been applied to the problem of integrating ferrites onto semiconductor platforms.[1] Other interesting platforms to which this technology may be applied are high-temperature superconductors, such as yttrium barium copper oxides, where YIG films may be used as tunable filters in monolithic microwave integrated circuit (MMIC) devices.

To achieve separation of the ferrite film, a sacrificial layer is created in a single-crystal magnetic garnet by means of energetic He ion implantation.[1] Most of the lattice damage from the implantation occurs towards the end of the ionic trajectories,[1] resulting in the formation of a deeply buried and strongly localized one-to-two micrometer-thick sacrificial layer. This process generates a high etch selectivity between the implantation layer and the rest of the sample, leading to layer detachment.

Since the quality factor (Q) of a ferrite filter is inversely proportional to the ferromagnetic resonance (FMR) linewidth of the ferrite, we investigated the X-band FMR and the dc magnetization of narrow linewidth GaYIG samples in various stages of the separation process. A 10.8 μm-thick epitaxial $Y_3Fe_{4.6}Ga_{0.4}O_{12}$ film was employed in this study, grown on <111> oriented gadolinium gallium garnet (GGG) single-crystal substrate by liquid phase epitaxy. At this Ga doping level, GaYIG has a $4\pi M_S$ that is nominally 1070 G reduced from 1750 G for pure YIG.

The samples were implanted with 3.8-MeV singly charged helium ions impinging normal to the surface without masking. The implant dose in all cases was 5×10^{16} ions/cm^2. Other details of the implantation procedure are discussed in Ref. 1. At the above energy, the implantation damage is concentrated approximately 9 μm beneath the surface, in a 1μm spread. The crystal ion slicing occurred entirely in the GaYIG film. Implanted but unseparated samples contain a ~2 μm thick region of GaYIG shielded from direct exposure to the He ions. 650 °C

Mat. Res. Soc. Symp. Proc. Vol. 603 © 2000 Materials Research Society

and 720 °C postimplantation rapid thermal anneals (RTA) were studied. The flash annealing was done in forming gas (5% hydrogen, 95% nitrogen) after the ion implantation but before any etching, for 35 seconds. After the separation process some of the detached GaYIG films were subjected to a variety of 1-to 3-hour second anneals in an attempt to reduce the FMR linewidth. Best results were obtained with argon-oxygen flow for 3h, although longer anneals have not yet been tested. To accommodate measurement, the detached and the post annealed samples were adhered to quartz or plastic substrates by a thin film epoxy. The detached samples were found to have optical quality finishes. We have not yet attempted to quantify the surface finish or the damage and stress incurred on remounting the detached samples.

FERROMAGNETIC RESONANCE AND MAGNETIZATION

Room temperature FMR was used to characterize the samples. All measurements are done at 9.45 GHz. Measurements were performed with the magnetic field parallel and perpendicular to the plane of the film. Samples that were implanted but not detached from the GGG substrate display two distinct FMR resonances with attendant magnetostatic modes. One FMR resonance has narrower linewidth and is attributed to the shielded bottom ~2-μm portion of the GaYIG film. The larger second resonance, with broader linewidth and shifted resonance field, is associated with the implanted portion of the GaYIG film. A summary of room temperature linewidth data is shown in Fig. 1. The unimplanted sample has an excellent 0.5-Oe linewidth. Implantation degrades the linewidth to 2.7 ± 0.3 Oe. Flash annealing prior to detachment at 650 °C does not improve the linewidth; however, a 750 °C flash anneal did improve the linewidth to 1 Oe. The linewidth increased upon detachment by approximately a factor of two relative to the 750 °C flash-annealed case, possibly as a result of bonding-induced strain. Post-detachment annealing for three hours in argon/oxygen at 400 °C reduces the room temperature linewidth to 0.7 Oe. Notice the trend towards narrower linewidth with anneal time in Ar/O_2. Further narrowing may result with longer annealing.

Ga YIG FMR Linewidths (9.45 GHz)

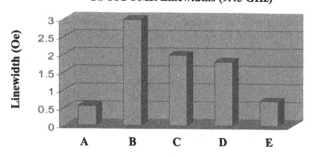

Fig. 1 Ferromagnetic resonance linewidth at various stages of the film fabrication process. A: unimplanted; B: implanted; C: detached; D: post-detachment anneal at 400 °C in Ar/O_2 for 1h; E: post-detachment anneal at 400 °C in Ar/O_2 for 3h.

If one assumes that the difference in the effective magnetization is due to variations in uniaxial anisotropy, one can solve for the gyromagnetic factor (or g-factor) and $4\pi M_{eff}$ from the parallel and perpendicular resonance relations:

$$2\pi f_{parallel} = \gamma \left(H(H + 4\pi M_{eff})\right)^{1/2},$$
$$2\pi f_{perpendicular} = \gamma (H - 4\pi M_{eff}),$$
$$4\pi M_{eff} = 4\pi M_S - H_A .$$

Here f is the microwave frequency, H is the applied magnetic field, γ is the gyromagnetic ratio, H_A is a constant uniaxial anisotropy field for easy axis normal to to the filmand $4\pi M_S$ is the saturation magnetization. We ignore the small crystal-field cubic anisotropy in writing down these relations. Concentrating on five samples: unimplanted, implanted unannealed, implanted and 720 °C flash-annealed, detached with no post-detachment anneal, and the post-detachment annealed sample (400 °C in Ar/O_2 for 3h), we plot the effective $4\pi M$ versus temperature from the in-plane and out of plane resonance field data. (Fig. 2) The apparent g value decreases roughly linearly with increasing temperature from a low temperature value of 2.10 at 5 °K to 2.035 at room temperature with a spread between samples of ± 0.8 %. Assuming that the magnitude and temperature dependence of $4\pi M_S$ is the same for all samples, then one concludes that the variations seen in $4\pi M_{eff}$ are entirely due to changes in H_A with sample processing and measurement temperature.

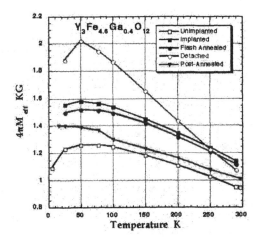

Fig. 2. Variation of $4\pi M_{eff}$ versus temperature for five samples in various stages of processing.

To further understand the variation of the magnetics on processing, we looked at these same samples with a vibrating sample (VSM) magnetometer at room temperature. From the sample volume and the magnetization at saturation we can find $4\pi M_S$. From the intercept at saturation of the in-plane and out-of-plane magnetization curves we can distinguish effective values of H_A and $4\pi M_{eff}$. In Fig. 3 we display the in-plane and perpendicular magnetization curves for the five samples discussed above. Particularly striking is the dramatic softening of the in-plane response upon implantation. The unimplanted sample apparently has a higher order anisotropy, requiring 60 to 70 Oe to bring the magnetization into the plane. Upon implantation the anisotropy is nearly eliminated. Some residual contribution remains from the lower portion of the sample shielded from the He ions. The higher-order uniaxial anisotropy in the original film is probably induced by lattice mismatch at the GGG/GaYIG interface. The softening upon implantation may be due to a decoupling of the implanted portion of the sample from the GGG substrate and the lattice mismatch at the GaYIG/GGG boundary. Varying the in-plane field angle we find a small 3 to 5 G variation in plane magnetization due crystal field effects (60° periodicity from the projection of {111} axis in the <111> plane) in both the unimplanted and implanted samples

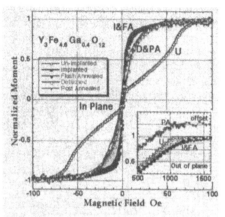

Figure 3. In-plane and out-of-plane magnetization curves for five GaYIG samples in various stages of processing. Letter designations are U: unimplanted; I: implanted; FA: flash annealed (720 °C); D: detached; PA: post-annealed (400 °C, Ar/O_2, 3 hrs). The moment has been normalized for clarity. The samples saturate at 950 G ± 80 G. The uncertainty arises from volume estimates.

CONCLUSIONS

We have used the crystal ion slicing technique to detach samples of narrow (0.5 Oe) linewidth Ga doped YIG. We find the implantation increases the FMR linewidth (3 Oe). Flash annealing partially reduces the linewidth (1 Oe) but does not fully restore it to the unimplanted state. Detached samples have increased linewidth (2 Oe) probably due to damage on detachment and remounting. Proper annealing after detachment can reduce the room temperature FMR linewidth to values approaching the unprocessed sample.

ACKNOWLEDGEMENTS

We wish to thank Drs. Peter Lubitz and James Krebs of NRL for useful discussions. M.L. and R.M.O. acknowledge support by MURI/DARPA under Contract No. F49620-96-1-0111 0111 and DARPA/FAME under Contract No. N00173-98-1-G014.

REFERENCES

[1] M. Levy, R.M. Osgood, Jr., A. Kumar, H. Bakhru, Appl. Phys. Lett. 71, 2617 (1997).

Frequency Agile Microwave Applications Using (Ba,Sr)TiO$_3$/Y$_3$Fe$_5$O$_{12}$ Multilayer Grown by Pulsed Laser Deposition

W.J. Kim*, W. Chang**, S.B. Qadri**, H.D. Wu*, J.M. Pond**, S.W. Kirchoefer**, H.S. Newman**, D.B. Chrisey** and J.S. Horwitz**
* SFA Inc.,1401 McCormick Dr., Largo, MD 20774
** Naval Research Laboratory, 4555 Overlook Ave., Washington D.C. 20375

ABSTRACT

(Ba,Sr)TiO$_3$ (BST) thin films have been deposited by pulsed laser deposition (PLD) onto single crystal Y$_3$Fe$_5$O$_{12}$ (YIG) substrates with/without an MgO buffer layer. The structure and microwave properties of BST films have been investigated as function of substrate orientation and O$_2$ deposition pressures (5-800mTorr). The orientation of BST film is varies with the deposition conditions. The dielectric constant, loss tangent, and change in dielectric constant with an applied electric field have been measured at room temperature using interdigitated capacitors at 0.1 – 20 GHz. Polycrystalline BST films have a high tunability (~40%) with a dc bias field of 67kV/cm and a dielectric Q (=1/tanδ) between 30 and 40, while (001) oriented BST films have a lower tunability (~20%) but higher dielectric Q (~50). A coplanar waveguide transmission line was fabricated from a (001) oriented BST film on (111)YIG which exhibited a 17° differential phase shift with an applied dc bias of 21 kV/cm. An equivalent differential phase shift was achieved with a magnetic field of 160Guass.

INTRODUCTION

Ferroelectric thin films are being used to develop a new class of tunable microwave devices.[1,2] The ferroelectric material (Ba$_{0.5}$Sr$_{0.5}$)TiO$_3$ (BST) exhibits a large electric field dependant dielectric constant, which can be used to produce a resonant frequency shift in a tunable oscillator or a time delay (phase shift) in a transmission line.

A concern in the development of tunable microwave circuits, such as a coplanar waveguide (CPW) transmission lines, is the large change in the characteristic impedance (Z_o) of the device, which occurs when the dielectric constant of the ferroelectric is reduced by a factor of four or more. A novel approach to this problem is to fabricate devices from a ferroelectric/ferrite multilayer.

In a coplanar waveguide (CPW) transmission line, the differential phase shift ($\Delta\phi$) of CPW transmission line between two bias states can be expressed by following,

$$\Delta\phi = \frac{2\pi f}{c} l \ (\sqrt{\varepsilon_{eff}^1 \mu_{eff}^1} - \sqrt{\varepsilon_{eff}^2 \mu_{eff}^2}\ ), \qquad (1)$$

where f is microwave frequency, l is the length of transmission line, c is light velocity in vacuum, and ε_{eff} and μ_{eff} are effective dielectric constant (permittivity) and permeabilty of device, respectively, and superscript 1 and 2 are indicating zero bias and an applied bias states, respectively. The characteristic impedance Z_o of the CPW transmission line is related as following equation,

$$Z_o \propto \frac{\sqrt{\mu_{eff}}}{\sqrt{\varepsilon_{eff}}}. \qquad (2)$$

Using eqs. (1) and (2), estimated $\Delta\phi$ and Z_o are shown in Table 1 and 2, respectively. The values are calculated at the operating frequency of 15GHz for an 1cm long transmission line, with a 2:1 change in ε_{eff} (μ_{eff}) from zero to non-zero electric (magnetic) bias field. The differential phase

Mat. Res. Soc. Symp. Proc. Vol. 603 © 2000 Materials Research Society

shift increases more with two bias fields than either an electric field or a magnetic field. The characteristic impedance of the transmission line behaviors differently; it changes a factor of $\sqrt{2}$ with either an electric field or a magnetic field by itself, while it remains at a constant value Z_o with both electric and magnetic fields.

In this paper, we report the growth and characterization of ferroelectric/ferrite multilayers to demonstrate independent control of the characteristics of the transmission line, and to lay the foundation for the development of a constant impedance device. The selected materials for the ferroelectric/ferrite multilayer stucture are $(Ba_{0.5}Sr_{0.5})TiO_3$ (BST) and $Y_3Fe_5O_{12}$ (YIG), respectively. The BST films show a large dielectric constant change with dc bias electric field and YIG shows a large permeability change with applied magnetic field, while both are reported as low dielectric loss materials at microwave frequencies.[1-3]

Table 1. Calculated differential phase shift.

$\Delta\phi$	$\varepsilon_{eff}= 2$	$\varepsilon_{eff}= 1$
$\mu_{eff}= 2$	0	75°
$\mu_{eff}= 1$	75°	180°

Table 2. Calculated impedance.

Z_o	$\varepsilon_{eff}= 2$	$\varepsilon_{eff}= 1$
$\mu_{eff}= 2$	Z_o	$\sqrt{2}\, Z_o$
$\mu_{eff}= 1$	$Z_o/\sqrt{2}$	Z_o

EXPERIMENT

Single phase BST films were deposited using pulsed laser deposition (PLD) onto LPE grown single crystal (001) or (111)YIG on single crystal $Gd_3Ga_5O_{12}$. A short pulse of Kr:F excimer laser (248nm, 30ns FWHM) was focused on the rotating BST target with energy density of ~ 2 J/cm^2 in flowing O_2 at pressures between 50 and 800 mTorr. PLD provides unique advantages for the deposition of multi-component oxide films, because it reproduces the stoichiometry of the target in the deposited film.[4] To grow (001) oriented epitaxial BST films on (001)YIG, a thin layer of MgO has been deposited on YIG prior to the BST deposition. The thickness of the MgO and BST were 100 and 500 nm, respectively. The structure of the multilayer films was investigated by an x-ray diffraction. Cross-sectional view of the multilayer films were investigated by transmission electron microscopy. Microwave properties of the BST films on YIG were measured at 0.1-20 GHz range by HP 8510C network analyzer using interdigitated capacitors and coplanar transmission line fabricated from depositing Ag electrodes with a thin Au layer by e-beam evaporation through a PMMA lift off mask. Dielectric constants were extracted using a modified conformal-mapping partial-capacitance method from measured capacitance and dimension of the capacitors.[5]

STRUCTURAL AND MICROWAVE PROPERTIES OF BST/YIG

The crystallographic orientation of deposited BST thin film depends on the crystallographic orientation of YIG substrates and the deposition conditions. Further, the deposition of a thin layer of MgO prior to the BST deposition also affects the crystallographic orientation of the BST films. We investigated the BST film growth on (001) and (111)YIG substrates. The substrate temperature for the BST film deposition onto YIG is fixed at 850°C to achieve a large dielectric constant change with applied dc electric field.

BST on (001) YIG

Figure 1 shows the x-ray diffraction patterns obtained for BST films deposited onto (001)YIG substrates with/without a thin MgO buffer layer. The deposited BST film on (001)YIG at 850°C without an MgO layer is a single phase and polycrystalline (Figure 1 (a)), even though there is a close match in the lattice parameters. The lattice mismatch is less than 3%

between 3 times of the bulk lattice parameter of BST (3.965Å) and that of YIG (12.380Å). It is worth to note that the lattice mismatch between BST and MgO is more than 6%, however, (001) oriented epitaxial BST films are readily observed on (001)MgO.

To grow an epitaxial BST film on YIG, a 1000Å MgO buffer layer was used. Figure 1 (b) and (c) show x-ray diffraction patterns of BST/MgO/YIG multilayers. The orientation of BST film deposited at 850°C on MgO/YIG show a strong dependence on the MgO deposition temperature. A strong $(002)_{BST}$ reflection is observed from the BST film on the MgO layer deposited at 750°C (figure 1 (b)), while a strong $(111)_{BST}$ reflection is found in

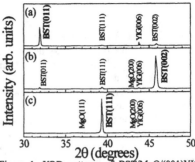

Figure 1. XRD patterns of BST/MgO/(001)YIG shows different orientations of BST. (a) (011)BST (850°C)/YIG, (b) (001)BST(850°C)/MgO(750°C) /YIG, and (c) (111)BST(850°C)/MgO(850°C)/YIG.

the BST film on the MgO layer deposited at 850°C (figure 1 (c)). High resolution x-ray diffraction measurements made using a 4-circle diffractometer shows that BST(850°C)/MgO(750°C)/YIG multilayers grows epitaxially; <001>BST // <001>MgO // <001>YIG with a ~18.5° rotation in in-plane direction between <010>BST // <010>MgO and <010>YIG, which relation is same as those reported from YBCO/MgO/YIG.[6] However, for the BST(850°C)/MgO(850°C)/YIG structure, there is no preferential in-plane orientation, while surface normal direction has a relation (<111>BST // <111>MgO // <001>YIG). From TEM we confirmed the orientational relation observed by x-ray diffraction. Bright field images of BST/MgO/YIG show columnar grains of BST films with ~50nm column width, which is typical for the films grown by PLD.

The measured microwave properties, quality factor Q (=1/tanδ), dielectric constant and %tuning (=$(\varepsilon_0-\varepsilon_b)/\varepsilon_0 \times 100$, where ε_0 and ε_b are dielectric constant at 0 and 67kV/cm), at 10GHz are summarized and shown in figure 2. The dielectric constant of the BST films ranges from 400 to 1200, which is comparable to those measured for BST films on dielectric substrates. The dielectric constant at zero bias field and %tuning (1200 and 40%, respectively) is a maximum for polycrystalline BST film on (001)YIG without an MgO buffer layer, and dielectric Q ranges 30-40. The (001) oriented BST film on MgO/YIG shows a high value for Q (~40 with zero bias field) and ~20% tuning. The BST film with the highest dielectric constant shows the largest dielectric constant change (~40%). The polycrystalline BST film shows a better overall figure of merit (= Q(0V) × %tuning), ~1100, than other two BST films, (001) epitaxial film and (111) textured film (~750 and ~870, respectively).

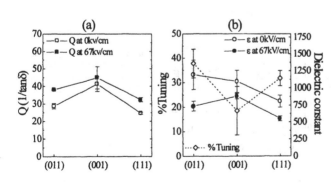

Figure 2. Microwave properties of BST on (001) YIG at 10GHz; (a) Quality factor Q (=1/tanδ) and (b) Dielectric constant and its % Tuning.

BST on (111) YIG

Figure 3 shows x-ray diffraction patterns for BST films grown on (111)YIG single crystal with different O_2 deposition pressures of 50, 350 and 800mTorr (from bottom to top) at deposition temperature of 850°C. An interesting relation between structure and microwave properties was observed for epitaxial BST films deposited on MgO using different O_2 deposition pressures.[7] In this study, a large tetragonal distortion was observed in the deposited films. The distortion was a minimum for films grown at 50mTorr, which also corresponded to films with a high dielectric constant and a large dielectric

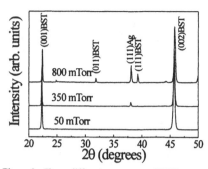

Figure 3. X-ray diffraction patterns of BST grown on (111)YIG with different O_2 deposition pressures. All BST films show a preferential growth along (001) direction.

constant change with applied dc bias field.[7] All three BST films grown on (111)YIG exhibit intense (001) and (002) reflections, indicating that the films are (001) oriented, however, no preferential in-plane relation between BST and YIG was observed. Therefore, the BST films are strongly textured but not epitaxial as in the case of BST on MgO and BST on MgO/(001)YIG. The x-ray diffraction pattern of BST film deposited in 800mTorr of O_2 pressure exhibits weak intensity for $(011)_{BST}$ and $(111)_{BST}$ reflections, suggesting that this film is less textured than those deposited in 50 and 350mTorr.

The measured microwave properties, quality factor Q, capacitance, and %tuning at 10GHz, for the BST films on (111)YIG are shown in figure 4. As the O_2 deposition pressure is increased, the Q increases, while dielectric constant and %tuning exhibit a maximum at 350mTorr. The figure of merit for the BST film deposited in 800mTorr is ~450, which is better than those for the BST films deposited in 50 and 350mTorr (100-300). Interestingly, the less textured BST film grown at 800mTorr on (111)YIG exhibits a better figure of merit than those deposited in low O_2 deposition pressures.

MICROWAVE PROPERTIES OF PHASE SHIFTERS

Coplanar waveguide (CPW) transmission lines were fabricated by PMMA lift off method from metal conductor (Au/Ag) on top of BST films deposited on different type of substrates,

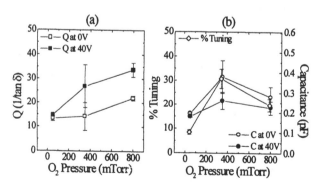

Figure 4. Microwave properties of BST on (111) YIG at 10GHz; (a) Quality factor Q (=1/tanδ) and (b) Dielectric constant and its %tuning.

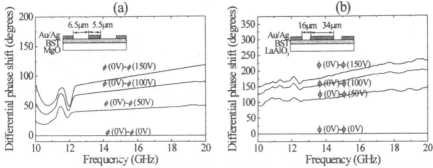

Figure 5. Measured differential phase shift ($\Delta\phi = \phi(0V) - \phi(bias)$) from coplanar transmission lines fabricated on ferroelectric BST films grown on dielectric materials; (a) BST/MgO with upto 270kV/cm dc bias field, and (b) BST/LaAlO$_3$ with upto 94kV/cm dc bias field.

MgO, LaAlO$_3$, and YIG. The differential phase shifts ($\Delta\phi = \phi$(zero bias)-ϕ(bias)) were measured as a function of the dc bias field. Magnetic and electric field tuning was achieved for the CPW transmission line for BST on YIG substrate.

Coplanar Transmission Line for Phase Shifter on BST on Dielectrics

Two CPW lines were fabricated from BST films deposited on MgO and LaAlO$_3$ single crystals, and the measured differential phase shifts are shown in figure 5 (a) and (b), respectively. The CPW line on BST/MgO has 5.5µm gap between center conductor and ground electrode and 6.4µm width of center conductor line, while that on BST/LaAlO$_3$ is 16µm and 34µm, respectively. The dc bias electric field was applied between center conductor and ground electrodes. The amount of differential phase shift on BST/LaAlO$_3$ is 160 - 240° at 10 - 20GHz with a dc bias field of 94kV/cm, while that on BST/MgO is 90 - 120° at 10 – 20GHz with 270kV/cm. Frequency dependent differential phase shift from BST films on dielectric substrates is proportional to the frequency.

Coplanar Transmission Line for Phase Shifter on BST on (111)YIG

To measure the magnetic and electric field effect at the same time, CPW transmission lines were fabricated from metal (Au/Ag) conductor on BST grown on (111)YIG. Figure 6 shows frequency dependent differential phase shift as a function of either electric and/or magnetic bias fields. The center conductor line (1cm long and 30µm width) is separated by 19µm from ground electrodes. The electric field is applied between a center conductor and two ground electrodes, while magnetic filed is applied along a center conductor line. The differential phase shift at 10GHz is 17° with a dc electric field of 21kV/cm. The same amount of differential phase shift at 10GHz is achieved with magnetic field of 160Gauss. This phase shift is equivalent with 4.7ps of time delay. With both 21 kV/cm of electric field and 160Guass of magnetic field, the measured differential phase shift at 10GHz is 32° (equivalent with 8.9ps of time delay). At 21kV/cm of electric bias field, estimated phase shifts from CPW transmission lines fabricated on BST/MgO and BST/LaAlO$_3$ are 12° and 70°, respectively, which are comparable with 17° and 32° phase shift observed on that of BST/YIG. At higher frequencies, the differential phase shift associated with electric field tuning increases monotonically, while that associated with magnetic field

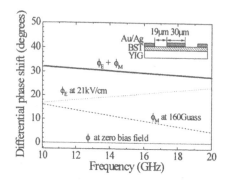

Figure 6. Measured differential phase shift ($\Delta\phi =$ ϕ(no bias) $-$ ϕ(bias)) from coplanar transmission lines fabricated on ferroelectric BST films grown on ferrite (111)YIG with electric dc bias field of 21kV/cm and/or magnetic field of 160Gauss.

decreases. The frequency dependence of the differential phase shift with magnetic field is proportional to $1/(f_o^2-f^2)$ which is similar to the magnetic susceptibility.

CONCLUSION

High quality single phase BST films were deposited on YIG substrates by PLD. The structure of BST film on YIG grown by PLD depends on the deposition temperature, buffer layer and the orientation of substrates. A (001) oriented BST film was grown at 850°C on (001)YIG with a thin buffer layer of MgO grown at 750°C. The polycrystalline BST film on (001)YIG exhibits better microwave properties among films grown at similar conditions.

A coplanar waveguide transmission line (1 cm long) fabricated from metal (Au/Ag) conductor on BST deposited onto (111)YIG exhibits a 17° differential phase shift with an applied electric field of 21kV/cm at 10GHz. An equivalent differential phase shift is achieved with a magnetic field of 160Gauss applied along center conductor line at 10GHz. For comparison, differential phase shifts with only electric dc bias field are measured from two coplanar transmission lines fabricated on BST grown on dielectric substrates, such as MgO and LaAlO$_3$. A transmission line on BST/MgO shows a 90° of phase shift with 270kV/cm at 10GHz, and that on BST/ LaAlO$_3$ shows a 160° of phase shift with 94kV/cm at 10GHz.

ACKNOWLEDGEMENT

The authors would like to thank Dr. D. Webb for a helpful discussion. This work was partially supported by DARPA.

REFERENCES

1. W. Chang, J.S. Horwitz, A.C. Carter, J.M. Pond, S.W. Kirchoefer, C.M. Gilmore and D.B. Chrisey, Appl. Phys. Lett. **74**, 1033 (1999).
2. H.D. Wu, and F.S. Barnes, Integrated Ferroelectrics **22**, 291 (1998).
3. D.E. Otaes, G.F. Dionne, D.H. Temme, and J.A. Weiss, IEEE Trans. Appl. Super. **7**, 2347 (1997)
4. *Pulsed Laser Deposition of Thin Films* edited by D.B. Chrisey and G.K. Hubler, John Wiley & Sons, Inc., New York (1994).
5. S.S. Gevorgian, T. Martinsson, P.I.J. Linner, and E.L. Kollberg, IEEE Trans. Microwave Theory Tech., **44**, 896 (1996).
6. Alberto Pique, Ph. D. Thesis in Physics, Univ. of Maryland (1996).
7. W.J. Kim, W. Chang, S.B. Qadri, J.M. Pond, S.W. Kirchoefer, D.B. Chrisey, and J.S. Horwitz, Appl. Phys. Lett. – Submitted.

High-Frequency Applications
of Ferroelectrics

PROPERTIES OF Zr-SUBSTITUTED (Ba, Sr)TiO$_3$ THIN FILMS FOR INTEGRATED CAPACITORS

G. T. STAUF*, P. S. CHEN*, W. PAW*, J. F. ROEDER*, T. AYGUAVIVES**, J.-P. MARIA**, A. I. KINGON**.
*Advanced Technology Materials, 7 Commerce Dr., Danbury, CT 06810, gstauf@atmi.com
** North Carolina State University Department of Materials Science and Engineering, Raleigh, NC 27695

ABSTRACT

There has been significant interest recently in use of BaSrTiO$_3$ (BST) thin films for integrated capacitors; these devices have benefits for high frequency operations, particularly when high levels of charge or energy storage are required. We discuss the electrical properties of BST thin films grown by metalorganic chemical vapor deposition (MOCVD) which make them suitable for these applications, as well as the impact of processing conditions such as growth temperature on specific film properties. We have also examined addition of Zr in amounts ranging up to 20% to the BST films. X-Ray diffraction indicates that the Zr is incorporated into the BST lattice. Voltage withstanding capability, leakage and dielectric constant of the thin films have been measured as functions of deposition temperature and Zr content. Addition of Zr to BST films increases breakdown voltages by as much as a factor of two, to approximately 2 MV/cm, raising their energy storage density values to levels approaching 30 J/cc. Charge storage densities of above 60 fF/μm^2 were also obtained.

INTRODUCTION

The market for personal communications devices such as cellular phones and computer notepads with built-in faxes and modems is expected to undergo explosive growth over the next decade. Circuits in these devices typically have many bypass and decoupling capacitors which operate at low voltages and that must occupy very little space. Due to the high cost of chip area for Si and GaAs active devices, an improvement over the charge storage density achievable with SiO$_2$ dielectrics is needed. Ba$_x$Sr$_{(1-x)}$TiO$_3$ (BST) offers this possibility. Although DRAM applications have been examined for this material, integrated capacitors in communications devices add several new challenges. The devices must operate at high frequencies (100 MHz to several GHz) and often at high temperatures, due to close packing of circuit components.

Several options are available to improve the basic properties of BST, including process modification, electrode modification (metals vs. conductive oxides) and doping or substitution. There have been reports that Zr addition can reduce relaxation and leakage currents in chemical solution deposited (CSD) thin films of BaTiO$_3$.[1] In this work we have investigated the effect of adding Zr as a substitutional cation on the B-site of the perovskite lattice, examining electrical properties in these thin films. Significant improvements in breakdown voltage, leakage properties, high temperature performance and storage density due to Zr incorporation were seen in metalorganic chemical vapor deposited (MOCVD) films in this work, and we will discuss applicability of the substituted films to charge storage and energy storage capacitors. To the best of our knowledge, this is the first systematic examination of which we are aware of Zr addition to BST thin films made by MOCVD.

Mat. Res. Soc. Symp. Proc. Vol. 603 © 2000 Materials Research Society

EXPERIMENT

Barium bis(2,2,6,6-tetramethyl-3,5-heptanedionate) (abbreviated (thd)) with Lewis base adduct (Ba(thd)$_2$)·pmdeta and (Sr(thd)$_2$)·pmdeta were used as the Group II metal source reagents, while titanium bis(isopropoxide)bis(thd) [Ti(OCH(CH$_3$)$_2$)$_2$(thd)$_2$] was chosen as the titanium precursor and Zr(thd)$_4$ as the Zr precursor.[2] These compounds were dissolved in N,N,N',N'',N''-pentamethyldiethylenetriamine (pmdeta)/solvent solution[3] and flash vaporized in an ATMI vaporizer.

A modified Watkins Johnson (WJ) Select MOCVD reactor with 6" wafer capability was used for BSZT deposition. Typical growth conditions have been reported previously.[4] Growth temperatures were measured by calibrating the susceptor with an instrumented wafer, and ranged from 540 to 660 °C. Reactor pressure was 750 mTorr, with growth rates of 3-6 nm/min and 500 sccm of O$_2$ and N$_2$O as oxidizers. Substrates were 6" wafers of Ir/SiO$_2$/Si with the Ir bottom electrode prepared by sputtering in a Varian M2I tool. A continuous thin film of BST or Zr-incorporated BST (BSZT) was then deposited over this by MOCVD, and Pt top electrodes were then e-beam evaporated over this through a shadow mask to define top electrodes. An HF etch was used to remove BST and expose an area of the bottom electrode for testing, a post-anneal for approximately ½ hour at 500 °C was performed. Measurements of capacitance and breakdown were made with a computer-controlled multifrequency HP 4192A Impedance Analyzer.

RESULTS

Two major variables were examined in this study, deposition temperature and level of Zr incorporation in films. Crystallinity of basic BST has a strong effect on dielectric constant; lower temperature film growth tends to form amorphous or microcrystalline BST, which can have dielectric constants as low as 20-30. By contrast, film growth above about 600 °C enhances grain growth, and BST in the grains has more long range order, giving dielectric constants between 300 and 500. The effect of large amounts of Zr incorporation on physical and electrical properties was examined since the large amounts of Zr used with some samples could cause second phase formation with concomitant change in microstructure. These might form weak links for breakdown, or reduce the dielectric constant of the Zr-incorporated BST (BSZT). The range examined for film deposition temperature was from 540 °C to 660 °C, while Zr incorporation ranged from zero up to approximately 20%.

Figure 1. Zr incorporation in BST film compared to fraction injected into reactor, showing that incorporation is linear with increasing Zr in the solution. Note that efficiency is lower than that of the other precursors.

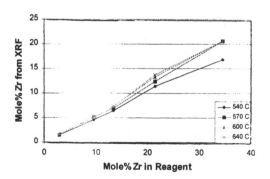

We first examined the incorporation efficiency of the Zr(thd)₄ precursor in the MOCVD process. Ideally one would like this to have a linear behavior, such that addition of Zr along with the BST precursors would lead to a predictable amount of incorporation into the film. As can be seen in Figure 1, this behavior was observed. It can be seen, though, that incorporation efficiency of Zr was lower than that of the Ba, Sr and Ti precursors. For example, 30% (by mole) of Zr(thd)₄ in the solution gives about 17% Zr in the film. Incorporation is relatively independent of deposition temperature, though some decrease compared to the other precursors can be seen at the lowest temperature examined (540 °C).

The physical effects of incorporating Zr into the BST lattice were studied via x-ray diffraction (XRD). At the highest growth temperature, 660 °C, BSZT films were primarily <110> oriented. With the replacement of Ti by Zr, lattice parameter steadily increased (as would be expected from the larger atomic radii), resulting in a shift of XRD peaks to lower 2θ angles. This is visible in Figure 2(a) showing shift of the (110) peaks of the BSZT XRD spectra with increasing Zr incorporation. Complete θ -2 θ XRD scans revealed no sign of secondary phases in the BSZT films. XRD spectra in Figure 2(b) for films grown at temperatures below about 600 °C show no peaks, indicating amorphous or very finely grained nanocrystalline material.

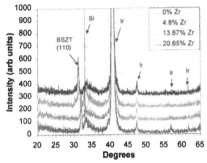 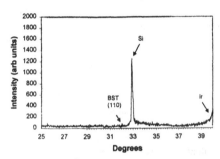

Figure 2. (a) XRD spectra of BSZT films approximately 500 Å thick, showing peak shift to larger lattice spacings for the (110) BSZT peak as Zr content in the film increases (on left) and (b) amorphous material resulting from growth below 600 °C (on right).

Atomic Force Microscopy (AFM) measurements made on films with and without Zr incorporation showed no significant difference in RMS film roughness due to addition of Zr.

We next examined electrical properties of BSZT. First, we observed that films with and without Zr grown at low temperatures (540-570 °C) have low dielectric constants, around 30-50, as shown in Figure 3. This is consistent with the degree of crystallinity for this material, as it is largely amorphous at these growth temperatures. The dielectric constant of this low temperature material was not affected significantly by Zr addition, though there was some decrease at the highest Zr levels. At an intermediate growth temperature of 600 °C, there was a more significant effect from Zr incorporation. Addition of 5% Zr lowered the dielectric constant from 200 to below 100. Finally, at a growth temperature of 660 °C the dielectric constant started out above 350 in unsubstituted BST, and decreased slowly and approximately linearly to approximately 140 at 20% Zr. This points to use of relatively low levels of Zr (2-10%) incorporation and high growth temperatures as being optimum for charge storage capacitor dielectrics, as relatively high ε values are retained. Scatter in the dielectric constant values with the high temperature films are believed to be due to a combination of possible interface layers at the growth interface (the reactor used did not have a load lock) and variations in Group II/(Ti+Zr) ratio in the films.

Figure 3. BSZT permittivity as a function of deposition temperature and Zr content in films approximately 500 Å thick. Low temperature growths have uniformly low ε, and Zr addition lowers ε in films grown at higher temperatures.

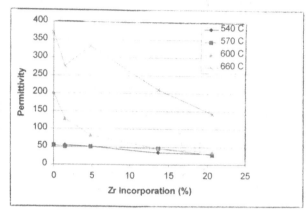

The other electrical property which has a strong influence on storage density for an integrated application is breakdown voltage. We have observed that leakage correlates well with trends in breakdown voltages, that is, lower leakage is seen for films which exhibit higher breakdown strength. As can be seen in Figure 4, addition of Zr increases breakdown voltage (V_b) in BSZT thin films by as much as a factor of two, from 1 MV/cm to over 2 MV/cm. This was reflected in lower measured leakage currents for Zr substituted material as well.

Figure 4. BSZT breakdown strength as a function of deposition temperature and Zr content in films approximately 500 Å thick. Low temperature growths have somewhat lower V_b, and Zr addition increases V_b in all films.

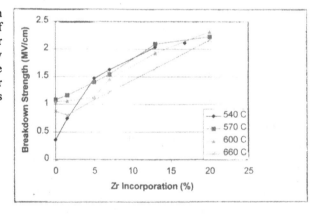

Proposed applications for thin film and integrated capacitor devices can be divided into two categories, charge storage, or capacitance per unit area (C/A), and energy storage, which is storage of the maximum possible amount of "reserve" electrical energy to handle surge demands from circuits or motors. Energy storage density is calculated by combining permittivity and breakdown strength to make a figure of merit according to the following equation;

$$U \text{ (J/cc)} = 1/2 K \varepsilon_o F^2 \tag{1}$$

where U is volumetric energy storage density, K is dielectric constant of the material, ε_0 is the permittivity of free space, and F refers to the electric field on the capacitor. This combined figure of merit for BSZT films based on the breakdown voltage and permittivity data of Figures 3 and 4 is shown in Figure 5. Energy density calculations are based on BST dielectric volumes only, and do not include the substrate or electrodes for this preliminary work since it has not yet been established what thicknesses of substrate and electrode are the minimums necessary for practical devices. For comparison, the best bulk ceramic capacitors typically reach 2 J/cc.

Finally, charge storage density, or capacitance per unit area (C/A), is perhaps the most important electrical parameter in order for BSZT to displace existing dielectric thin films. For integrated capacitor applications, we seek high charge storage density so that compact structures can be made on ICs, which are very expensive per unit area. The standard material for these applications, SiON, can achieve about 2.2 fF/μm^2, or 220 nF/cm^2. Ta$_2$O$_5$ has also been examined for embedded DRAMs, but gains only about an order of magnitude, to about 20 fF/μm^2. Charge storage density for BSZT films is shown in Figure 5, and reaches over 60 fF/μm^2.

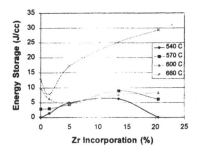 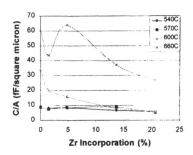

Figure 5. (left) BSZT energy storage density (J/cc), showing storage densities above 10 J/cc, with higher Zr content pushing density up to almost 30 J/cc. (right) BSZT charge storage density, C/A, showing charge storage above 60 fF/μm^2. High temperature film growth is clearly important in achieving high charge and energy storage, but optimum Zr content is different for the two applications.

While both the energy storage and charge storage applications are maximized for the highest growth temperatures examined for BSZT, 660 °C, different amounts of Zr incorporation are optimum for each application. Due to the strong sensitivity of energy storage to electric field (it appears as a squared term), higher Zr contents are optimum for high energy storage. Reduction of dielectric constant with increasing Zr, on the other hand, point to lower (0-5%) Zr content being optimum for high levels of charge storage. Some evidence was also seen that low levels of Zr reduce leakage currents at high measurement temperatures (up to 300 °C), which may point to improved long term degradation behavior for Zr-incorporated BST. This effect will require further examination.

We also examined tunability and loss in a film grown on a Pt bottom electrode with Zr incorporation at a level of 1.5%. Addition of Zr reduced tunability, from approximately 2:1 in the unsubstituted film to 1.2:1, as shown in Figure 6. Losses at 10 kHz were measured, and were approximately the same as in a similar unsubstituted film, tan δ = 0.005. We intend to carry out further examination of BSZT films with varying levels of Group II/(Ti+Zr), with particular emphasis on loss reduction at high (RF and microwave) frequencies.

Figure 6. BSZT C-V curve, showing tuning of dielectric constant with applied voltage (film approximately 700 Å thick). Tuning of 1.2:1 was seen, with tan δ = 0.005 at 10 kHz. In contrast with other films in this study, Pt was used as the bottom electrode.

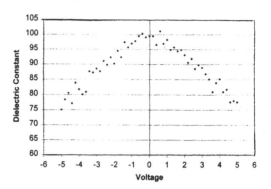

CONCLUSIONS

An MOCVD process to deposit high quality BSZT films has been developed. A range of Zr content from 0 to 20% was examined, and film deposition temperatures ranging from 540 °C to 660 °C. These materials prove to have useful properties, in particular high levels of charge storage density (C/A) and energy storage density (J/cc). This data shows that Zr incorporation can significantly enhance the properties of BST for use in embedded capacitor applications. Optimum film growth temperature for both applications was 660 °C, while low Zr content was optimum for charge storage and high Zr content was optimum for energy storage.

ACKNOWLEDGEMENTS

This work has been funded by Wright Patterson Air Force Base (Contract # F33615-98-C-2850), and by DARPA (Contract # F33615-98-C-5411).

REFERENCES

1. "Dielectric Properties, Leakage Behaviour, and Resistance Degradation of Thin Films of the Solid Solution Series $Ba(Ti_{1-y}Zr_y)O_3$", S. Hoffmann and R. M. Waser, Integrated Ferroelectrics **17**, p. 141 (1997).

2. "Nuclear Magnetic Resonance and Infrared Spectral Studies on Labile cis-Dialkoxy-bis(acetylacetonato) titanium(IV) Compounds.", D.C. Bradley and C.E. Holloway, J. Chem. Soc. (A), p. 282 (1969).

3. "Advances in Precursor Development for CVD of Barium-Containing Materials," B.A. Vaartstra, R.A. Gardiner, D.C. Gordon, R.L. Ostrander and A.L. Rheingold. Presented at the Fall MRS Meeting, Boston, 1993.

4. "MOCVD of $BaSrTiO_3$ for ULSI DRAMs", P. Kirlin, S. Bilodeau, P. Van Buskirk, Integrated Ferroelectrics **7**(2), p. 307 (1995).

BISMUTH PYROCHLORE FILMS FOR DIELECTRIC APPLICATIONS

Wei REN, Ryan THAYER, Clive A. RANDALL, Thomas R. SHROUT, and Susan TROLIER-McKINSTRY

Materials Research Laboratory, Pennsylvania State University, University Park, PA 16802-4801

ABSTRACT

Bismuth pyrochlore ceramics have modest temperature coefficients of capacitance, good microwave properties, and can be prepared at relatively modest temperatures (~900 – 1100 °C). This work focuses on the preparation and characterization of thin films in this family for the first time. A sol-gel procedure using bismuth acetate in acetic acid and pyridine, in combination with zinc acetate dihydrate and niobium ethoxide in 2-methoxyethanol was developed. The solution chemistry was adjusted to prepare $(Bi_{1.5}Zn_{0.5})(Zn_{0.5}Nb_{1.5})O_7$ and $Bi_2(Zn_{1/3}Nb_{2/3})_2O_7$ films. Solutions were spin-coated onto platinized Si substrates and crystallized by rapid thermal annealing. In both cases, crystallization occurred by 550 °C into the cubic pyrochlore structure. $(Bi_{1.5}Zn_{0.5})(Zn_{0.5}Nb_{1.5})O_7$ films remained in the cubic phase up to crystallization temperatures of 750 °C, while the structure of the $Bi_2(Zn_{1/3}Nb_{2/3})_2O_7$ thin films is dependent of the firing temperature: cubic below 650 °C and orthorhombic above 750 °C. A mixture of cubic and orthorhombic structures is found at 700 °C. The resulting BZN films are dense, uniform, and smooth (rms roughness of < 5 nm). Cubic bismuth zinc niobate films show dielectric constants up to 150, a negative temperature coefficient of capacitance, TCC, (~ - 400 ppm/°C), tan δ < 0.01, and a field tunable dielectric constant. Orthorhombic films showed smaller dielectric constants (~80), low tan δ (< 0.01), positive TCC, and field independent dielectric constants. TCC could be adjusted to new 0 ppm/°C using a mixture of orthorhombic and cubic material.

INTRODUCTION

The rapid development of communication technologies, especially mobile communication systems, is facilitated by miniaturization of devices. Dielectric materials for these applications must possess a large dielectric constant, low loss tangent (tan δ), high dielectric quality factor Q, and a small temperature coefficient of resonator frequency for resonators. The Bi_2O_3-ZnO-Nb_2O_5 (BZN) pyrochlore ceramic system with medium dielectric constants and high Q values has been developed for low firing multilayer capacitors [1-4]. Recent studies showed that some members of the BZN system also exhibit excellent microwave properties [5, 6]. BZN ceramics with dielectric constants >100, temperature coefficient of resonant frequency $|T_f|$ < 10 ppm/°C and Q-f > 5000 GHz have been reported [6]. The good dielectric properties shown by the BZN system suggest that thin films of this composition are potential materials for integrated microwave resonators and decoupling capacitors. Thin films may have the advantage of lower crystallization temperatures and smaller device size than bulk ceramics and can be integrated in microelectronic devices. To our knowledge, no BZN thin films have been reported previously.

In this study, we report on the fabrication and properties of BZN thin films prepared by a sol-gel process. The sol-gel process has been widely used to deposit dielectric and ferroelectric thin films due to low processing temperatures, precise composition control, uniform deposition over large area substrate and low cost.

EXPERIMENTAL PROCEDURE

Two compositions of BZN thin films were investigated in this study: $(Bi_{1.5}Zn_{0.5})(Zn_{0.5}Nb_{1.5})O_7$ and $Bi_2(Zn_{1/3}Nb_{2/3})_2O_7$. These two compositions are members of the general family $(Bi_{3x}Zn_{2-3x})(Zn_xNb_{2-x})O_7$.

137

The starting materials in the sol-gel process were bismuth acetate, zinc acetate dihydrate and niobium ethoxide (Aldrich). 2-methoxyethanol, pyridine and acetic acid (Aldrich) were selected as solvents. Zinc acetate dihydrate was first mixed with 2-methoxyethanol and vacuum distilled at 110 °C to expel the water of hydration and to prevent niobium ethoxide from hydrolyzing. Niobium ethoxide was then added into the solution and refluxed at 120 °C for 1 h to form a (Zn, Nb) complex precursor, followed by vacuum distillation of by-products. The solution was cooled down below 80 °C. In a separate flask, bismuth acetate was mixed with pyridine and stirred for 0.5 h. 30 vol% of acetic acid was added into the solution and stirred for 1 h until the solution became totally clear. The bismuth acetate solution was then added to the (Zn, Nb) precursor and the solution was refluxed at 120 °C for 0.5 h. After vacuum distilling off by-products, the final precursor solution was diluted using 2-methoxyethanol to a concentration of 0.3 M.

To prepare films, the precursor solution was spin coated on platinum-coated Si wafers Pt/Ti/SiO$_2$/Si (Nova Electronic Materials, Inc., Richardson, TX) at a speed of 3000 rpm for 30 s. The as-deposited films were pyrolyzed on a hot plate at a temperature of 350 °C for 1 min to remove the organics. The coating-pyrolysis procedure was repeated until the desired thickness was reached. The film was then crystallized in air either using a preheated tube furnace for 5 min. or a rapid thermal annealer with a heating rate of 100 °C/s and a soak time of 60 s. The thickness of the films prepared was 0.4 to 0.5 μm.

The crystallinity of the BZN thin films was characterized with a Scintag DMC-105 X-ray diffractometer using Cu K_α radiation. A Digital Instruments Dimension 3100 atomic force microscope was used to investigate the surface morphology and roughness of the thin films. Tapping mode was adopted in this work. To examine the electrical properties, platinum dots of 0.5 mm or 1.5 mm in diameter were sputtered onto the films as top electrodes to form a Pt/BZN/Pt sandwich configuration. The dielectric properties of the BZN films were measured with a Hewlett Packard 4284A multi-frequency LCR meter with a test signal of 0.03 V rms. Two temperature measurement systems were employed to measure the temperature dependence of the dielectric properties. For the temperature range between 200 and –175 °C, a computer-controlled Delta 9023 temperature oven with a temperature cooling ramp of 2 °C/min was used. For temperatures between 25 and -269 °C, a second set-up was employed (again with a ramp rate of 2 °C/min). The measurement frequencies were between 20 Hz and 1 MHz. The field dependence of the dielectric properties was measured using a DC bias voltage from a HP4284A LCR meter.

RESULTS AND DISCUSSION

Bulk (Bi$_{1.5}$Zn$_{0.5}$)(Zn$_{0.5}$Nb$_{1.5}$)O$_7$ ceramics have a cubic pyrochlore structure with a = 1.056 nm [2, 4]. The recent study by Wang et al. suggests that Bi$_2$(Zn$_{1/3}$Nb$_{2/3}$)$_2$O$_7$ has a distorted pyrochlore phase with an orthorhombic structure with a = 0.7202 nm, b = 0.7603 nm and c = 1.064 nm [4]. The structure of Bi$_2$(Zn$_{1/3}$Nb$_{2/3}$)$_2$O$_7$ ceramics is firing temperature dependent and the material converts to a cubic phase at a higher firing temperature (~ 1100 °C) [7].

The XRD patterns of (Bi$_{1.5}$Zn$_{0.5}$)(Zn$_{0.5}$Nb$_{1.5}$)O$_7$ films fired at different temperatures are shown in Fig. 1. The films were amorphous at temperatures of 500 °C and below: no sharp diffraction peaks were observed in the film. Films fired at 550 °C were crystalline and showed a cubic pyrochlore structure. With increasing firing temperature, the intensities of the diffraction peaks strengthened, but the films maintained the cubic pyrochlore structure. No preferred orientation was found in the films. The lattice constant calculated from films fired at 750 °C is 1.055 nm, which is close to the bulk value.

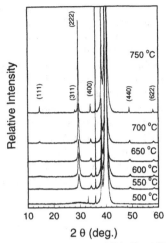

Figure 1: X-ray diffraction patterns of $(Bi_{1.5}Zn_{0.5})(Zn_{0.5}Nb_{1.5})O_7$ films as a function of crystallization temperature.

For $Bi_2(Zn_{1/3}Nb_{2/3})_2O_7$ solutions, after the 500 °C anneal, the film was amorphous. Between 550 °C and 650 °C, the films appeared to have a cubic structure with a main peak (222) at 2θ = 29.3°. For the film fired at 750 °C, the diffraction pattern found was indexed to an orthorhombic phase with lattice constants of a = 0.718 nm, b = 0.759 nm, and c = 1.058 nm. A mixture of cubic and orthorhombic structures was observed in films fired at 700 °C. $Bi_2(Zn_{1/3}Nb_{2/3})_2O_7$ films fired at 800 °C still maintained an orthorhombic structure and no cubic phase was detected from XRD analysis. All films were polycrystalline.

AFM was used to measure the surface morphologies of the BZN thin films as a function of firing temperature. Featureless AFM scans were found in the BZN films fired at 500 °C, which were X-ray amorphous. When the temperature was increased to 550 °C, very fine grains with a surface grain size of about 30 nm were observed in the films. The surface grain size increased with firing temperature. At temperatures of 750 °C, the grain size was ~100 nm. The BZN films exhibited a dense microstructure with no cracks or defects. RMS values of the surface roughness for $Bi_2(Zn_{1/3}Nb_{2/3})_2O_7$ films were very small. At firing temperature of 550 °C, the average roughness was 0.4 nm. The roughness increased with temperature and was 4.2 nm at 750 °C. AFM data for $(Bi_{1.5}Zn_{0.5})(Zn_{0.5}Nb_{1.5})O_7$ films showed similar features.

Table I gives the dielectric properties of $(Bi_{1.5}Zn_{0.5})(Zn_{0.5}Nb_{1.5})O_7$ films as a function of firing temperature. The dielectric constants of the films increased monotonically with crystallization temperature and reached 150 at 750 °C, which was comparable to the values of bulk ceramics with the same composition (140 – 170) [2, 4]. All films exhibited very low dielectric losses (tanδ < 0.01). Between -50 and 100 °C, the dielectric constants changed linearly with temperature. The results are shown in Table II. The TCC of the $(Bi_{1.5}Zn_{0.5})(Zn_{0.5}Nb_{1.5})O_7$ films fired at 750 °C was comparable to the reported values of bulk ceramics [2, 4].

The dielectric properties of $Bi_2(Zn_{1/3}Nb_{2/3})_2O_7$ films versus firing temperature are given in Table I. The dielectric constant of the $Bi_2(Zn_{1/3}Nb_{2/3})_2O_7$ films was a function of firing temperature and had a maximum at 650 °C. The dielectric constant at 750 °C was 80, which was comparable to the bulk value [2, 4]. XRD indicated that $Bi_2(Zn_{1/3}Nb_{2/3})_2O_7$ films had an orthorhombic phase at firing temperatures of 750 °C. Based on the above results, we can conclude that dielectric constant of orthorhombic BZN was smaller than cubic films. Because of the cubic phase formed in $Bi_2(Zn_{1/3}Nb_{2/3})_2O_7$ films at the lower firing temperatures range of 650 – 700 °C, the dielectric constant increased with decreasing firing temperature. Further lowering the firing temperature lead to a decrease in the dielectric constant due to weak crystallization of the films. The loss tangent of the films was lower than 0.01. The temperature dependence of dielectric constant of the $Bi_2(Zn_{1/3}Nb_{2/3})_2O_7$ films is shown in Table II. At 750 °C, TCC of the films was 150 ppm/°C, which was also comparable to the bulk values [2, 4]. Cubic BZN films had a negative TCC (-400 ppm/°C), while orthorhombic films had a positive TCC (150 ppm/°C). As the firing temperature decreased from 750 to 650 °C, the TCC of $Bi_2(Zn_{1/3}Nb_{2/3})_2O_7$ films decreased due to the cubic phase which formed at low temperatures. A mixture of cubic and orthorhombic of $Bi_2(Zn_{1/3}Nb_{2/3})_2O_7$ films fired at 700 °C give a very small TCC (~20 ppm/°C).

Table I: Dielectric constant for $(Bi_{1.5}Zn_{0.5})(Zn_{0.5}Nb_{1.5})O_7$ (BZN1) and $Bi_2(Zn_{1/3}Nb_{2/3})_2O_7$ (BZN2) films as a function of crystallization temperature.

Crystallization Temperature (°C)	BZN1	BZN2
550	90	80
600	105	95
650	110	100
700	130	85
750	150	80

Table II: Temperature coefficient of capacitance for $(Bi_{1.5}Zn_{0.5})(Zn_{0.5}Nb_{1.5})O_7$ (BZN1) and $Bi_2(Zn_{1/3}Nb_{2/3})_2O_7$ (BZN2) films as a function of crystallization temperature

Crystallization Temperature (°C)	BZN1 (ppm/°C)	BZN2 (ppm/°C)
550	0.1	75
600	-70	-0.1
650	-150	-60
700	-350	20
750	-400	150

The dielectric constant and loss measured down to -269 °C of the $(Bi_{1.5}Zn_{0.5})(Zn_{0.5}Nb_{1.5})O_7$ films fired at 750 °C are given in Fig. 2. A frequency dispersion in the dielectric properties was observed at low temperature. The dielectric constant maxima decreased and shifted to higher temperatures with increasing measuring frequency. The corresponding maxima in the loss tangent increased and shifted to higher temperature with increasing frequency. The frequency dispersion of $(Bi_{1.5}Zn_{0.5})(Zn_{0.5}Nb_{1.5})O_7$ films are similar to those of dipole glasses [8-10].

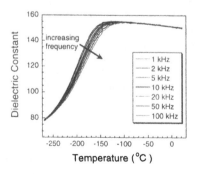 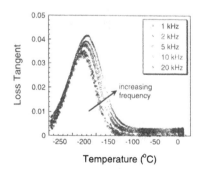

Figure 2: Low temperature relaxation in cubic BZN films.

An Arrhenius equation was used to fit the dispersion data of $(Bi_{1.5}Zn_{0.5})(Zn_{0.5}Nb_{1.5})O_7$ films:

$$1/\tau_c = v_0 \exp[-E_a/k_B T] \qquad (1)$$

where τ_c is the characteristic relaxation time, v_0 is the attempt jump frequency, E_a is the activation energy and k_B is Boltzmann constant. v_0 and E_a are temperature independent. The logarithmic measuring frequency had a linear relationship with the reciprocal of the temperature at which the loss tangent was maximum. v_0 and E_a were 2×10^{13} Hz and 0.13 eV respectively. The jump frequency v_0 of $(Bi_{1.5}Zn_{0.5})(Zn_{0.5}Nb_{1.5})O_7$ films had the same order as typical ionic vibrations of the lattice (~ 10^{13} Hz), and was comparable to that of the K(Br, CN) system [10]. Attempts to fit to a Vogel-Fulcher equation were unsuccessful.

The electric field dependence of dielectric properties of the BZN films was investigated by applying DC bias voltage (± 40 V max.) during dielectric property measurements. Fig. 3 gives the dielectric constant and loss for $(Bi_{1.5}Zn_{0.5})(Zn_{0.5}Nb_{1.5})O_7$ films fired at 750 °C as a function of the bias electric field at the measuring frequency of 10 kHz. It can be seen that dielectric constant of the cubic BZN films changed with the bias field, while the loss tangent was constant. The dielectric constant decreased with increasing bias field and changed by 10% with a bias of 830 kV/cm. The curve was symmetric with respect to zero-bias and had no hysteresis. The voltage variable dielectric constant makes cubic BZN thin films candidates for tunable microwave device applications.

The tunability of $(Bi_{1.5}Zn_{0.5})(Zn_{0.5}Nb_{1.5})O_7$ films was investigated at the different measuring temperatures. Fig. 4 shows the normalized tunability of the films fired at 750 °C at measuring temperatures of 25 °C, -100 °C, -130 °C and -170 °C. It can be seen that all films demonstrated the same amount of relative tunability.

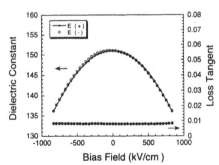

Figure 3: Tunability of $(Bi_{1.5}Zn_{0.5})(Zn_{0.5}Nb_{1.5})O_7$ film.

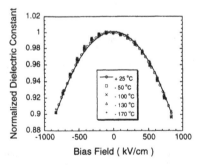

Figure 4: Dependence of dielectric constant on dc bias field for cubic BZN films as a function of temperature.

The bias electric field dependence of the dielectric constant and loss for $Bi_2(Zn_{1/3}Nb_{2/3})_2O_7$ films fired at 750 °C is small. For the $Bi_2(Zn_{1/3}Nb_{2/3})_2O_7$ films fired at 650 °C and 700 °C, their dielectric constants can be tuned by a bias field. This tunability is attributed to the cubic phase existing in the $Bi_2(Zn_{1/3}Nb_{2/3})_2O_7$ films fired at lower temperatures.

The dielectric measurements demonstrate that the dielectric constants, TCC and tunability of the BZN films depended not only on their composition, but also on the firing temperature and phase content. BZN films with the cubic phase have larger dielectric constants, negative TCC values and larger tunability. Films with the orthorhombic phase have smaller dielectric constants, positive TCC and low tunability. The above results suggest that the dielectric properties of the BZN films can be tailored to a given application by adjusting composition, phase and firing temperature.

CONCLUSIONS

$(Bi_{1.5}Zn_{0.5})(Zn_{0.5}Nb_{1.5})O_7$ and $Bi_2(Zn_{1/3}Nb_{2/3})_2O_7$ thin films were successfully prepared by the sol-gel process. $(Bi_{1.5}Zn_{0.5})(Zn_{0.5}Nb_{1.5})O_7$ films showed the cubic pyrochlore structure, while the structure of $Bi_2(Zn_{1/3}Nb_{2/3})_2O_7$ thin films depended on the firing temperature. Such films were cubic in the temperature range of 550-650 °C and orthorhombic when crystallized at 750 °C. A mixture of cubic and orthorhombic pyrochlore structures was formed at 700 °C. The $(Bi_{1.5}Zn_{0.5})(Zn_{0.5}Nb_{1.5})O_7$ films fired at 750 °C have a dielectric constant of ~150 and a negative TCC. $Bi_2(Zn_{1/3}Nb_{2/3})_2O_7$ thin films fired at 750 °C had a smaller dielectric constant of ~80 and a positive TCC. Loss tangents of both types of BZN films are below 0.01 at 10 kHz. The cubic $(Bi_{1.5}Zn_{0.5})(Zn_{0.5}Nb_{1.5})O_7$ films fired at 750 °C had a large tunability and the orthorhombic $Bi_2(Zn_{1/3}Nb_{2/3})_2O_7$ films had a low tunability.

ACKNOWLEDGEMENTS

Significant support for this work was given by the TDK Corporation and Intel. We are grateful to Dr. L.E. Cross for helpful discussions. We would like to thank Dr. X. X. Xi for use of low temperature facilities.

REFERENCES

1. M. F. Yan, H. C. Ling, and W. W. Rhodes, *J. Am. Ceram. Soc.*, **73**, 1106 (1990).

2. D. Liu, Y. Liu, S. Huang, and X. Yao, *J. Am. Ceram. Soc.*, **76**, 2129 (1993).

3. D. P. Cann, C. A. Randall, and T. R. Shrout, *Solid State Comm.*, **100**, 529 (1996).

4. X. Wang, H. Wang, and X. Yao, *J. Am. Ceram. Soc.*, **80**, 2745 (1997).

5. H. Kagata, T. Inoue, J. Kato, and I. Kameyama, *Jpn. J. Appl. Phys.*, **31**, 3152 (1992).

6. S. L. Swartz, and T. R. Shrout, "Ceramic compositions for BZN dielectric resonators," U. S. Patent, No. 5449652 (1995).

7. D. P. Cann, "Bismuth pyrochlores for high frequency dielectric applications," MS Thesis, The Pennsylvania State University (1993).

8. E. Courtens, *Phys. Rev. Lett*, **52**, 69 (1984).

9. U. T. Höchli, *Phys. Rev. Lett.*, **48**, 1494 (1982).

10. S. Bhattacharya, S. R. Nagel, L. Fleishman and S. Susman, *Phys. Rev. Lett.*, **18**, 1267 (1982).

11. D. Huser, L.E. Wenger, A. J. van Duyneveldt, and J. A. Mydosh, *Phys. Rev.* **B 27**, 3100 (1983).

STRUCTURAL AND DIELECTRIC PROPERTIES OF PULSED LASER DEPOSITED Pb[Yb$_{1/2}$Nb$_{1/2}$]O$_3$-PbTiO$_3$ THIN FILMS

Véronique BORNAND* and Susan TROLIER-McKINSTRY
The Pennsylvania State University, Department of Materials Science and Engineering, Materials Research Laboratory, University Park, PA 16802-4801
*current address : Laboratoire de Physicochimie de la Matière Condensée UMR 5617, C.C. 003, Place Eugène Bataillon, 34095 MONTPELLIER Cédex 5, France

ABSTRACT

Heterostructures consisting of (001) LaAlO$_3$ or (111) SrTiO$_3$ substrates, SrRuO$_3$ metallic oxide bottom electrodes and Pb[Yb$_{1/2}$Nb$_{1/2}$]O$_3$-PbTiO$_3$ ferroelectric films were deposited by pulsed laser deposition. The combination of oxidic perovskite-type materials results in highly <001>- or <111>-heteroepitaxial capacitors with well-defined and homogeneous columnar microstructures. Most of the films show room temperature dielectric constant greater than 1500 associated with low dielectric loss (tgδ < 4%) and exhibit saturated hysteresis loops with remanent polarizations up to P_r ~ 50μC.cm^{-2}. It was found that the ferroelectric characteristics and, in particular, the fatigue phenomena greatly depend on the orientation and crystalline quality of the as-grown films. The stabilization of the <001>-orientation enhances the fatigue resistance and a good endurance up to 10^{11} cycles has been determined for highly <001>-textured samples. In contrast, fatigue tests performed on strongly <111>-oriented capacitors have revealed poor fatigue performances, with a progressive decrease in the switchable polarization by ac voltage cycling.

INTRODUCTION

The last decade has seen a surge of interest in ferroelectric thin films due to their potential integration in a wide range of capacitive applications and microelectromechanical systems (MEMS).[1,2] In particular, considerable attention has been recently paid to the preparation of high-quality epitaxial relaxor ferroelectric thin films to take full advantage of their anisotropic properties. Indeed, the demonstration of high strains, high piezoelectric coefficients and high electromechanical coupling constants in relaxor ferroelectric single crystals of the Pb[B'B'']O$_3$-PbTiO$_3$ family suggests that the development of judiciously-oriented lead-based relaxor-PbTiO$_3$ thin films could improve the reliability and performances of existing devices.[3-6]

Of the reported relaxor ferroelectric-PbTiO$_3$ solid solutions, $(1-x)$ Pb[Yb$_{1/2}$Nb$_{1/2}$]O$_3$ $-$ x PbTiO$_3$ (PYbN-PT) has the highest transition temperature (~360°C) at the morphotropic phase boundary (MPB x~0.5).[7,8] This should greatly improve the high temperature capabilities of piezoelectric driven MEMS relative to lead magnesium niobate-based compounds. Thus, this study focuses on the preparation and characterization of heteroepitaxial PYbN-PT thin films with 50:50 and 60:40 compositions.

EXPERIMENTAL PROCEDURE

PYbN-PT thin films with SrRuO$_3$ (SRO) bottom electrodes were sequentially deposited by the pulsed laser deposition (PLD) process onto pseudo-cubic <001>-oriented LaAlO$_3$ (LAO) and <111>-oriented SrTiO$_3$ (STO) substrates. Details of the specific processing conditions and device optimizations for both materials can be found elsewhere.[9,10] The out put of a short pulsed KrF excimer laser operating at 248nm was focused to a spot size of 10mm^2 and an energy density of 2.5J.cm^{-2} onto the desired rotating targets. Ablations of the two materials were performed in a 200mTorr (SRO) $-$ 300mTorr (PYbN-PT) 10%O$_3$-90%O$_2$ partial pressure ambient, which was also maintained during the cool down. Heteroepitaxial SRO layers

(~350nm-thick) were first deposited at 680°C from a stoichiometric SrRuO₃ target (Target Materials International). The subsequent ablations of PYbN-PT were realized from ceramic targets including 25w% excess PbO to compensate for the high volatility of Pb species and obtain nearly stoichiometric films with reasonable thickness uniformity profiles and good crystallinity. The vaporized material was deposited onto a heated substrate approximately 5.5cm away from the target. Pt top electrodes (∅=0.3mm) were then sputtered at room temperature, through a shadow mask, onto the surface of the film to complete the capacitor structures.

The phase identification and texture quantification of the as-grown heterostructures were analyzed by 4-circle X-ray diffraction (XRD) using the K_α emission of a Cu anode. The morphologies and microstructures were observed by scanning electron microscopy (SEM) and atomic force microscopy (AFM). Low and high field electrical properties were measured using a HP47240 LCR meter and a Radiant Technology RT66A test system.

RESULTS AND DISCUSSION

1. Structural Analyses

The influence of deposition parameters such as the laser frequency, the chamber pressure, the substrate temperature and the target composition on the film properties was previously investigated.[10,11] It was demonstrated that high crystalline quality materials can be grown in the 560°C-660°C temperature range, with a dynamic O_3/O_2 pressure of 300mTorr and high laser repetition rates around 16Hz. For the lowest temperatures, an intermediate pyrochlore phase preceding and/or accompanying the formation of the bulk-like perovskite PYbN-PT one is hard to avoid totally. Nevertheless, by increasing the temperature, *i.e.* enhancing both atomic mobility and reactivity at the surface of the substrate, improved crystalline quality and purity can be achieved. The reasonable lattice mismatch and good chemical compatibility between SRO and PYbN-PT allow the elaboration of highly <001>- and <111>-textured heterostructures according to the orientation of the substrate. Rocking curves and azimuthal scans confirmed the strong out- and in-plane orientation with, in both cases, a "pseudo-cube" on "pseudo-cube" heteroepitaxial relationship between the growing film and the bottom electrode. Figure 1 shows the 4-circle X-ray data for an optimized PYbN-PT (60/40) thin film deposited on a (001) SRO/LAO substrate.

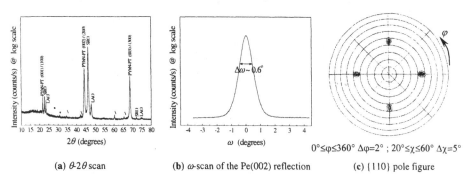

(a) θ-2θ scan	(b) ω-scan of the Pe(002) reflection	(c) {110} pole figure

Figure 1 : XRD patterns of a PYbN-PT (60/40) thin film deposited at 655°C on a SRO/LAO substrate
*→Perovskite (Pe)   Δ→Pyrochlore (Py), PbO_x

Both <001>- and <111>-oriented heterostructures show dense and homogeneous columnar microstructures consistent with the PLD process. However, for a given deposition temperature, <001>-oriented materials tend to develop much coarser microstructures (Fig. 2a) than <111>-

oriented ones (Fig. 2b). This coarser appearance can be associated with the higher roughness and lateral grain size of the SRO bottom electrode deposited on (001) LAO substrates. Indeed, AFM studies suggest that the morphology and cleanliness of SRO is very sensitive to the orientation and nature of the underlying material. Films grown on (111) STO substrates exhibit a regular and atomically smooth surface with an average roughness around 10nm (Fig. 3b). This value is, at least, 6 times better than the one obtained for <001>-oriented SRO (Fig. 3a).

 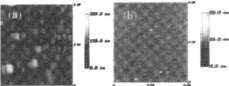

Figure 2 : SEM of the cross-section of a
(a) 1.8μm-thick <001> PYbN-PT (60/40) thin film
(b) 1.2μm-thick <111> PYbN-PT (60/40) thin film
①→substrate ②→350nm-thick SRO ③→PYbN-PT

Figure 3 : AFM of the surface of a 350nm-thick
SRO film deposited on
(a) (001) LAO   and   (b) (111) STO

2. Dielectric Properties

As illustrated in Figure 4 for a <001>-oriented PYbN-PT (50/50) 'thin film, changes in the dielectric constant and loss as a function of frequency and temperature are consistent with a relaxor ferroelectric behavior. The relative permittivity exhibits a peak (MPB) near 370°C at 1kHz, shifted towards 380°C at 100kHz. Moreover, dielectric loss measurements reveal modest values through the entire frequency and temperature range. Optimized <001>-grown samples show losses below 5% up to 250°C.

Electric properties are enhanced with PT content and PYbN-PT (50/50) thin films engineered at the MPB exhibit higher dielectric constants with modest dielectric losses at room temperature. The better crystalline quality and purity of the films deposited at the higher temperatures are well effective to obtain optimal dielectric characteristics (Fig. 5). For a given film thickness, the relative permittivity increases with the deposition temperature, *i.e.* with the degree of orientation and perovskite content of the films. The lower properties observed in <111>-oriented films can be associated, at least, in part, with the finer microstructures producing more grain boundaries.[12,13]

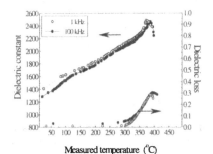 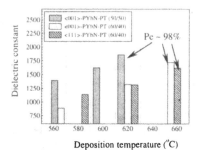

Figure 4 : Temperature and frequency dependence
of ε_r and tgδ of a <001> PYbN-PT (50/50) thin film
(0.04kV.mm⁻¹)

Figure 5 : Room temperature ε_r
of a 800nm-thick PYbN-PT thin films
(100kHz ; 0.04kV.mm⁻¹)

The thickness dependence of the dielectric constant of a <001> PYbN-PT (50/50) thin film grown at 580°C is shown in Figure 6a. ε_r initially increases with the film and levels off for thicknesses between 1.5µm-1.6µm. If we assume that the bulk dielectric constant remains unchanged with thickness, the observed trend may be explained by the fact that the interfacial dominance of ε_r which occurs for thin films is overtaken by bulk effects as the thickness increases. This seems reasonable because a plot of the reciprocal capacitance versus thickness yields a straight line. In Figure 6b, the data are plotted in the relationship between $\varepsilon_0 S/C$ and d, where ε_0 is the dielectric constant of the free space, S the capacitance area, C the measured capacitance and d the total thickness.

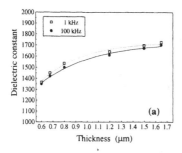 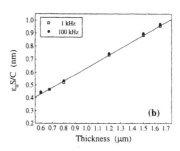

Figure 6 : Thickness dependence of **(a)** ε_r and **(b)** $\varepsilon_0 S/C$ of a <001> PYbN-PT (50/50) thin film (deposition temperature : 580°C ; 0.04kV.mm^{-1})

These results are consistent with the previous structural analyses. Indeed, grazing incidence X-ray diffraction experiments have pointed out that the patchy nucleation of some pyrochlore material, for the lowest temperatures, meanly occurs in the initial stages of the deposition process.[11] If we assume that such a low-permittivity interfacial layer (ε_{Py}, d_{Py}) exists in series with the bulk of the high-permittivity perovskite film (ε_{Fe}, d_{Fe}), the reciprocal capacitance appears as a linear function of the film thickness ($d = d_{Fe} + d_{Py}$) :

$$\frac{\varepsilon_o S}{C} = \frac{d_{Fe}}{\varepsilon_{Fe}} + \frac{d_{Py}}{\varepsilon_{Py}} = \frac{d}{\varepsilon_{Fe}} + d_{Py}\left(\frac{\varepsilon_{Fe} - \varepsilon_{Py}}{\varepsilon_{Fe}\varepsilon_{Py}}\right) \qquad \text{Eq. (1)}$$

Extrapolation at d=0, giving a positive value (0.128nm) of $\varepsilon_0 S/C$, truly indicates that an interfacial layer is present in addition to the ferroelectric layer. From equation (1), ε_{Fe} was calculated from the slope of the fitted line as 1960. This should be compared with the 1950 bulk value of PYbN-PT (50/50) ceramics.[8]

3. Ferroelectric characteristics

Optimized samples exhibit well-developed hysteresis loops at room temperature with remanent polarizations as high as 50µC.cm^{-2} (Fig. 7). The mismatch-induced lattice deformation to an appropriate degree is quite effective to enhance ionic displacements and obtain large spontaneous polarizations. Indeed, especially in the case of <001>-textured films, the lattice parameter along the growing direction was found larger than the one expected for powder samples and also increases when the degree of orientation along the thickness direction increases.[10] These ferroelectric properties are significantly better than those reported previously for PLD-grown PMN-PT thin films,[14] making PYbN-PT materials promising candidates for the next generation components in the microelectronic arena.

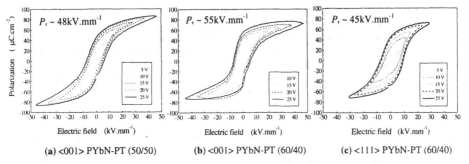

$P_r \sim 48kV.mm^{-1}$

$P_r \sim 55kV.mm^{-1}$

$P_r \sim 45kV.mm^{-1}$

(a) <001> PYbN-PT (50/50)   (b) <001> PYbN-PT (60/40)   (c) <111> PYbN-PT (60/40)

Figure 7 : Hysteresis loops of 1μm-thick heteroepitaxial PYbN-PT thin films

Even though many improved device properties stem from the large induced polarization exhibited by ferroelectric materials, the polarization itself can be subjected to various forms of degradation such as fatigue. There are several proposals that attempt to explain the decrease of the remanent polarization with bipolar switching in ferroelectric compounds. These proposals are all related to defects that may be trapped at the film/electrode interfaces or domain boundaries. Oxygen vacancies and electron/hole injection have been suggested as possible mechanisms for both surface and domain wall pinning.[15,16] Of particular interest in the present study is the significant impact on fatigue resistance that can be expected from the proper control of orientation of the as-grown films.[17] Whereas <001>-textured capacitors withstand negligible fatigue up to 10^{11} cycles, <111>-oriented heterostructures obviously fatigue and show more than 40% decay after 10^{11} cycles (Fig. 7). Such an orientation dependence of the fatigue behavior has also been reported in PZN-PT single crystals.[18]

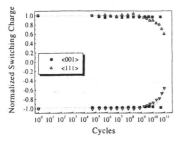

Figure 7 : Fatigue characteristics of heteroepitaxial PYbN-PT (60/40) thin films (14kV.mm⁻¹)

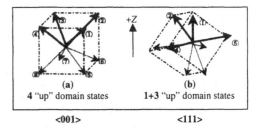

(a) 4 "up" domain states   (b) 1+3 "up" domain states

<001>   <111>

Figure 8 : "Up" domain states in (a) <001> and (b) <111>-oriented thin films

The interpretation of fatigue effects and the complete elucidation of the related mechanisms in both <001>- and <111>-oriented materials remain unclear. However, it seems reasonable to interpret the observed difference in fatigue resistance as a difference in domain wall switching dynamics. In the core of the strongly oriented grains, domains are formed and reversible domain wall motions are expected. These will give the largest contribution to the polarization. As shown in Figure 8, there are 4 possible polar directions which have a component of the polarization along the +Z direction (*i.e.* lying out of the plane of the film). In the case of <001>-oriented capacitors, the contribution of these 4 components to the polarization is identical and there is no driving force for wall motion between the 4 equivalent directions. Domain switching is feasible

147

and only limited by the intrinsic velocity of the walls. Thus, very little –if any- fatigue degradation is expected. In contrast, 3 directions have a smaller contributions in <111>-oriented films if we assume that non-180° domains are mostly inactive and do not contribute to the macroscopic behavior. This anisotropic distribution offers more opportunity for the domains to entangle and interact with each other. It becomes much more difficult for the domain walls to overcome the back stress of the geometric confinement. This long-range friction might result in pinning force and fatigue upon ac voltage cycling.

CONCLUSION

To the best of the authors' knowledge, it is the first time that the growth of heteroepitaxial PYbN-PT thin films is reported. The good structural and chemical compatibility between SRO and PYbN-PT allow the elaboration of highly oriented and textured thin films on (001) LAO and (111) STO substrates. X-ray diffraction patterns display a high crystalline order of both SRO and PYbN-PT. Such heterostructures exhibit good dielectric and ferroelectric properties. In particular, the massive field-induced polarizations ($P_r \sim 50 \mu C.cm^{-2}$) should lead to high energy storage capabilities. Although a complete description of the fatigue process has not been obtained to date, it has been clearly demonstrated that a significant orientation effect on the fatigue performances can be expected for ferroelectric materials. <001>-textured thin films offer the advantages of improved fatigue resistance. This orientation dependence is believed to result from a difference in the domain configuration and switching process.

ACKNOWLEDGEMENTS

This work has been supported by the Office of Naval Research (U.S.) under the project number N00014-98-1-0527. One of the author (V.B.) also acknowledges the DRET/DGA (France) for its financial support under the grant No 93811113.

REFERENCES

1. J.F. Scott and C.A. Paz de Araujo, Science 246, 1400 (1989)
2. R. Ramesh, Thin Film Ferroelectric Materials and Devices (Kluwer Academic Publisher, 1997)
3. L.E. Cross, Ferroelectrics 76, 241 (1987)
4. T.R. Shrout and J. Fielding, Proc. of the 1990 IEEE Ultrasonics Symp. 7115 (1990)
5. Y. Yamashita and N. Ichinose, Proc. of the 10th IEEE Int. Symp. Appl. Ferr. 1, 71 (1996)
6. S.E. Park and T.R. Shrout, Presentation at the 1997 Williamsburg Workshop on Ferroelectrics (1997)
7. T. Yamamoto and S, Ohashi, Jpn J. Appl. Phys. 34, 5349 (1995)
8. H. Lim, H.J. Lim and W.K. Choo, Jpn J. Appl. Phys. 34, 5449 (1995)
9. J.P. Maria, S. Trolier-McKinstry, D.G. Schlom, M.E. Hawley and G.W. Brown, J. Appl. Phys. 83, 4373 (1998)
10. V. Bornand and S. Trolier-McKinstry, submitted to Thin Solid Films
11. V. Bornand and S. Trolier-McKinstry, submitted to J. Appl. Phys.
12. Y. Park, K.M. Knowles and K. Cho, J. Appl. Phys. 83, 5702 (1998)
13. F. Xu, PhD Thesis, The Pennsylvania State University (1999)
14. J.P. Maria, W. Hackenberger and S. Trolier-McKinstry, J. Appl. Phys. 84, 5147 (1998)
15. I.K. Yoo and S.D. Desu, Mat. Sci. Eng. B13, 319 (1992)
16. W.L. Warren, D. Dimos, B.A. Tuttle and D.M. Smyth, J. Am. Ceram. Soc. 77, 2753)1994)
17. V. Bornand and S. Trolier-McKinstry, submitted to J. Mat. Res.
18. K. Takemura, M. Ozgul, V. Bornand, S. Trolier-McKinstry and C.A. Randall, Proc. 9th US-Japan Workshop on Dielectric and Piezoelectric Materials, Okinawa, Nov. 3th-5th (1999)

PULSED LASER DEPOSITED $Na_{0.5}K_{0.5}NbO_3$ THIN FILMS

CHOONG-RAE CHO, ALEX GRISHIN, BYUNG-MOO MOON *

Department of Condensed Matter Physics, Royal Institute of Technology,
SE-100 44 Stockholm, SWEDEN;
* Department of Electrical Engineering, Korea University, Seoul, KOREA.

ABSTRACT

$Na_{0.5}K_{0.5}NbO_3$ films have been grown onto polycrystalline $Pt_{80}Ir_{20}$ substrates at two different regimes: high ($\sim$ 400 mTorr) and low ($\sim$ 10 mTorr) oxygen pressure. Films grown at high oxygen pressure have been found to be single phase and highly c-axis oriented. The concept of *discriminated thermalization* has been developed to explain the dynamics of the laser ablation process and to find reliable pulsed laser deposition (PLD) processing conditions. The phenomenon of *self assembling* of $Na_{0.5}K_{0.5}NbO_3$ films along [001] direction has been observed. On the other hand, films grown at low oxygen pressure have been found to be mixed phase of ferroelectric $Na_{0.5}K_{0.5}NbO_3$ and paraelectric potassium niobates. Superparaelectric behavior has been observed in these films: 5% tunability at electric field of 100 kV/cm, losses as low as 0.003 and excellent stability to the temperature and frequency changes.

INTRODUCTION

Recently several considerably interesting characteristics, both from fundamental and technological points of view, of perovskite $Na_{0.5}K_{0.5}NbO_3$ (hereafter NKN) thin films have been reported. [1,2] Strong electric field dependence of dielectric permittivity with very low loss at room temperature and high piezoelectric coefficient with moderate dielectric constant promise vast variety of emerging applications. We report on the pulsed laser deposited (PLD) NKN thin films onto $Pt_{80}Ir_{20}$ (here after Pt) substrates at different oxygen ambient gas pressure.

EXPERIMENTAL PROCEDURES

A KrF excimer laser (Lambda Physik-300, pulse width of 20 ns) has been used to ablate a stoichiometric NKN ceramic target. Slightly c-axis textured polycrystalline $Pt_{80}Ir_{20}$ substrates have been used. Two different PLD processes have been implemented (see the Table): deposition at **high oxygen ambient pressure** (hereafter

TABLE. Processing conditions for HP- and LP-$Na_{0.5}K_{0.5}NbO_3$ films

PARAMETER	HP	LP
Temperature (°C)	575 ~ 640	670
Oxygen pressure (mTorr)	~ 400	10
Laser energy density (J/cm²)	4 ~ 5	3 ~ 4
Annealing condition (°C, min)	650, 30	400, 30

149

HP) and *low pressure* (hereafter LP). To make dielectric characterization, Au upper electrodes with a diameter of 0.55 mm have been thermally evaporated on the top of NKN layer at room temperature. The dielectric, ferroelectric, and crystalline quality characterization techniques used are described elsewhere. [2]

RESULTS AND DISCUSSION

Composition and crystalline properties of the fabricated NKN films have been proved to be crucially dependent on ambient gas pressure. Figs.1 show x-ray diffraction (XRD) θ-2θ scans for NKN LP-film (**a**), HP-film (**b**), and rocking curves of the NKN-002 and Pt-002 reflections for HP-film (**c**).

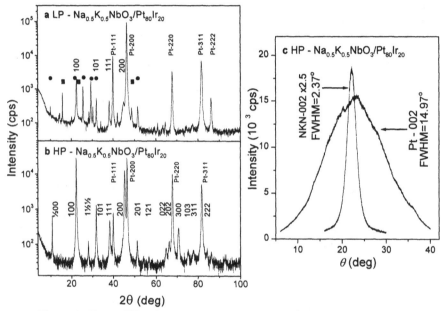

Figure 1. X-ray diffraction (XRD) scans in CuK_α radiation for $Na_{0.5}K_{0.5}NbO_3$ films deposited on $Pt_{80}Ir_{20}$ substrates: **a** - θ-2θ scan for 0.47 μm thick NKN film deposited in 10 mTorr O_2 (LP-film) Symbol ● corresponds to $K_2Nb_8O_{21}$ phase, while ■ to $K_2Nb_4O_{11}$. **b** - θ-2θ scan for 1.5 μm thick NKN film deposited in 400 mTorr O_2 (HP-film). *hkl* Bragg peaks indicate the $Na_{0.5}K_{0.5}NbO_3$ reflections. **c** - XRD rocking curves (ω-scans) of the NKN-002 and Pt-002 reflections from the same film shown in Fig. 1b.

The XRD spectrum of LP-film (Fig. 1a) consists of NKN, $K_2Nb_8O_{21}$, and $K_2Nb_4O_{11}$ phases. Significant loss of Na is ascribed to the discriminated thermalization of Na (which is the lightest cation species of atomic mass 23) in the plume expansion process, while the transport of heavier K (atomic mass 39) and Nb (atomic mass 92) species could still be ballistic at low oxygen pressure regime. Sharp Bragg peaks of three phases indicate high crystallinity and the absence of crystal intergrowth between different phases. One can expect that LP film should exhibit relaxor behavior, since relaxors are highly inhomogeneous materials with locally homogeneous ferroelectric clusters on a nano-meter scale. [3]

Fig.1b presents θ-2θ XRD pattern for HP-film. In contrary to the XRD pattern of LP-film, it is clearly seen in logarithmic scale that HP-film has been grown as a single NKN phase in spite of volatile Na and K constituents. A high degree of preferential c-axis orientation is observed. The intensities of NKN-001 and -011 reflections are in the ratio of $I_{NKN-001}/I_{NKN-011} = 60$. Congruent material transfer from the target to the substrate at high ambient pressure can be explained by thermalization of all constituents and forming shock waves at sufficiently high oxygen pressure of 400 mTorr. The observed strong preferential c-axis orientation of NKN films is also suggested by the surprisingly narrow rocking curve of NKN-002 reflection shown in Fig.1 c. Its full width at half maximum (FWHM) is $2.37°$, which is more than six times narrower than FWHM = $14.97°$ of the rocking curve from the neighboring substrate Pt-002 reflection. This result reveals the phenomenon of *self-assembling* of NKN films and clearly demonstrates feasibility of growing highly crystalline NKN film onto polycrystalline substrates by PLD.

Fig. 2 shows the ferroelectric P-E loops of HP-NKN films deposited on Pt at different laser repetition rates, f_{rep}, and temperatures.

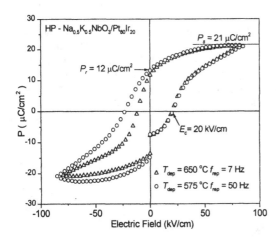

Figure 2. P-E hysteresis loops of HP-NKN/Pt films fabricated at different deposition temperatures T_{dep}, and laser repetition rates, f_{rep}, measured at room temperature. Symbols Δ show the P-E loop for T_{dep} of 650 °C and f_{rep} of 7 Hz, while the symbol O corresponds to 575 °C and 50Hz.

Although high-repetition rate f_{rep} = 50 Hz film has been deposited at temperatures 75 °C lower than the low-repetition rate f_{rep} = 7 Hz film, their P-E hysteresis loops are very similar to each other: remnant polarization P_r is 12 $\mu C/cm^2$ and saturated polarization P_s is 21 $\mu C/cm^2$ at the field of 80 kV/cm. This result implies that the increment of momentum transfer from high energy species in the plume per unit time could raise the adatom mobility. We reported that the preferential orientation and self-assembling phenomenon are also affected by the minority of highly energetic Nb ions. [2]

Fig. 3 Shows the frequency dependencies of the dielectric permittivity ε' and dissipation factor *tan* δ of HP and LP films measured at room temperature.

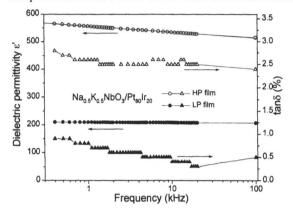

Figure 3. Frequency dependencies of the dielectric permittivity ε' and *tan* δ of HP-film (open symbols) and LP-film (solid symbols).

LP-film exhibits superior dielectric performance at room temperature: the dielectric permittivity ε' is almost frequency independent varying from 210.9 at 400 Hz to 207.3 at 100 kHz (1.7 % change), loss tangent does not exceed 0.009 and decreases as low as 0.003 with frequency increase from 400 Hz to 20 kHz. The dielectric loss has been found to be lower than that of bulk ceramics and even lower than that of epitaxial NKN films grown on oxide substrates. [1] One of the reasons for the very low loss of the LP film could be attributed to the absence of ferroelectric domains, since LP films showed no ferroelectricity at room temperature. Dielectric permittivity of HP-film can be straightened in $\log f$ scale: $\varepsilon' = 556.56 - 21.685 \cdot \log (f / 1 \text{ kHz})$. ε' of HP-film is a little bit higher than what has been observed for polycrystalline bulk NKN of 290 ~ 420 [4], while in LP-film is lower. For HP-film the dissipation factor *tan* δ ~ 0.025 is nearly invariable to frequency change in 400 Hz to 100 kHz range. The dielectric permittivity of LP film was also measured as a function of temperature (77 - 415K) at different frequencies and electric fields.

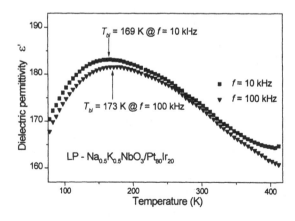

Figure 4. Temperature dependence of ε' of LP-NKN film measured at two frequencies: 10 and 100 kHz.

Fig. 4 shows typical relaxor behavior as follows: i). broaden maximum of dielectric permittivity around "blocking" temperature T_{bl}; ii). T_{bl} shifts to higher temperatures with the frequency increase (for 10 and 100 kHz T_{bl} has been found to be 166 and 173 K respectively; iii). frequency dispersion of the dielectric permittivity is more predominant at temperatures around and below T_{bl}. As temperature decreases, the density of the polar clusters always increases while interactions between them also increases, hence some of the large polar clusters cannot follow high frequency signal. [5] All the features of the curves can be explained by above mechanism and it appears natural that T_{bl} increases with the frequency.

Also, the electric field dependence of dielectric permittivity has been measured in the wide range of temperatures 77 K to 415 K, as shown Fig. 5. The measurements have been

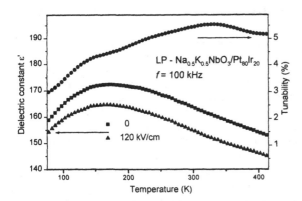

Figure 5. Dielectric permittivity of LP-film measured at 100 kHz at zero bias (symbols ■) and electric field of 120 kV/cm (symbols ▲). Electric field tunability $(1 - \varepsilon'_{120 \text{ kV/cm}}/\varepsilon'_0)$ shown by symbols ● reaches the maximum value of 5.5 %.

performed at zero and 6 V (E = 120 kV/cm) external bias, at the frequency of 100 kHz. All over the measured temperature range, the biased film shows lower permittivity. The tunability $1 - \varepsilon'_{120 \text{ kV/cm}} / \varepsilon'_0$, calculated from above results is shown in the same figure. The tunability increases rapidly with temperature increase, reaches the maximum value of 5.54 % at 327 K and then decreases slightly. This also can be qualitatively explained by the superparaelectric model. It is based on well-known fact that incremental polarizability in ferroelectric state is much lower compared with the polarizability of the paraelectric state. At temperatures much lower than the maximum of permittivity most of the polar clusters are "frozen" in ferroelectric state since thermally activated fluctuations of cluster dipole are not sufficient for frequent switching between two equivalent polar states. Thus, only small amounts of clusters having small "activation" volume remain in the paraelectric state and can contribute to the polarizability. As temperature goes up, more and more clusters "melt" beginning to fluctuate between two equivalent polar directions. Therefore, the polarizability increases. When the temperature rise exhausts all the ferroelectric clusters and makes dipole thermal fluctuations dominate then polarizability goes down $\propto 1/T$ according to the Curie-Weiss law. Frequency dependence of the dielectric permittivity of LP-film can be qualitatively explained by the expression for "escape" frequency

$$f = v_{att} \cdot e^{-\frac{U \cdot V_{act}}{kT}}$$

where v_{att} is the attempt frequency, U is the height of potential barrier, and V_{act} is the activation (polar cluster) volume. According to this equation, "blocking" temperature, which corresponds to the maximum of permittivity, logarithmically increases with measuring frequency increase. Also, shortening the measuring time $2\pi/f$ reduces the probability to find polar cluster flipping between two polar directions, thus results in polarizability decrease with the frequency increases.

SUMMARY AND CONCLUSIONS

The effects of oxygen ambient pressure on the growth of the NKN thin films have been investigated. It has been demonstrated that stoichiometric NKN films can be prepared by the ablation of stoichiometric ceramic target through thermalization process at sufficiently *high oxygen pressure*. The minority of high-energy species in laser plume plays an important role in the formation of high performance ferroelectric film. HP-film showed preferential *c*-axis orientation and the effect of strong *self-assembling*. High remnant and spontaneous polarization, high resistivity and low loss *tan* δ, small dielectric dispersion and *self-assembling* characteristics promises a vast variety of applications in ferroelectric devices.

On the other hand, NKN films fabricated at *low oxygen pressure* showed considerable Na deficit because of discriminated thermalization of the species with different masses, hence highly inhomogeneous material. These films possess characteristics which are the typical features of relaxor material: broadened dielectric maximum and shift of this maximum to higher temperatures at higher frequency. We demonstrate the possibility to engineer and manipulate relaxor behavior by controlling the size and concentration of ferroelectric NKN clusters. Although the dielectric performance of fabricated NKN films have been studied at low frequencies, we expect low loss, low frequency dispersion, high electric tunability, and temperature stability at higher frequencies ranging to the microwave band.

ACKNOWLEDGMENTS

This research was funded by the Swedish Agency NUTEK.

REFERENCES

1. X. Wang, U. Helmersson, S. Olafsson, S. Rudner, L. Wernlund, S. Gevorgian, Appl. Phys. Lett., **73**, 927 (1998).
2. C.-R. Cho, Alex Grishin, Appl. Phys. Lett., **75**, 268 (1999).
3. S. Li, J.A. Eastman, R.E. Newnham, and L.E. Cross, Phys. Rev. B. **55**, 12067 (1997).
4. L. Egerton, D.M. Dillon, J.Am.Ceram.Soc. **42**, 438 (1959).
5. Z.-Y. Cheng, R.S. Kaiyar, X. Yao, A.S. Bhalla, Phys. Rev. **57**, 8166 (1998).

Ferroelectrics

EFFECT OF THE OXYGEN PARTIAL PRESSURE ON THE MICROSTRUCTURE AND PROPERTIES OF BARIUM STRONTIUM TITANATE THIN FILMS SYNTHESIZED BY PULSED LASER DEPOSITION

C.G. Fountzoulas, C. W. Hubbard , Eric H. Ngo, J. D. Kleinmeyer , J. D. Demaree, P. C. Joshi and M. W. Cole
Army Research Laboratory, Weapons Materials Directorate, APG, MD, 21005-5069.

ABSTRACT

Barium strontium titanate (BSTO) films were synthesized by the pulsed laser deposition technique (PLD) on silicon substrates at room temperature. The BSTO film synthesis took place at constant laser energy, 500 mJ, and partial oxygen pressure of 3, 15, 30, 45, 100 mTorr. All films were post annealed at 750 °C in a tube furnace in an oxygen atmosphere. The microstructure, crystallinity and lattice constant of the BSTO films were studied with the aid of scanning electron microscopy (SEM), photon tunneling microscopy (PTM) and Glancing Angle X-ray Diffraction analysis (GAXRD). The hardness and modulus of elasticity of the films were studied with the aid of a nanohardness indenter. The film stoichiometry was determined with the aid of Rutherford Back Scattering (RBS). The results of this research will be combined with the results of our previous work [1] on the effect of substrate temperature on the microstructure and mechanical properties of the BSTO films in order to construct a structural zone model (SZM) of the BSTO films synthesized by PLD.

INTRODUCTION

Thin films of novel barium strontium titanate (BSTO), deposited by the pulsed laser deposition (PLD) technique exhibit excellent electronic properties including tunable dielectric constants and low electronic loss. The dielectric constant of the BSTO depends on the applied electric field. This variable dielectric constant results in a change in phase velocity in the device allowing it to be tuned in real time for a particular application. The dielectric requirements for tunable BSTO thin film are (a) loss less than 0.01; (b) high tunability; (c) dielectric constant between 30 and 100; (d) low leakage current; and (e) good frequency and temperature stability of dielectric properties [1].

A tunable BSTO film must be a single phase and crystalline. It must also have a smooth, defect and crack free surface, uniform microstructure and exhibit good thermal stability with the substrate. The microstructure of the film influences the electronic, and mechanical properties (internal stresses and adhesion), important factors affecting the mechanical integrity and reliability of a device made of these thin films, which in turn influences the performance of the film.

The concept of electronic thin film deposition at lower temperatures is desirable because it enables the integration of several device fabrication steps. However, the inherent material properties required for any application have to be maintained at the lower deposition temperatures. In the case of electronic thin films used for tunable dielectric applications, these properties include dielectric constant, voltage tunability, and the electronic loss tangent. However, the mechanical integrity of the thin film (i.e., the adhesion and cohesion) is just as important.

It is well known that the substrate temperature Ts, the ambient gas pressure, and the energy of any incoming ions influence the growth conditions and therefore, the film structure produced under low-pressure conditions [2-4]. A structural classification system that has gained the broad acceptance for thin films produced by physical vapor deposition (PVD) process has been presented by Movchan and Demchishin [2]. They proposed three zones to describe the microstructures than can develop in deposits produced by vacuum evaporation as a function of T_s/T_m, where T_s is the absolute substrate temperature and T_m is the absolute melting temperature of the deposited material. Thornton [3] elaborated on the approach of Movchan and Demchishin extending it to typical sputtering conditions.

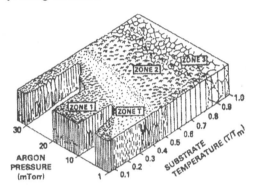

Fig. 1. Thornton's structure zone model for coatings produced by sputtering. [3]

Thornton also concluded that the structure and physical properties of films produced by sputtering could be represented as a function of T_s/T_m, where T_m is the absolute melting temperature of the film material, in terms of four zones as shown in Fig.1, each with its own characteristic structure and physical properties. The general features of Thornton's model were based on the examination of 25- to 250-μm-thick coatings deposited at argon pressures 1.33 X 10^{-4} (1 mTorr) to 3.9 X 10^{-3} Pa (30 mTorr) using cylindrical-post and hollow cathode magnetron sputtering sources. The pulsed laser deposition technique resembles sputtering.

We have initiated an investigation of the crystallinity, microstructure, mechanical and adhesive properties of dielectric thin film of barium strontium titanium oxide (BSTO) deposited on silicon substrate by the pulsed laser deposition method (PLD) [5-7]. This paper presents the initial results of this study of the properties of BSTO thin films as a function of oxygen partial pressure.

EXPERIMENTAL

The experimental apparatus consists of a pulsed laser deposition chamber equipped for optical diagnostics. The 248-nm output of an excimer laser (Lamda Physik, EMG 150 MSC) was directed through a 50-cm focal length lens and focused at 45° near the stoichiometric $Ba_{0.6}Sr_{0.4}TiO_3$ target (BSTO), which was mounted inside a stainless-steel chamber on a high-vacuum, rotatable holder. The details of this deposition technique are given elsewhere [7]. In this work thin films of BSTO were deposited by the PLD technique on 6 to 9 cm^2 (110) silicon substrates at room temperature, 500 mJ laser energy, 10 Hz repetition rate, and 3, 15, 30, 45 and

100 mTorr partial oxygen pressures [1 mTorr = 1.33 X 10^{-4} Pa]. Prior to film synthesis the silicon substrates were degreased, cleaned for 5 minutes in warm (70-75 °C) methanol and ethanol baths and rinsed in warm (70-75 °C) distilled water for another 5 minutes. Subsequently the native oxide of the silicon substrate was removed in a dilute HF aqueous solution. All films were post annealed at 750°C for 45 minutes in a tube furnace in a continuous oxygen flow.

The microstructure of the films was observed by scanning electron microscopy (SEM) and photon tunneling microscopy (PTM); the crystallinity of the films was determined by Glancing Angle X-ray Diffraction (GAXRD); the stoichiometry of the films was determined by Rutherford Back Scattering (RBS); and the Modulus of Elasticity (Young's Modulus) and the nanohardness of the films were determined with the aid of a nanoindenter.

RESULTS AND DISCUSSION

Adhesion and Thickness

The adhesion of the films was determined by the standard "Scotch-Tape" pull test. The films synthesized at 3, 15 and 30 mTorr passed the adhesion test. However, the adhesion of the film synthesized at 45 mTorr was poor. We were not able to obtain a film at 100 mTorr oxygen pressure. The thickness of the films, determined with the aid of a Tencor profilometer, ranged from 200 nm to 600 nm for all pressures throughout the substrate.

Microstructure

The microstructure and morphology of the BSTO films, at various partial oxygen pressures, was determined by metallographic microscopy, and SEM and PTM photomicrographs (Fig. 2). All films observed by metallographic microscopy exhibited red and green elliptical striations, indicative of the thickness variation throughout the substrate, an inherited disadvantage of the PLD technique.

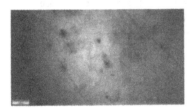 

a   b

Figure 2. Photon tunneling photomicrographs of the BSTO film synthesized at 3 mTorr (a) and 45 mTorr (b) oxygen pressure.

The surfaces of the films were relatively smooth on a macroscopic scale, as observed in the SEM, in spite of thermal grooving which occurred in all films. Thermal grooving occurs when a material is hot enough to generate considerable atomic migration. The groove forms where the grain boundary meets the surface in order to reduce the grain boundary area and hence free energy [8]. However, particulates were evident on the surface of the films. As became evident by the SEM and PTM analysis the density of the particulates increases with increasing oxygen pressure. Previous study of the particulates [9] has been shown that they are expelled parts of the target. In addition, the surface of the film examined in 10,000x SEM and PTM showed microvoids and cracks at oxygen pressures higher than 15 mTorr. The film synthesized at 3 mTorr oxygen pressure showed lines of brittle fracture, not present at any of the other films synthesized at higher oxygen pressure.

RBS Results

The RBS spectrum of the BSTO films synthesized at 3 mTorr is shown in Fig. 3. The dotted line is the measured data for the RBS spectrum and the solid line is the RUMP fit to data.

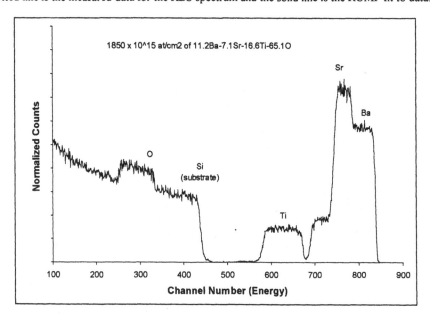

Figure 3. RBS spectrum of the BSTO film synthesized at 3 mTorr oxygen pressure.

The stoichiometry of the film synthesized at 3 and 30 mTorr was $Ba_{0.632}Sr_{0.4}TiO_{356}$.

Glancing Angle X-ray Diffraction

The GAXRD showed that all films were crystalline for all oxygen partial pressures. Figure 4 shows the GARXD pattern for BSTO deposited at 45 mTorr oxygen pressure. For analysis purposes the BSTO films were initially treated as orthorhombic. The lattice constants of the BSTO films are shown in the Table I. It becomes evident from the lattice constants in Table I that the unit cell of all film approaches a cubic unit cell with an average lattice constant **a** equal to 3.96 Å. This average lattice constant is in very good agreement with the published values elsewhere [10].

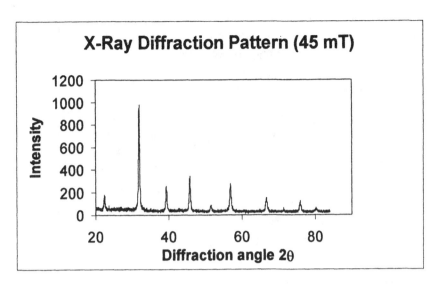

Figure 4. Glancing Angle X-ray Diffraction pattern for BSTO synthesized at 45 mTorr oxygen pressure.

Oxygen Pressure (mT)	a (Å)	b (Å)	c (Å)
3	3.9580	3.9685	3.9649
15	3.9568	3.9588	3.9612
30	3.9614	3.9638	3.9692
45	3.9600	3.9610	3.9631

Table I. Lattice constants of the BSTO films as a function of the oxygen partial pressure.

Nanohardness and Modulus of Elasticity

The nanohardness and the Modulus of Elasticity decrease as a function of the oxygen pressure (Fig. 5). This may be attributed to the increased defect density of the BSTO film with increased partial oxygen pressure.

CONCLUSIONS

The crystallinity, surface morphology, stoichiometry, adhesion, modulus of elasticity and nanohardness of BSTO films synthesized on silicon substrates by the PLD technique were studied as a function of the the oxygen partial pressure. All films were crystalline after post annealing at 750°C for 45 minutes, as it was shown by GAXRD analysis. The adhesion of the films to the substrate, Modulus of Elasticity and nanohardness decreased with increased oxygen pressure. The defect density, microvoids, microcracks and particulates increased with increased oxygen pressure.

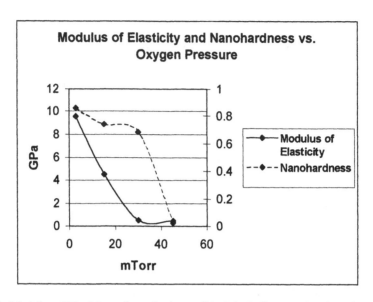

Figure 5. Modulus of Elasticity and nanohardness of the BSTO films as a function of the oxygen partial pressure.

REFERENCES

1. "Mechanical Properties of Ferroelectric Composite Thin Films ", C. G. Fountzoulas* and Somnath Sengupta, Thin Films-Stresses and Mechanical Properties VII, MRS Vol. 505 (1998)
2. B. A. Movchan and S. V. Demchishin, Phys. Met. Metallogr., **28**, 653 (1969).
3. J. A. Thornton, Annual. Rev. Mater. Sci., **7**, 239 (1977).
4. C. Fountzoulas and W. B. Nowak, J. Vac. Sci. Technol. **A 9 (4)**, 2128 (Jul/Aug 1991).
5. S. Sengupta, L. C. Sengupta, D. P. Vijay and S. B. Desu, J. of Integrated Ferroelctrics ,**13**, 239-245 (1996).
6. C. J. Brennan, Integrated Ferroelectrics, **2**, 73 (1992).
7. J. Lee, R. Ramesh and V. G. Keramidas, MRS Symposium Proceedings, **361**, 67 (1195).
8. W. W. Mullins, J. Appl. Phys., **28**, 333 (1957)
9. S. Sengupta, L. C. Sengupta, J. D. Demaree, and W. Kosik, Technical report, ARL-TR-654 (1994).
10. B. A. Baumert, L. H. Chang, A. T. Matsuda, T. L. Tsai, C. J. Tracy, D. J. Taylor, T. Otsuki, E Fujii, a. Hayashi and K. Suu, J. Appl. Phys., **82**, (5), 3558-3565 (1997)

MICROSTRUCTURE OF YBCO AND YBCO / SrTiO₃ / YBCO* PLD THIN FILMS ON SAPPHIRE FOR MICROWAVE APPLICATIONS

M. LORENZ [1], H. HOCHMUTH [1], D. NATUSCH [1], T. THÄRIGEN [1], V. L. SVETCHNIKOV [2], H. W. ZANDBERGEN [2], C. SCHÄFER [3], G. KÄSTNER [3], D. HESSE [3]

[1] Universität Leipzig, Fakultät für Physik und Geowissenschaften, D-04103 Leipzig, Germany, mlorenz@physik.uni-leipzig.de
[2] National Center for High Resolution Electron Microscopy, 2628 AL Delft, The Netherlands
[3] Max-Planck-Institut für Mikrostrukturphysik, D-06120 Halle/Saale, Germany

ABSTRACT

A large-area pulsed laser deposition process for high-quality YBa₂Cu₃O₇₋δ (YBCO) thin films on both sides of R-plane sapphire substrates with CeO₂ buffer layer is used routinely to optimize planar microwave filters for satellite and mobile communication systems. With the experience of more than 700 double-sided 3-inch diam. YBCO:Ag films a high degree of reproducibility of j_c values above 3.5 MA/cm² and of state of the art R_s values is reached. TEM cross sections of the large-area and double-sided PLD-YBCO:Ag thin films on R-plane sapphire with CeO₂ buffer layers show typical defects like stress modulation, stacking faults, a-axis oriented grains, precipitates and interdiffusion layers. YBCO films on SrTiO₃ / YBCO* / CeO₂ film systems on R-plane sapphire wafers have more growth defects compared to bare CeO₂ buffers on sapphire but show as microwave resonators encouraging electrical tunability.

INTRODUCTION

High-T_c superconducting (HTS) Y₁Ba₂Cu₃O₇₋δ (YBCO) and dielectric or ferroelectric thin films on low dielectric loss substrates are suitable candidates for applications as passive microwave devices in future communication systems. There is a huge number of activities in many countries to develop microwave devices using HTS thin films, because in this field real market applications of HTS subsystems seem to be possible in the nearest future. Very recently, the physical and technological aspects of HTS thin films at microwave frequencies were summarized in the excellent book by Hein [1]. To develop electrically tunable microwave filters and devices, combinations of HTS and dielectric or ferroelectric thin films are necessary.

A highly reproducible PLD process for large-area 3-inch diameter and double-sided YBCO films on sapphire substrates for microwave applications is developed and continuously improved by Lorenz et al [2 – 5]. The electrical and microwave performance of this PLD-YBCO films is comparable to high-quality films deposited by other techniques as e. g. thermal coevaporation developed by Kinder et al [6]. Reproducible deposition processes are base technologies for the successful application of HTS microwave devices in commercial products. Several processes are now ready to fulfill the demands of the device developing community.

This paper describes the state of the art of a highly reproducible large-area PLD technique for HTS-YBCO and dielectric SrTiO₃ and ferroelectric Sr(Ba)TiO₃ thin films on buffered sapphire wafers by comparing TEM cross sections and maps of the critical current density j_c.

EXPERIMENTAL

YBCO:Ag thin films, CeO$_2$ buffer layers, and recently also dielectric SrTiO$_3$, and ferroelectric Sr(Ba)TiO$_3$ films were deposited by PLD at very similar conditions, using a KrF excimer laser operating at 248 nm wavelength, on both sides of 3-inch diameter R-plane sapphire wafers of 430 µm thickness as described in more detail by Lorenz et al [2 - 4]. The double-sided films are deposited subsequently at substrate temperatures around 760°C. During YBCO deposition on several hundreds of 3-inch diameter sapphire wafers a high degree of YBCO film quality and reproducibility is reached.

Most of the PLD-YBCO:Ag films are used for development of passive microwave filters for the advanced satellite and communication technique at Robert BOSCH GmbH Stuttgart. Mapping of j$_c$ at 77 K, employing the inductive and side-selective method by Hochmuth and Lorenz [7], is used as a simple and fast routine control of the stability of PLD process and of the resulting YBCO film quality. R$_s$-mapping at 145 GHz [8] and R$_s$ measurements in dependence on the microwave surface magnetic field at 8.5 GHz and 77 K [9] were used for further optimization of the large-area PLD process. TEM cross sections were prepared at National Center for HREM in Delft, Netherlands [10] and at Max-Planck-Institute for Microstructure Physics in Halle/Saale [11].

ELECTRICAL AND MICROWAVE PERFORMANCE

Fig. 1 compares scans of the critical current densities (j$_c$) of double-sided 3-inch diam. YBCO films directly on CeO$_2$ buffered R-plane sapphire (top) to the j$_c$ of YBCO films on 300 nm thick SrTiO$_3$ (100) films on thin YBCO* seed layers on ceria buffered sapphire (sides two on the bottom, only right j$_c$-map of every sample). Figure 1 demonstrates the very good lateral homogeneity and reproducibility of the critical current density of 4 to 5.5 MA/cm^2 at 77 K with a YBCO thickness of about 250 nm for YBCO on bare ceria buffer on sapphire (top and sides 1 on the bottom). Much lower j$_c$ values of about 1.0 – 2.5 MA/cm^2 are obtained up to now for YBCO on SrTiO$_3$ and the microstructural reasons for this drop in j$_c$ will be explained in the following by TEM cross sections.

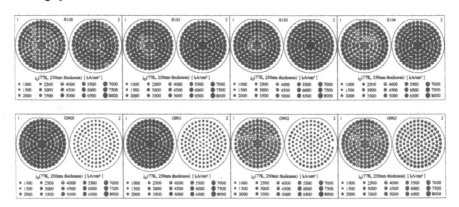

Fig. 1. Comparison of: (top) j$_c$ – maps of 4 consecutively deposited double-sided 3-inch diam. YBCO:Ag thin films on CeO$_2$ buffered 3-inch diam. sapphire and (bottom) of wafers with YBCO / SrTiO$_3$ / YBCO* / CeO$_2$ film system on side 2. The thickness of the YBCO* seed layer in between CeO$_2$ and SrTiO$_3$ increases from about 40 nm (left) to 130 nm (right).

More important than electrical film properties at low frequencies of some kHz is the film performance at microwave frequencies of some GHz. Therefore, R_s maps of typical double-sided 3-inch diam. YBCO films on CeO_2 at 145 GHz and 77 K as measured at Research Center Karlsruhe are a usefull control of homogeneity of large-area films. In general, the R_s maps demonstrate very good lateral homogeneity of R_s over single 3-inch diam. films which is within the statistical variation of R_s values due to the measuring system. However, from side one to side two of the same sample and from sample to sample there are higher differences of the levels of R_s which are due to the sequential deposition of both YBCO:Ag films on side 1 and 2. However, the planar stripline structures of the microwave filters are patterned always on side 2 whereas side 1 which is the ground plane affects less sensitively the total filter performance. Indeed, C-band filters with very good bandpass performance which is suitable for satellite communication subsystems have been structured from PLD-YBCO:Ag thin films at Robert BOSCH GmbH Stuttgart. The microwave power handling capabilities of selected PLD - YBCO:Ag films on CeO_2 show similar $R_s(B_s)$ characteristics and a good reproducibility. The R_s values at 8.5 GHz remain constant below $400\mu\Omega$ up to a microwave surface magnetic field of about 10 mT as shown in [5]. R_s measurements of YBCO thin films on $SrTiO_3$ films are in progress.

TEM CROSS SECTIONS OF YBCO / CeO$_2$ FILMS

To get further insight into microstructure of double-sided HTS - YBCO thin films on 3-inch diam. sapphire substrates with CeO_2 buffer layers selected TEM cross sections are shown in Figures 2 – 4. In general, the interface between YBCO and the CeO_2 buffer layer was found to be atomically flat despite some imperfectness of the substrate. At the substrate/CeO_2 interface, an amorphous or polycrystalline reaction layer was found occasionally. Film No. G 701 side 2 shown in Fig. 2 has a smooth surface. Inside the film both a deformed and a strain-free structure is present. The latter is limited to domains of about 100 nm x 200 nm. It was found that the precipitates at the CeO_2 / substrate interface have no influence on the surface quality.

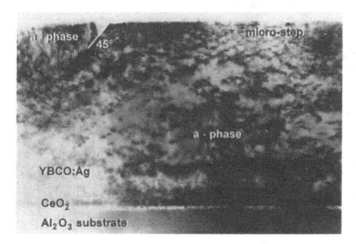

Fig. 2. TEM cross section of sample No. G 701 (side 2, 28 mm from the wafer rim). Note that the contrast of the layer located in between the CeO_2 and Al_2O_3 varies along the interface. This layer corresponds either to amorphous material (left corner), or to a crystalline compound (the rest of the layer).

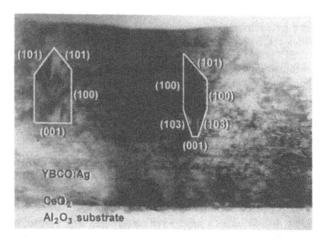

Fig. 3. TEM cross section of film No. G 704 (side 1, 20 mm from the rim) showing the a-axis oriented grains inside the c-oriented YBCO film. These grains are faceted with low-index crystallographic planes {001}/{101}/{103}/{100}.

As Figures 2 and 3 show, a-axis orientated grains nucleate almost near the film surface but it does not affect the smoothness of the surface (Fig. 2) or a-axis orientation starts to grow at approximately 1/3 of the film thickness (Fig. 3). It is interesting to note that in the investigated YBCO films on CeO_2 / sapphire a-axis oriented grains never grew over the level of the film surface (see Fig. 2), but usually formed a concavity on the film surface. The growth rate ratio in YBCO $v_{c\text{-phase}}$ / $v_{a\text{-phase}}$ varies from 1:8 to 1:3 depending on $p(O_2)$ pressure [12]. In our case high $p(O_2)$ in PLD provides a low growth rate ratio and makes it possible to confine the a-phase to the limit of the film thickness (see fig. 3).

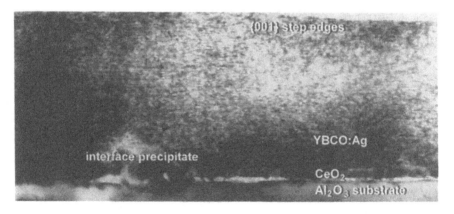

Fig. 4. TEM cross section of film No. G 702 (side 2, 28 mm from the wafer rim). The surface of the film has unit-cell high singular micro-steps. Close to the interface with the substrate the CeO_2 buffer layer is locally broken which facilitates the formation of a precipitate but does not affect the quality of the film surface.

Sample No. G 702 shows at both wafer sides a „wavy" TEM contrast with uniform stress distribution in the (001) planes (Fig. 4). The amorphous layer in G 702 does not cover the whole substrate as it was in G 701 (Fig. 2) and is found to be rather crystalline than amorphous. The broken CeO_2 buffer layer facilitates the formation of interface precipitates which, however, do not introduce any considerable film surface roughness. On the film surface unit-cell high singular micro-steps are present (compare Figures 2 and 4).

YBCO / SrTiO₃ FILMS FOR TUNEABLE MICROWAVE DEVICES

In order to develop electrically tuneable microwave filters and phase shifters there is a need for combinations of dielectric $SrTiO_3$ or ferroelectric $Sr_{1-x}Ba_xTiO_3$ films at one hand and of HTS-YBCO films on the other hand. For example, several microstrip-based phase shifter designs were proposed by van Keuls [13]. In order to deposit YBCO / $SrTiO_3$ bilayers on R-plane sapphire substrates, there are necessary additional CeO_2 buffer and YBCO* seed layers in between the sapphire substrate and the $SrTiO_3$ film as proposed e.g. by Boikov et al [14]. Only by introducing a more than 30 nm thick c-axis oriented YBCO* film on the CeO_2 buffer layer we found a nearly perfect (100)-texture of the $SrTiO_3$ film as proved by XRD phi-scans. Fig. 5 shows a TEM cross section of one of the first deposited 4-layer film systems on sapphire. All films were deposited in-situ in one deposition run which appears as a main advantage of the flexible PLD technique. Obvious in Fig. 5 are the a-axis oriented grains in the top YBCO film on $SrTiO_3$ which are found with much higher probability compared to YBCO on CeO_2 buffer layers (see Figures 2 and 3).

Because of this structural imperfectness, the j_c at 77 K of the YBCO films on $SrTiO_3$ is only about 2 MA/cm² up to now, which is about a factor of two smaller than the j_c of PLD-YBCO films on CeO_2 buffered sapphire as shown in Fig. 1. Correlation of microstructure and microwave properties of the PLD - YBCO films were discussed for example by Kästner et al [11] and Lorenz et al [4, 5]. A-axis oriented grains are known to degrade both j_c and R_s. However, test resonators made from the described YBCO / $SrTiO_3$ / YBCO* / CeO_2 film system on sapphire wafers showed encouraging tunability of bandpass frequency.

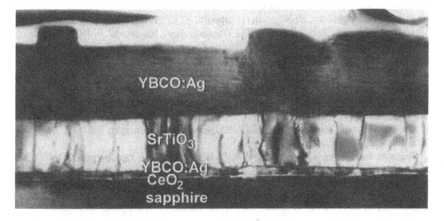

Fig. 5. TEM cross-section of sample No. G 877 with YBCO / $SrTiO_3$ / YBCO* / CeO_2 films on R-plane sapphire showing two a-axis oriented grains within the c-axis oriented YBCO:Ag film on top. The right a-axis oriented grain nucleates at a step of the dielectric $SrTiO_3$ layer.

CONCLUSIONS

A PLD technique is presented which allows the fully reproducible double-sided coating of 3-inch diameter sapphire wafers by thin YBCO, $SrTiO_3$, and CeO_2 films with laterally homogeneous j_c and R_s values. Hundreds of PLD-YBCO:Ag films were already used as microwave bandpass filters for future communication systems by Robert BOSCH GmbH Stuttgart, Germany. The flexible PLD-technique seems to be advantageous compared to other deposition techniques, particularly if more complicated combinations of HTS and dielectric films as e. g. YBCO / $SrTiO_3$ films on sapphire substrates with CeO_2 buffer and YBCO* seed layer for use as electrically tuneable microwave filters have to be deposited in-situ.

ACKNOWLEDGMENTS

We thank M. Klauda, C. Neumann, T. Kässer, F. Schnell and R. Schmidt from Robert BOSCH GmbH for the long-term friendly cooperation. We are indebted to the German BMBF and to Robert BOSCH GmbH for continuous support within „Leitprojekt: Supraleiter und neuartige Keramiken für die Kommunikationstechnik der Zukunft", and to the Saxonian Ministry of Science and Art.

REFERENCES

1. M. Hein, *High-Temperature-Superconductor Thin Films at Microwave Frequencies*, Springer Tracts in Modern Physics **155** (Springer, Berlin, Heidelberg, New York, 1999).
2. M. Lorenz, H. Hochmuth, D. Natusch, H. Börner, G. Lippold, K. Kreher and W. Schmitz, Appl. Phys. Lett. **68**, 3332 (1996).
3. M. Lorenz, H. Hochmuth, D. Natusch, H. Börner, T. Thärigen, D. G. Patrikarakos, J. Frey, K. Kreher, S. Senz, G. Kästner, D. Hesse, M. Steins, W. Schmitz, IEEE Transact. Appl. Supercond. **7**, 1240 (1997).
4. M. Lorenz, H. Hochmuth, J. Frey, H. Börner, J. Lenzner, G. Lippold, T. Kaiser, M. Hein, G. Müller, Inst. Phys. Conf. Series **158**, 283 (1997).
5. M. Lorenz, H. Hochmuth, D. Natusch, G. Lippold, V. L. Svetchnikov, T. Kaiser, M. Hein, R. Schwab, R. Heidinger, IEEE Transact. Appl. Supercond. **9**, 1936 (1999).
6. H. Kinder, P. Berberich, W. Prusseit, S. Rieder-Zecha, R. Semerad, B. Utz, Physica C **282 – 287**, 107 (1997).
7. H. Hochmuth and M. Lorenz, Physica C **265**, 335 (1996).
8. R. Schwab, R. Heidinger, J. Geerk, F. Ratzel, M. Lorenz, H. Hochmuth, Proc. 23rd Int. Conf. on Infrared and Millimeter Waves, University of Essex, Colchester, U.K., September 7 - 11 (1998).
9. T. Kaiser, C. Bauer, W. Diete, M. Hein, J. Kallscheuer, G. Müller, H. Piel Inst. Phys. Conf. Series **158**, 45 (1997).
10. C. Traeholt, J. G. Wen, V. Svetchnikov, A. Delsing, and H. W. Zandbergen, Physica C **206**, 318 (1993).
11. G. Kästner, C. Schäfer, S. Senz, T. Kaiser, M. Hein, M. Lorenz, H. Hochmuth, D. Hesse, Supercond. Sci. Technol. **12**, 366 (1999).
12. P. J. M. van Beutum, F. A. J. M. Driessen, J. W. Gerritsen, H. van Kempen, and L. W. M. Schreurs, J. Crystal Growth **98**, 551 (1989).
13. F.W. Van Keuls, R.R. Romanovsky, F.A. Miranda, Integrated Ferroelectrics **22**, 373 (1998).
14. Y. A. Boikov, Z. G. Ivanov, A. N. Kiselev, E. Olsson, T. Claeson, J. Appl. Phys. **78**, 4591 (1995).

MOCVD GROWTH AND CHARACTERIZATION OF $(Ba_xSr_{1-x})Ti_{1+y}O_{3+z}$ THIN FILMS FOR HIGH FREQUENCY DEVICES

P.K. BAUMANN[1], D.Y. KAUFMAN[2], S.K. STREIFFER[1], J. IM[1], O. AUCIELLO[1], R.A. ERCK[2], J. GIUMARRA[2], J. ZEBROWSKI[3], P. BALDO[4] AND A. McCORMICK[4]
[1]Argonne National Laboratory, Materials Science Division, Argonne, IL 60439
[2]Argonne National Laboratory, Energy Technologies Division, Argonne, IL 60439
[3]New Brunswick Laboratory, Argonne, IL 60439
[4]Argonne National Laboratory, Materials Science Division, Argonne, IL 60439

ABSTRACT

We have investigated the structural and electrical characteristics of $(Ba_xSr_{1-x})Ti_{1+y}O_{3+z}$ (BST) thin films. The BST thin films were deposited at 650°C on platinized silicon with good thickness and composition uniformity using a large area, vertical liquid-delivery metalorganic chemical vapor deposition (MOCVD) system. The (Ba+Sr)/Ti ratio of the BST films was varied from 0.96 to 1.05 at a fixed Ba/Sr ratio of 70/30, as determined using x-ray fluorescence spectroscopy (XRF) and Rutherford backscattering spectrometry (RBS). Patterned Pt top electrodes were deposited onto the BST films at 350°C through a shadow mask using electron beam evaporation. Annealing the entire capacitor structure in air at 700°C after deposition of top electrodes resulted in a substantial reduction of the dielectric loss. Useful dielectric tunability as high as 2.3:1 was measured.

INTRODUCTION

Nonlinear dielectrics such as barium strontium titanate (BST) exhibit a large variation in permittivity, ε, as a function of changes in the electric field applied to the material. For this reason, BST is a suitable candidate for high-frequency tunable phase-shifters [1-4]. To optimize the performance of such devices it is critical to maximize the dielectric tunability, i.e., the ratio of the permittivity at zero field to the permittivity at a defined field, and minimize the dielectric loss (tan δ) in the device operational frequency range. Fabrication of BST thin film devices by metalorganic chemical vapor deposition (MOCVD) provides high compositional control, superior thickness and composition uniformity, high deposition rates, excellent conformality for films grown on high-aspect ratio structures, and the ability to scale film growth to large area substrates.

EXPERIMENTAL APPROACH

$Ba_{1-x}Sr_xTi_{1+y}O_{3+z}$ thin films were synthesized using a large area vertical MOCVD system. A schematic diagram of the deposition system is shown in Fig.1. Metalorganic precursors of $Ba(thd)_2$, $Sr(thd)_2$, and Ti $(O-iPr)_2$-$(thd)_2$ with polyamine adducts were introduced using high purity nitrogen as a carrier gas into the MOCVD reactor, via a temperature-controlled flash-vaporizer and a computer-controlled liquid delivery system (ATMI LDS-300B) that provides good composition control and reproducibility of the delivered precursor mixture. The temperature of the delivery lines was carefully controlled to avoid condensation or premature reaction of the precursors prior to introduction into the MOCVD reactor. The precursors were thoroughly mixed with high purity reactive gases (O_2 and N_2) in a showerhead designed to provide deposition of BST films with uniform composition and thickness over large area substrates. Table I summarizes the film deposition and processing conditions.

169

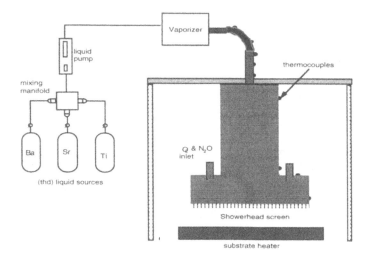

Figure 1: Schematic diagram of large area vertical MOCVD reactor with liquid precursor delivery system.

The Ba/Sr ratio of the BST thin films was kept at 70/30 while the (Ba+Sr)/Ti ratios were varied between 0.96 and 1.05. The composition and thickness of the BST films were analyzed by x-ray fluorescence spectroscopy (XRF) and Rutherford backscattering spectrometry (RBS), while x-ray diffraction (XRD) was used to determine their crystallographic orientation. Patterned Pt top electrodes were deposited by electron beam evaporation onto the BST films at 350°C through a shadow mask. Subsequent to top electrode deposition, whole capacitors were annealed in air at 550°C for 30 min followed by a 700°C anneal for 60 min, to improve the structure and chemistry of both the bulk BST film and the top electrode/BST interface, with the aim of reducing bulk and interfacial contributions to the dielectric loss tangent. Earlier experiments performed using *in-situ* surface analytical techniques [5] indicate that impurities at, and the structure of the top electrode/BST interface are strongly correlated with the capacitor losses. Measurement of the capacitance and loss tangent were performed using an HP4192A impedance analyzer at a frequency of 1MHz and 0.1 Vrms oscillation level. Measurements were performed between two top electrodes, resulting in the electrical characterization of two capacitors in series. This top electrode to top electrode geometry is consistent with device design for which these samples are targeted.

Table I. Deposition and processing parameters for MOCVD synthesis of BST thin films.

Substrates:	Pt(1000Å)/SiO$_2$(1000Å)/Si
Substrate heater temperature:	650°C
Reactive gases:	O$_2$ and N$_2$O
Reactive gas flow rate:	250-1000 SCCM
Reactor pressure:	1.5-2.7 Torr
Top electrodes:	e-beam evaporated Pt (1000Å)
Post electrode anneal:	550°C for 0.5 hrs
Electrical characterization:	HP4192A at 1 MHz and 0.1 V rms

RESULTS AND DISCUSSION

XRD analysis demonstrated that the thin BST films were polycrystalline. Fig. 2 shows XRD spectra for BST films deposited with 500 sccm and 1000 sccm total reactive gas flow (O$_2$ and N$_2$O). Strong (100) fiber texture was observed in films grown with a total reactive gas flow rate of 500 sccm.

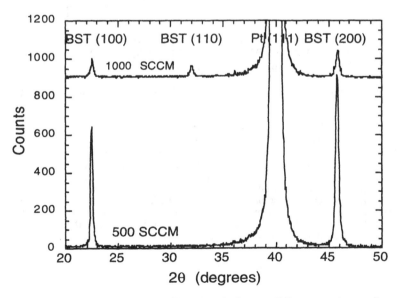

Figure 2:. XRD scans of BST thin films deposited at two different reactive gas flow rates.

Excellent film uniformity was obtained over a 4-inch wafer. Initial growth rates were ≈ 12 Å/min, but higher rates of ≈ 50Å/min were achieved after optimization of the reactor showerhead configuration and vaporizer temperature uniformity (± 5 °C). With the control achieved in the deposition process, we produced BST films with (Ba+Sr)/Ti ratios between 0.96 and 1.05. For different (Ba+Sr)/Ti ratios slightly different characteristics for the permittivity and dielectric loss were observed as a function of applied field. A comparison of these curves is shown in Fig. 3. It is well-established that the electrical properties depend on the (Ba+Sr)/Ti ratio [6-9]. A broader range of (Ba+Sr)/Ti ratios is still under investigation. Since Ti rich films generally display lower leakage, all further electrical data presented in this paper are from Ti rich samples ((Ba+Sr)/Ti = 0.96).

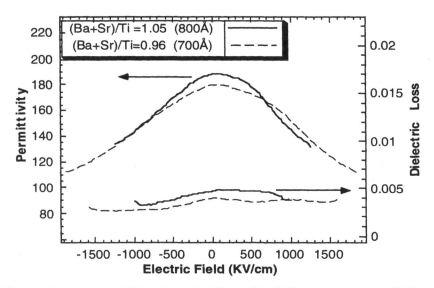

Figure 3: Permittivity and dielectric loss for different (Ba+Sr)/Ti ratios measured at 1MHz.

It has also previously been reported that post-growth annealing of BST films can enhance the microstructure and improve electrical properties [10, 11]. Annealing the capacitor at 700°C for 1 hour in air resulted in a substantial decrease of the dielectric loss, as shown in Fig. 4. However, a broadening of the permittivity-electric field curves was observed. Dielectric losses as low as 0.003 were measured at room temperature and zero field for 80 nm thick films with useful tunabilities of 1.5:1. Useful tunability is defined as the tunability at a field achieved before leakage causes the loss to increase above the zero bias. Useful tunabilities of 2.3:1 were observed at the expense of a somewhat higher loss (<0.005) as shown in Fig. 5. Tunabilities of 3.6:1 were achieved, but at the expense of substantially increased dielectric loss at the voltage extremes.

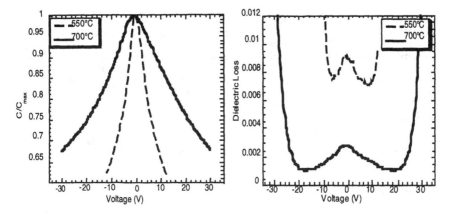

Figure 4: Relative capacitance and dielectric loss following a 550°C post electrode anneal for 30 min in air and a subsequent 700°C anneal for 1h in air. Measured at 1MHz.

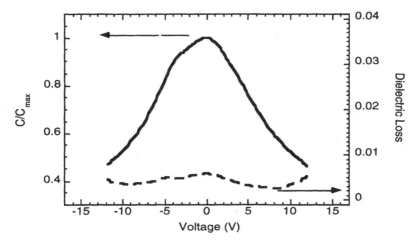

Figure 5: Relative capacitance and dielectric loss following a 550°C post electrode anneal for 30 min in air and a subsequent 700°C anneal for 1h in air. Measured at 1MHz.

Further work is underway in our laboratory to elucidate the role of composition, thickness, and interfacial quality on the electrical properties of BST capacitors.

CONCLUSIONS

We have grown polycrystalline, {001} fiber-textured BST thin films with good thickness and composition control on platinized silicon, using a large area vertical MOCVD system. The (Ba+Sr)/Ti ratio content of the BST films was varied from 0.96 to 1.05. Dielectric losses as low as 0.003 were measured at room temperature, zero field and 1 MHz for 80nm thick films. Higher tunabilities of 2.3:1 could be achieved at the expense of increased loss. A reduction of the dielectric loss could be achieved by annealing the BST films at 700°C in air for 1 hr after top electrode deposition.

ACKNOWLEDGMENTS

This work has been supported by the U.S. Department of Energy, BES-Material Sciences and Office of Transportation Technologies, under Contract W-31-109-ENG-38, and by DARPA, contract # 978040.

REFERENCES

1. D. Flaviis, N.G. Alexopolous, and M. Staffsudd, IEEE Trans. Microwave. Tech. **45**, 963 (1997).
2. V.K. Varadan, D.K. Ghodgaonker, V.V. Varadan, J.F. Kelly and P. Glikerdas, Microwave Journal, **35**, 116 (1992).
3. J.M. Ponds, S.W. Kirchoefer, W. Chang, J.S. Horwitz, and D.B. Chrisey, Integr. Ferroelectr., **22**, 317 (1998).
4. F.A. Miranda, F.W. Van Keuls, R.R. Romanofsky, and G. Subramanyam, Integr. Ferroelectr. **22**, 269 (1998).
5. J. Im, O. Auciello, A.R. Krauss, P.K. Baumann, and S.K Streiffer (APL, to be published).
6. S. Yamamichi, H. Yabuta, T. Sakuma, and Y. Miyasaka, Appl. Phys. Lett. **64**, 1644 (1994).
7. P.C. van Buskirk, J.F. Roeder, and S. Bilodeau, Integr. Ferroelectr. **10**, 9 (1995).
8. C. Basceri, S.K. Streiffer, A.I. Kingon, R. Waser, J. Appl. Phys. **82**, 2497 (1997).
9. S.K. Streiffer. C. Basceri, C.B. Parker, S.E. Lash, and A.I. Kingon, J. Appl. Phys. **86**, 4565 (1999).
10. L.A. Knauss, J.M. Pond, J.S. Horwitz, D.B. Chrisey, C.H. Mueller, and R. Treece, Appl. Phys. Lett., **69**, 25 (1996).
11. H.D. Wu, and F.S. Barnes, Integr. Ferroelectr. **22**, 291 (1998).

DIELECTRIC PROPERTIES OF Ba$_{1-x}$S$_x$TiO$_3$ FILMS GROWN ON LaAlO$_3$ SUBSTRATES

Y. GIM, T. HUDSON, Y. FAN, A. T. FINDIKOGLU, C. KWON*, B. J. GIBBONS, AND Q. X. JIA
Superconductivity Technology Center, MS-K763, Los Alamos National Laboratory, NM 87545
*Department of Physics and Astronomy, California State University-Long Beach, CA 90840

ABSTRACT

We report the crystal structures and dielectric properties of barium strontium titanate, Ba$_{1-x}$Sr$_x$TiO$_3$ (BST), films deposited on LaAlO$_3$ substrates using pulsed laser deposition, where x = 0.1 to 0.9 at an interval of 0.1. We have found that when x < 0.4 the c-axis is parallel to the plane of the substrate but normal as x approaches 1. Temperature-dependent capacitance measurements at 1 MHz show that the capacitance has a peak and that the peak temperature decreases with increasing x. We have found that the peak temperatures of the films are about 70 °C higher than those of bulk BSTs when x < 0.4. From room-temperature capacitance (C) vs applied voltage (V) measurements, we have found that the C-V curves of the BST films exhibit hysteresis except for x = 0.9 and that the peak voltage at which the capacitance becomes maximum decreases with increasing x. At room temperature, the Ba$_{0.6}$Sr$_{0.4}$TiO$_3$ film exhibits the largest capacitance tunability ($\approx$ 37%) with an applied electric field of 40 kV/cm.

INTRODUCTION

In recent years, there have been intense research efforts to develop various microwave devices, such as electrically tunable filters and high-Q resonators, by exploiting a wide range of the dielectric properties of barium strontium titanate, Ba$_{1-x}$Sr$_x$TiO$_3$ (BST).[1-3] Barium titanate (BTO) is a ferroelectric material with a bulk Curie temperature of 120 °C and has a very high bulk dielectric constant up to a few thousands at room temperature. On the other hand, strontium titanate (STO) is not ferroelectric down to very low temperatures but has a great structural stability over a wide range of temperatures.

BTO and STO can form solid solutions with each other at any Ba/Sr composition ratio because of their similar crystal structures and ionic radii of Ba^{2+} and Sr^{2+}. Therefore, by making a mixture of dielectric BTO and STO, it is possible to produce a solid solution of BST whose dielectric properties can meet specific applications. One example is the dependence of the Curie temperature of bulk BST on x; one can precisely predict the Curie temperature based on the Ba/Sr composition ratio.[4]

When an epitaxial BST film is used as an integral part of a microwave device, the crystal structure and dielectric properties of the film can be different from those of bulk BST because the BST film is usually grown on a single crystal substrate at a relatively high deposition temperature. Due to the lattice mismatch and difference in the thermal expansion between the film and the substrate, the crystal structure of the BST film can be distorted from that of bulk BST. And, in most oxide films with a layered structure, such structural distortion usually results in changes in the electrical properties of the films. As a result, for the development of BST thin-film microwave devices, it is important to characterize the structure and electrical properties of the BST film in order to find an optimum Ba/Sr composition ratio for each specific application.

Here we report a systematic study of the crystal structures and dielectric properties of BST films grown on LAO substrates, where x = 0.1 to 0.9 at an interval of 0.1. X-ray analysis shows that the c-axis, the longest unit-cell axis (c > a $\approx$ b), is parallel to the plane of the substrate when x < 0.4 but becomes normal as x approaches 1. Even though the direction of the c-axis is dependent on x, the c-axis lattice constants of the BST films appear to be larger for all x values than those of bulk BSTs. Temperature-dependent capacitance measurements at 1 MHz show that the capacitance has a peak and that the peak temperature decreases with increasing x. When x $\leq$ 0.4, the peak temperatures of the BST films are about 70 °C higher than those of bulk BSTs. But, as x increases over 0.4, the peak temperatures of the BST films and bulk BSTs have a very similar temperature dependence on x. Capacitance vs applied voltage (C-V) measurements at room temperature show

175

that the C-V curves of the BST films exhibit hysteresis except for x = 0.9 and that the peak voltage at which the capacitance becomes maximum decreases with increasing x. Among the BST films, the $Ba_{0.6}Sr_{0.4}TiO_3$ film exhibits a maximum capacitance tunability of 37% at room temperature with an applied dc electric field of 40 kV/cm.

EXPERIMENT

We have used standard pulsed laser ablation technique to grow 300 nm thick BST films on LAO substrates. To grow films with various Ba/Sr ratios, we have used a series of sintered stoichiometric ceramic targets of $Ba_{1-x}Sr_xTiO_3$, where x = 0.0 to 1.0 at an interval of 0.1. The BTO (x = 0.0) and STO (x = 1.0) are used only for structural analyses. We have deposited all the films at the same deposition temperature (800 °C) and oxygen pressure (200 mTorr), which was used for the growth of highly epitaxial BST films for tunable microwave devices.[1] After the deposition, the films are cooled in an oxygen pressure of 300 Torr.

After the deposition, We have performed x-ray θ-2θ scans to determine the crystal structures of the BST films. For each x, we have calculated the out-of-plane and in-plane lattice constants from normal ($\chi \approx 90°$) and tilted ($\chi \approx 45°$) scans, respectively. To characterize the dielectric properties of the BST films, we have made capacitors by depositing gold on tope of film surface. Each gold electrode is 2 mm long and 200 μm wide and the gap between two gold electrodes for each capacitor is 10 μm. We have measured the capacitance over a wide range of temperatures (from 20 K to 500 K) in order to find the capacitance peak temperature. In addition, we have studied the dependence of the capacitance on applied electric field at room temperature.

RESULTS

Filled circles and open squares in Fig.1 represent the out-of-plane and in-plane lattice constants of BST films as a function of x. For comparison, we also plot the c-axis (solid line) and

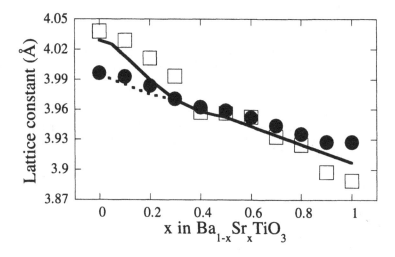

Fig.1 The out-of-plane (solid circles) and in-plane (open squares) lattice constants of the BST films. The a-axis (dotted curve) and c-axis (solid curve) lattice constants of bulk BSTs are from Ref. [5].

176

a-axis (dotted line) lattice constants of bulk counterparts;[5] bulk BST is tetragonal (c > a = b) when x < 0.3 while cubic when x ≥ 0.3. As shown in Fig. 1, when x < 0.4 the out-of-plane lattice constants of the BST films are very close to the a-axis lattice constants of bulk BSTs and the in-plane lattice constants to the c-axis lattice constants. As x increases over 0.4, the in-plane and out-of-plane lattice constants of the BST films are very close to each other until the out-of-plane lattice constants become larger than the in-plane lattice constants near x ≈ 1.

The unit-cell lattice constant of a pseudocubic LAO substrate is 3.795 Å, which is smaller than any other unit-cell dimensions for bulk BST (see solid and dotted curves in Fig. 1). In such case, the in-plane lattices of an epitaxial film are expected to be shrunken and the out-of-plane to be elongated under a compressive stress; this will make the c-axis (the longest unit-cell axis) normal to the plane of the substrate. However, the BST films of x < 0.4 show that it is not true. Instead, the c-axis appears to be in the plane of the substrate. In addition, the c-axis lattice constants of the BST films are larger than those of bulk BSTs when x < 0.4.

As x increases from 0.4 to 0.8, both the in-plane lattice and out-of-plane lattice constants are very close to each other. But, for x values of 0.9 and 1.0, the out-of-plane lattice is elongated and the in-plane lattice is shortened. Because bulk BST is cubic when x ≥ 0.3, the elongation of the out-of-plane lattice at x ≈ 1 can be considered as a result of the in-plane compressive stress.[6] Following the definition of the c-axis, the out-of-plane lattice becomes the c-axis and again elongated. In other words, the c-axis seems to be always elongated although the orientation of the c-axis with respect to the plane of the substrate is parallel when x < 0.4 but normal when x ≈ 1.

To study the dependence of the capacitance peak temperatures of the BST films on x, we have measured the capacitance over a wide range of temperatures, from 20 K to 500 K. Compared with bulk BSTs, the films exhibit a relatively broad capacitance peak. Filled circles in Fig.2 represent the peak temperatures of the BST films. For comparison, we also plot the peak temperatures of bulk BSTs, represented by open squares in Fig. 2.[4] As shown, both film and bulk have a similar temperature dependence on x: the peak temperature decreases with increasing x. However, when x ≤ 0.4 the peak temperatures of the BST films are about 70 °C higher compared with those of bulk BSTs.

In the case of a single crystal BTO, it has been known that a structural distortion under a mechanical pressure can increase or decrease the peak temperature.[7-9] For example, if the BTO is

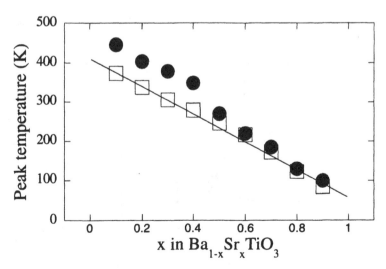

Fig. 2 The capacitance peak temperatures of the BST films (filled circles) and bulk BSTs (open squares). Bulk data are from Ref. [4].

under a hydrostatic pressure so that all of its unit-cell dimensions are shrunken, then the peak temperature decreases. On the other hand, if one of the unit-cell dimensions is allowed to expand under a two-dimensional pressure, the peak temperature increases. Although such detailed study is not available for other Ba/Sr composition ratios, we may, in a first approximation, assume that the trend would be similar although the change in the peak temperature is expected to depend on x. In the case of an epitaxial film, a stress in the film can produce a similar structural distortion as the mechanical pressure does to the single crystal. From the previous x-ray analysis, we learn that the c-axis is always elongated although their direction with respect to the plane of the substrate depends on x. Following the case of the single crystal BTO, we expect the peak temperature to increase for all x values. Indeed, the peak temperatures of the BST films have increased over those of bulk BSTs when $x \leq 0.4$, as shown in Fig. 2. However, the increase in the peak temperature is not significant when $x > 0.4$. A possible explanation is that the effect of a structural distortion on the peak temperature is very minor when the peak temperature of bulk BST is already significantly decreased as x increases over 0.4 (see a solid line in Fig. 2).

To investigate the dependence of the capacitance on electric field, we have applied a slowly varying electric field across two gold electrodes on top of film surface and measured the capacitance at a probing frequency of 1 MHz. Fig. 3(a) shows a plot of the capacitance (C) vs applied voltage (V) of the $Ba_{0.5}Sr_{0.5}TiO_3$ film at room temperature; here the $Ba_{0.5}Sr_{0.5}TiO_3$ film is chosen as an example and the applied electric field (V/cm) corresponds to 10^3 x V. Dotted and solid curves correspond to capacitance values measured while the applied voltage is ramping up and down, respectively. Except for $x = 0.9$ (the $Ba_{0.1}Sr_{0.9}TiO_3$ film), the C-V curves exhibit hysteresis and have two capacitance peak voltages corresponding to V_p^+ and V_p^- (see Fig. 3(a)). We have observed that the magnitudes of the C-V hysteresis increase as x decreases from 0.9. Especially when $x < 0.4$, we have found that the C-V curves become very asymmetric and have multiple loops. Considering that the peak temperatures of the BST films are higher than room temperature when $x \leq 0.4$, we believe that such characteristics are related to ferroelectricitiy in the films.

Fig. 3(b) shows a plot of the V_p^+ (filled circles) and V_p^- (open squares) vs x for the BST films. We can see that there are three different trends in the dependence of the V_p^+ and V_p^- on x. At $x = 0.9$, the V_p^+ and V_p^- are very close to 0 (little hysteresis in the C-V curve). As x decreases from 0.9 to 0.5, the V_p^+ and V_p^- remain at ±2.5 V. However, as x decreases further from 0.4, we see that the magnitudes of the V_p^+ and V_p^- start to increase and reach almost 8 - 9 V. Again, as the

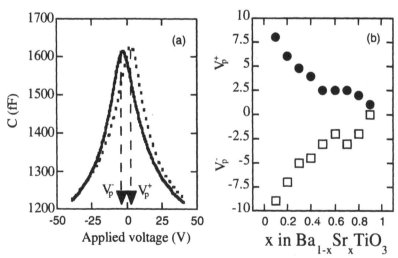

Fig. 3 (a) Capacitance vs applied voltage of the $Ba_{0.5}Sr_{0.5}TiO_3$ film. (b) Positive (filled circles) and negative peak (open squares) voltages vs x in $Ba_{1-x}Sr_xTiO_3$.

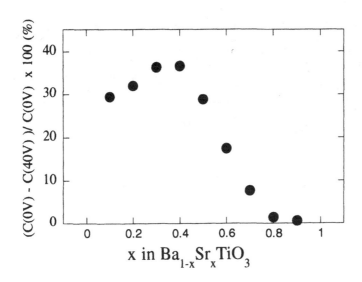

Fig. 4 Capacitance tunability vs x in $Ba_{1-x}Sr_xTiO_3$ at room temperature.

cases of the orientation of the c-axis and the capacitance peak temperature, there is a change in the dependence of the C-V hysteresis on x as x increases over 0.4.

From the previous room-temperature C-V measurements, we have measured the capacitance tunability of the BST films with an applied dc electric field of 40 kV/cm: (C(0V) - C(40V))/C(0V) x 100%. Fig. 4 is a plot of the capacitance tunability vs x in $Ba_{1-x}Sr_xTiO_3$. We can also observe a change in the dependence of the capacitance tunability on x as x increases over 0.4; the capacitance tunability is relatively high and exhibit a weak dependence on x when x ≤ 0.4 but starts to decrease rapidly as x approaches 1. Among the BST films, the $Ba_{0.6}Sr_{0.4}TiO_3$ film exhibits the highest capacitance tunability (≈ 37%) at room temperature.

In conclusion, we have investigated the crystal structures and dielectric properties of $Ba_{1-x}Sr_xTiO_3$ films grown by laser ablation on LAO substrates, where x = 0.1 and 0.9 at an interval of 0.1. We have found that there is a consistent change in the dependence of the orientation of the c-axis and the dielectric properties on x. When x < 0.4, the c-axis is parallel to the plane of the substrate but becomes normal as x approaches 1. The temperature-dependent capacitance measurements show that the capacitance peak temperatures of the BST films decrease with increasing x as those of bulk BSTs do. But, we have found that when x ≤ 0.4 the BST films have about 70 °C higher peak temperatures compared with bulk counterparts. The C-V measurements show that the C-V hystereses of the BST films are very strong when x ≤ 0.4 but become weak as x approaches 1. Especially when x = 0.9, the C-V curve shows little hysteresis. Our results indicate that there is a strong correlation among the chemical composition (Ba/Sr ratio), the orientation of the c-axis, the dielectric properties of the BST films. From the room-temperature capacitance tunability measurements, the $Ba_{0.6}Sr_{0.4}TiO_3$ film exhibits the highest tunability (≈ 37%) among the BST films grown by laser ablation in this work.

REFERENCES

[1]A. T. Findikoglu, Q. X. Jia, D. W. Reagor, and X. D. Wu, Microwave and Optical Technology Letters **9**, 306(1995).
[2]R. A. Chakalov, Z. G. Ivanov, Yu. A. Boikov, P. Larsson, E. Carlsson, S. Gevorgian, and T. Claeson, Physica C **308**, 279 (1998).

[3]C. L. Chen, H. H. Feng, Z. Zhang, A. Brazdeikis, Z. J. Huang, W. K. Chu, C. W. Chu, F. A. Miranda, F. W. Van Keuls, R. R. Romanofsky, and Y. Liou, Appl. Phys. Lett. **75**, 412 (1999).
[4]G. A. Smolenskii and K. I. Rozgachev, Zh. Tekh. Fiz. **24**, 1751 (1954).
[5]M. McQuarrie, J. of the American Ceramic Society **38**, 444 (1955).
[6]L. Ryen, E. Olsson, L. D. Madsen, X. Wang, C. N. L. Edvardsson, S. N. Jacobsen, U. Helmersson, S. Rudner, L. -D. Wernlund, J. of Appl. Phys. **83**, 4884, (1998).
[7]Franco Jona and G. Shirane, Ferroelectric Crystals, Ch. IV, Dover Publications, Inc., New York (1993).
[8]Forsbbergh, P. W., Jr., Phys. Rev. **93**, 686 (1954).
[9]W. J. Merz, Phys. Rev. **77**, 52 (1950).

MICROSTRUCTURAL ARCHITECTURE OF (Ba,Sr)TiO₃ THIN FILMS FOR TUNABLE MICROWAVE APPLICATIONS

WONTAE CHANG, JAMES S. HORWITZ, WON-JEONG KIM, CHARLES M. GILMORE, JEFFREY M. POND, STEVEN W. KIRCHOEFER, AND DOUGLAS B. CHRISEY
Naval Research Laboratory, 4555 Overlook Ave., SW, Washington, D.C. 20375

ABSTRACT

$Ba_xSr_{1-x}TiO_3$ (BST, x=0.5 and 0.6) thin films have been deposited onto (100) MgO single crystal substrates by pulsed laser deposition (PLD). The room temperature capacitance and dielectric quality factor (Q=1/tanδ) have been measured as a function of electric field (≤ 100 kV/cm) at microwave frequencies (1 to 20 GHz) using silver interdigitated electrodes deposited on top of the BST film. It has been observed that the dielectric constant of the film and its change with electric field are closely related to film phase (amorphous to crystalline phase) and film strain which affects the ionic polarization of the film. Amorphous BST films show high dielectric Q (> 100) with low dielectric constant (~30–200) and low dielectric tuning (< 1%), presumably due to small ionic polarization. Crystalline films have a higher dielectric constant (~1000–3000) and a higher dielectric tuning (~ 65%) but a lower dielectric Q (~20). As an optimal microstructure of the film for tunable microwave applications, strain-relieved large-grained (~5000 Å) randomly oriented polycrystalline films were deposited using a thin amorphous buffer layer of BST (~50 Å). Very large grains (size up to a few microns) were observed in BST films prepared using a thicker amorphous buffer layer (~500 Å). We will present results on how careful control of microstructure can lead to films with optimal dielectric properties for the tunable microwave devices.

INTRODUCTION

$Ba_xSr_{1-x}TiO_3$ (BST) is a solid solution ferroelectric material that exhibits an electric field dependent dielectric constant and (Ba,Sr) composition dependent Curie temperature [1]. These properties are currently being used to develop high frequency (1-20 GHz) tunable microwave devices for room temperature applications.

Since the early stage of this research, the ideal form of ferroelectric thin films for tunable microwave devices was believed to be a highly oriented single crystalline phase. However, we have observed that these materials exhibited either high tuning or high Q but not both at the same time, as in Figure 1(a). The general trend indicates that the epitaxial single crystalline materials may not be the ideal form of BST films for this application. The inverse relationship between dielectric tuning and dielectric Q shown in Figure 1(a) is due to the Kramer–Krönig relationship, film strain, and film crystallinity. In this paper, we will discuss the ideal microstructure of BST thin films for tunable microwave applications.

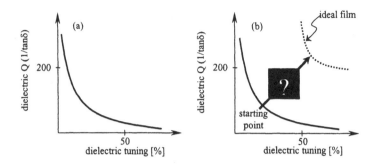

Figure 1   (a) A general trend in the dielectric properties of highly oriented and single crystalline phase BST thin films and (b) an ideal form of BST thin films providing both high tuning and high Q.

EXPERIMENTAL

BST (x=0.5 and 0.6) thin films (~0.5 μm thick) were grown on (100) MgO single crystal substrates at 750 °C in an oxygen ambient gas by pulsed laser deposition (PLD).   The output of a short pulsed (30 ns FWHM) excimer laser operating with KrF (λ=248 nm) at 5 Hz was focused to a spot size of ~0.1 cm^2 and an energy density of ~1.9 J/cm^2 onto a single phase BST (x=0.5 and 0.6) target.   BST films were also deposited using 2-step growth techniques.   First, a thin (~50 to 1000 Å) amorphous layer of BST was deposited at room temperature in an oxygen ambient pressure of 350 mTorr.   The substrate temperature was then increased to 750 °C and a second BST film (~0.5 μm thick) was deposited.   Films deposited by both procedures were characterized for structure and morphology using X-ray diffraction (XRD) and scanning electron microscopy (SEM).   Interdigitated capacitors with gaps from 5 to 12 μm were deposited on top of the BST films through a polymethylmethaacrylate (PMMA) lift off mask by e-beam evaporation of 1-2 μm thick Ag and a protective thin layer of Au [2].   Temperature dependent measurements were performed at 1 MHz with DC bias changes (0-40 V) using a HP 4284A Precision LCR Meter.   Microwave input reflection coefficient S$_{11}$ measurements were made on an HP 8510C network analyzer at room temperature.   The data are fitted to a parallel resistor-capacitor model to determine capacitance and dielectric Q (=1/tanδ) [2].   Dielectric constants were calculated from the device dimensions.

RESULTS AND DISCUSSION

Factors causing an Inverse relationship between Dielectric tuning and Dielectric Q

1. Kramer–Krönig relationship
        Figure 2 shows a temperature dependence of dielectric properties of a BST (x=0.5) film, which is well represented by the Kramer–Krönig relationship which states the dielectric loss follows similar temperature dependent behavior to the dielectric constant resulting in an inverse

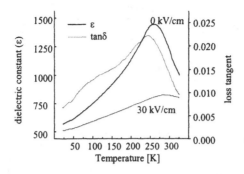

Figure 2   Temperature dependent dielectric properties of BST (x=0.5) film on MgO at 1 MHz.

relationship between dielectric constant (or its tuning) and dielectric Q.   More work causes more energy loss and consequently it can be applicable to any form of ferroelectric materials (i.e., film or bulk, strained or unstrained, amorphous phase or crystalline phase).   Even the ideal case BST films will keep the inverse relationship (Figure 1(b)).

2. Film Strain

Previously, we reported on the effect of film strain on the microwave (1–20 GHz) dielectric properties of epitaxial single crystalline BST films deposited on MgO and LAO substrates [3,4].   For strained BST films, a significant change in the dielectric constant and dielectric Q was observed depending on the type of strain, i.e., compressive or tensile strain [3,4].   For compressive strain, the dielectric constant and its tuning were decreased and for tensional strain, the opposite effect was observed.   Film strain is one of factors enhancing the inverse relationship between dielectric tuning and dielectric Q and should be relieved to improve the dielectric properties (Figure 1(b)).

3. Amorphous–Crystalline Ferroelectric Phase

Since M.E. Lines [5] and A.M. Glass *et. al.* [6] reported on the observation of ferroelectricity of amorphous ferroelectric materials in 1977 the understanding of amorphous ferroelectric materials has not been established completely, probably due to the microstructural disorder of amorphous phase with short range order causing various interatomic potentials. Amorphous ferroelectric materials were believed not to exhibit ferroelectricity because they did not have crystalline grains and domains.   Presumably, the short-range order of the amorphous phase may provide enough ionic dipoles to exhibit ferroelectricity.   Table I shows a general trend of 1–20 GHz dielectric properties of amorphous BST films and highly oriented crystalline BST films.   Film phase, i.e., from amorphous to crystalline phase, is a factor which leads to the inverse relationship between dielectric tuning and dielectric Q and could be optimized using either an amorphous–crystalline mixed phase or an optimized range of ordering between short range and long range (i.e. nanocrystalline phase).

Table I. Dielectric properties of amorphous and highly oriented crystalline BST (x=0.5) film.

1–20 GHz dielectric properties	amorphous phase	epitaxial crystalline phase
dielectric constant	30~100	1000~3000
dielectric tuning at 67 kV/cm	< 1%	~ 65%
dielectric Q	> 100	10~20

Film Strain Relief

A thin amorphous BST buffer layer was reported to relieve the strain in the subsequently deposited crystalline BST film [7]. Any BST phase from amorphous to randomly oriented crystalline phase can be used as the buffer layer if the buffer layer can effectively eliminate problems associated with mismatch between film and substrate such as lattice size and thermal expansion (Figure 3). The strain-relieved films using this method are also free to be annealed from any substrate effect (i.e., film strain due to thermal expansion mismatch). We found that a 500 Å-thick buffer layer could effectively relieve the film strain so that large grains (up to several microns) could grow in the strain-relieved film (Figure 4).

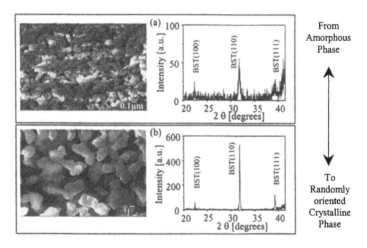

Figure 3   A possible BST buffer layer (a) amorphous and randomly oriented crystalline BST deposited at room temperature and then heated at 750°C for 1min. (b) randomly oriented polycrystalline BST deposited at room temperature and annealed at 900°C for 6 hrs.

Film Phase Optimization

A mixed phase material is a compromise starting point between amorphous and highly oriented crystalline phase (Figure 1(b)) and is schematically presented in Figure 5 (a). In this approach, the following critical variables need to be optimized for a better dielectric property; size of crystalline phase, distance between crystalline phases, and density and porosity of the amorphous phase. After a careful study of heat treatment effect on the film phase at the boundary between amorphous and crystalline phase (Figure 6), a sample close to the mixed

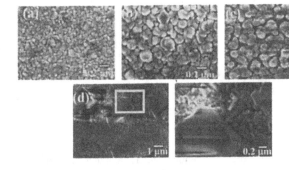

Figure 4 Strain-relieved BST (x=0.6) films using three different thickness of amorphous BST buffer layer (a) 100 Å, (b) and (d) 500 Å, and (c) 1000 Å. (a), (b), and (c) were annealed at 800°C for 12 hrs and (d) was annealed at 1000°C for 12hrs in O₂ flowing.

phase material was prepared by depositing a ~500 Å thick BST (x=0.6) at 750°C and another ~2000 Å thick BST at 400°C and then annealed at 400°C for 12 hours (Figure 5(b)). Even though the annealing temperature is below the crystallization temperature of 450°C, large crystalline grains grow on the crystalline seed layer deposited at 750°C. Also, Figure 6 shows that the annealing temperature for crystallization of BST amorphous films depend on which temperature the amorphous film is deposited. For example, the annealing temperature of 450°C could crystallize the BST amorphous film deposited at 400°C but not the film deposited at 200°C. The film deposited at 200°C and annealed 450°C shows an amorphous phase only.

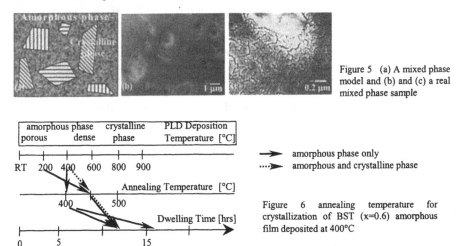

Figure 5 (a) A mixed phase model and (b) and (c) a real mixed phase sample

Figure 6 annealing temperature for crystallization of BST (x=0.6) amorphous film deposited at 400°C

Microwave Dielectric Properties

Previously, we reported that (Ba,Sr) compensation to the PLD target increased both dielectric constant and dielectric Q [8]. We also observed that donor/acceptor doping such as Mn, Fe, and W increased the dielectric Q significantly [8]. Therefore, once strain-relieved and/or phase-optimized films are obtained, these techniques (i.e., compensation and doping) and post heat treatment can be used to improve the dielectric properties of the films. Figure 7 shows 1–10 GHz dielectric properties of 3 mol.% Ba and 2 mol.% Sr compensated and 1 mol.% W doped strain-relieved BST (x=0.6) film using a 50 Å thick BST buffer layer. The film showed a

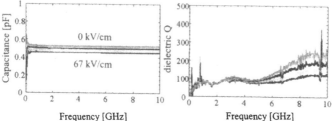

Figure 7 Dielectric properties of strain-relieved BST (x=0.6) film using a BST buffer layer

relatively high dielectric Q (~100) with a 20 % dielectric tuning, which is a higher figure of merit K factor (= dielectric $Q_{0V}\times$dielectric tuning) than strained BST (x=0.6) film with the same compensation and doping condition (K factor <1000). Preliminary measurements on a mixed phase (amorphous and crystalline) of BST (x=0.6) films showed ~50 in the dielectric Q with ~3% tuning at 67 kV/cm.

CONCLUSIONS

The dielectric properties of pulsed laser deposited $Ba_{1-x}Sr_xTiO_3$ (x=0.5 and 0.6) ferroelectric films were measured at microwave frequencies (1–20 GHz). Film strain (i.e., compressive and tensile) and film phase (amorphous and oriented crystalline) have a significant effect on dielectric properties of the film. As an ideal microstructure of the film for tunable microwave applications, strain-relieved and phase-optimized BST films were prepared using a thin BST buffer layer and post heat treatment. The dielectric properties of the films showed a potential possibility of the application. A further study on the mixed phased films in terms of size of crystalline phase, distance between crystalline phases, and density (i.e., range of ordering) and porosity of amorphous phase is needed to optimize its dielectric properties.

ACKNOWLEDGMENTS

Support of this research has been provided by Office of Naval Research and SPAWAR.

REFERENCES

1. L. Davis, J_R., and L.G. Rubin, J. Appl. Phys. **24**, 1194 (1953).
2. S.W. Kirchoefer, J.M. Pond, A.C. Carter, W. Chang, K.K. Agarwal, J.S. Horwitz and D.B. Chrisey, Microwave Opt. Technol. Lett. **18**, 168 (1998).
3. W. Chang, J.S. Horwitz, A.C. Carter, J.M. Pond, S.W. Kirchoefer, C.M. Gilmore, and D.B. Chrisey, Appl. Phys. Lett. **74**, 1033 (1999).
4. W. Chang, J.S. Horwitz, W.J. Kim, J.M. Pond, S.W. Kirchoefer, C.M. Gilmore, S.B. Qadri, and D.B. Chrisey, J. of Appl. Phys. **87**, 3044 (2000).
5. M.E. Lines Phys. Rev. **B 15**, 388 (1977).
6. A.M. Glass, M.E. Lines, K. Nassau, and J.W. Shiever Appl. Phys. Lett. **31**, 249 (1977).
7. W. Chang, J.S. Horwitz, W.J. Kim, J.M. Pond, S.W. Kirchoefer, C.M. Gilmore, and D.B. Chrisey, Materials Research Society Symposium Proceedings, **541**, 693 (1999).
8. W. Chang, J.S. Horwitz, W.J. Kim, J.M. Pond, S.W. Kirchoefer, and D.B. Chrisey, Materials Research Society Symposium Proceedings, **541**, 699 (1999).

ROLE OF SrRuO$_3$ BUFFER LAYERS IN ENHANCING DIELECTRIC PROPERTIES OF Ba$_{0.5}$Sr$_{0.5}$TiO$_3$ TUNABLE CAPACITORS

S.J. PARK*, J. SOK*, E.H. LEE*, J.S. LEE**
*Microelectronics Laboratory, Samsung Advanced Institute of Technology, P.O.Bos 111, Suwon, Korea 440-600
**Analytic Engineering Laboratory, Samsung Advanced Institute of Technology, P.O.Bos 111, Suwon, Korea 440-600

ABSTRACT

Ba$_{0.5}$Sr$_{0.5}$TiO$_3$ (BST) is a first candidate material for the development of voltage-tunable microwave devices, such as, filter, phase-shifter and VCO. In this work, crystal structures and dielectric properties of BST film are investigated with and without SrRuO$_3$ (SRO) buffer layers. BST and SRO thin films are sequentially prepared by pulsed laser deposition and Au/Ti metal electrodes are fabricated by a DC magnetron sputtering system. The capacitance of the capacitors has been measured as a function of bias voltages at room temperature using a low frequency LCR meter. For the high frequency characteristics (~2GHz), a microstrip resonator with ~2GHz resonance frequency and the center coupling design is fabricated. Using flip-chip BST capacitor attached at the position of the center coupling on the microstrip resonator, Its dielectric loss and tunability were obtained. The microwave loss was obviously enhanced in the film with SRO buffer layer. .

INTRODUCTION

Ferroelectric (FE) thin films with large dielectric nonlinearity and very low dielectric loss are increasingly considered to be promising for voltage-tunable microwave devices, such as resonators, filters, antennas and delay lines with reduced length scales since the fabrication technique of FE thin films have been impressively developed. [1]
Electronic tunability of the operation frequency is related to nonlinear dielectric constant, which can be tuned by an applied bias voltage. Dielectric loss in the FE thin film plays a crucial role in determining the performance of electrically tunable devices.[2] The interface effect of the capacitance hysteresis on voltage-capacitance characteristics was already investigated between Ag/YBCO/BST structure and Cu/Cr/BST structure. [3]
Therefore, in this study, the dielectric properties of the FE Ba$_{0.5}$Sr$_{0.5}$TiO$_3$ (BST) film deposited on the LaAlO$_3$ (LAO) substrates were investigated at the range of microwave frequency. The measurements were performed using a resonator incorporated with a flip-chip structure. For the resonator design, a conventional end-coupled approach was used. The resonator was basically designed to have different capacitance between the poles which is determined by the separation gap of each end-coupled resonator section.
Microwave properties, capacitance (C) and dielectric loss ($tan\delta$) of the BST thin film with or without SRO buffer layer were described as a function of bias voltages. It seems to help to understand interfacial effect between the FE thin film and the electrode.

EXPERIMENT

Testing chip capacitors were fabricated on LaAlO₃ substrate. BST films were deposited using a pulsed laser deposition with 248nm KrF excimer laser(Lambda Physik 300)., which is widely used for the growth of high temperature superconductors and dielectric oxide films. Deposition of BST was performed in the pure oxygen pressure of 50mTorr and substrate temperature of 750_ yielded the film deposition rate of 0.35 Å /pulse. The thickness of the BST film is about 0.7 μm. Subsequently the SRO buffer electrode of 15 nm thickness was grown onto the BST film The substrate temperature was 780°C with 200 mTorr oxygen pressure. All PLD has done with a laser pulse rate of 5Hz, the distance between substrate and BST target was 60mm, and energy density of about 2 J/cm² . The film was immediately cooled to 550°C and then annealed for 30 min under 500 Torr of pure oxygen pressure. The top electrode Au/Ti multilayer film were prepared with multi-target d.c. magnetron sputtering system from Au and Ti target. For adhesion between Au and LAO substrate Ti interlayer is deposited about 300Å on the substrate in d.c. magnetron sputtering system. The gaps in Au electrodes layer were formed by ion-milling.

Metal microstrip resonator was fabricated on Al₂O₃ substrate. Gold thin film as a resonator was DC-sputtered on the alumina substrate and the gold ground plane was also deposited on the opposite side of the substrate. Gold films are patterned by a standard photolithographic method and etched by an argon ion-milling system. The silver paste is used to contact the ground plane to the package and the 50 Ω input and output feed lines to the SMA connectors.

Our resonator and capacitor design is depicted in Fig.1. The gap is centered in the   microstrip resonator thus raising the odd-ordered modes frequencies relative to the resonant frequencies without a gap. The even mode frequencies are not shifted in frequency by the gap capacitor due to a current node for these modes located at the gap. The total gap capacitance obtained from the odd-order resonances, and loss information can be calculated from their Q's.

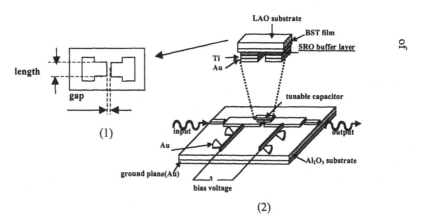

Fig.1 (1) Top views of a typical ferroelectric BST planar capacitor, (2) Tunable gold microstrip resonator with the planar flip-chip BST planar capacitor located at the break point of the microstrip.

RESULTS

Crystal Structures

The crystallinity of the BST films was investigated by XRD diffraction. A typical θ-2θ-scan X-ray diffraction of the double layer structures is shown in Fig.2. Only the (l00) peaks of BST film and the (020), (040) peaks of SRO film appear in the θ-2θ scan which is evidence for the high quality crystal lattice with c-axis perpendicular to the substrate surface. The in-plane epitaxy of BST film was also checked by φ–scan XRD. A 90° periodicity of the peaks from the film was observed.

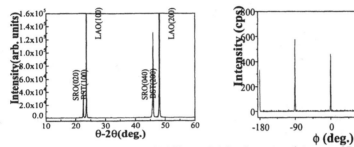

Fig. 2 XRD θ-2θ scan data of SRO/BST/ film on LAO substrate and φ scan of BST film on LAO substrate

Microwave properties of BST Ferroelectric film at 1MHz

Microwave properties of BST ferroelectric film at 1MHz was obtained from measurements of planar capacitors. The geometry of capacitor is 0.6mm x 4_m planar electrode. Using HP 4284A LCR meter, The capacitance as a function of applied voltage(0V~40V) is measured at 1MHz. Fig.3 shows the change of capacitance with bias voltage at R.T. in the films with and without SRO buffer layer. One electrode structure is (Au,Ti)/BST/LAO and the other is (Au,Ti)/SRO/BST/LAO. The 45% tunability of the film with SRO buffer layer is higher than 40% tunability of the film without SRO buffer layer. Considering the tunability at the bias voltage of 10V, the films with SRO buffer layer exhibits about 25% tunability, which is higher than 15% in the films without SRO buffer layer.

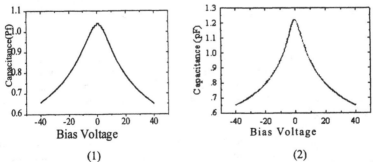

Fig. 3 Microwave properties of BST film with bias voltage at 1MHz (1) without and (2) with SRO buffer layer

Microwave Responses of BST Ferroelectric film at 2GHz

Microwave responses of the resonator were performed using a Hewlett-Packard (HP) 8510C network analyzer, simulated by HFSS (high frequency structure simulator). Fig 4 is simulation data of 2GHz resonator integrated with the capacitor. The parameter in simulation data was changes of capacitance of the resonator by the influence of flip-chip ferroelectric BST capacitor.

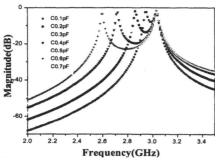

Fig.4 S_{21} Simulation of tunable microstrip resonator

Possibility to determine the more accurate value of the capacitance from the measurement of the resonance frequency is shown in Fig. 5.

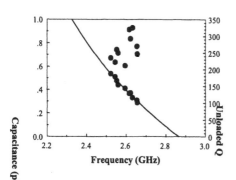

Fig.5 calibration sheet based on microwave measurements of standard capacitors(without FE film).

The use of standard capacitors with different dimensions would give calibration sheet and fitting of capacitance versus movable resonance frequency would be possible. The movable resonance frequency (ω_r) and the capacitance (C) are given by the following relations [4]

$$C = - \frac{\tan\varphi}{2Z_0\omega_r} - C_0 , \qquad \varphi = \pi\frac{\omega_r}{\omega_2} \qquad (1)$$

Where Z_0 and C_0 are the characteristic impedance and the parasitic capacitance of the microstrip line, respectively. C_0 was chosen to be temperature and bias-voltage independent, because the electric field in the substrate of both the capacitor and the microstrip line was assumed to be independent. The ω_2 is the immovable resonance frequency. Z_0 and C_0 values were obtained 23.56 Ω, 0.05 pF respectively.

Using S_{21} measurements (Fig.6) and calibration sheet (Fig.5) we obtained bias voltage dependencies of capacitance and calculated effective loss tangent by Eqs.3. We demonstrate the bias voltage dependence of capacitance and dielectric losses for FE capacitors produced with resonator.

190

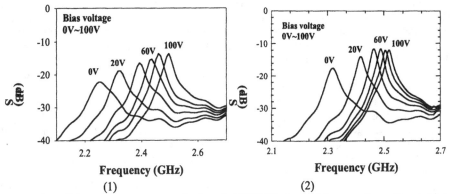

Fig. 6 S$_{21}$ of the gold resonator employing the FE BST planar chip capacitor without (1) and with (2) SRO buffer layer with various dc-bias voltages(0V~100V) at room temperature.

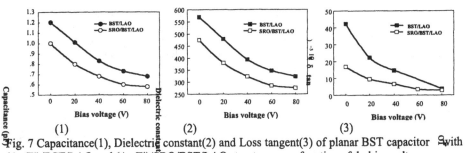

Fig. 7 Capacitance(1), Dielectric constant(2) and Loss tangent(3) of planar BST capacitor with (Au,Ti)/BST/LAO and (Au,Ti)/SRO/BST/LAO structures as a function of dc-bias voltage

Fig. 7 (1) shows bias voltage dependence of the capacitance at about 2GHz. The c apacitance values of the FE BST capacitor were extracted from the resonant frequencies in Fig. 6 using a calibration sheet shown in Fig. 5. The values were curve-fitted. With increasing bias voltage, there is a change in the capacitance. The tunability values in the film with and without SRO buffer layer at 2GHz frequency were almost matched with the values calculated at 1MHz from Fig.3. The films with SRO buffer layer exhibits higher tunability than in the films without SRO buffer layer at low bias voltage.

Fig. 7 (2) shows bias voltage dependence of the dielectric constant at about 2GHz In the planar ferroelectric capacitor, the estimation ε values were obtained using the following modified formula [5]

$$\varepsilon = 4C(\ln 2 + \pi W/4H)/ \pi \varepsilon_0 L \qquad (2)$$

Where W and L are the width and length of the capacitor structure gap, respectively, and H is the ferroelectric film thickness.

In order to calculate the loss tangent of the BST film, the following equations were used.

$$\tan \delta = \xi^{-1}(Q_u^{-1} - Q_0^{-1})$$

$$\xi = \frac{2}{1 - 2\varphi/\sin 2\varphi} , \qquad \varphi = \pi \frac{\omega_r}{\omega_2} \qquad (3)$$

Where Q_u is the unloaded Q-factor of the resonator with the BST capacitor and Q_0 is the unloaded Q-factor of the resonator with a capacitor without dielectric losses (standard capacitor). ξ is the inclusion coefficient of the capacitance. It is the main parameter that characterizes the influence of the capacitor on the Q-factor and on the frequency controllability of the resonator.

Fig. 7 (3) illustrated the loss tangent values which have the order of less than $3X\ 10^{-3}$ in the range of more than 80V with SRO buffer layer. The lower value was obtained with the electrode structure, which has SRO buffer layer.

Shown in this figure, the use of SRO layer as an electrode apparently leads to improvement in the dielectric loss and the tunability at the low bias voltage. It is thought that the SRO layer seems to serve as a good barrier for the degradation of oxygen vacancies in the BST thin film. It also works as a protecting layer for chemical reaction between metal (Au,Ti) electrode and BST thin film [6]. Therefore, the dielectric loss in the FE film is electrode-dependent. For both electrode structures, the dielectric loss decreases as the bias voltage is increased. However, the decreasing slope has more sharp without the SRO layer than that with the SRO layer. The low loss $tan\delta$ of $3X10^{-3}$ at 80 bias voltage is obtained by using the SRO layer at about 2GHz.

CONCLUSIONS

Microwave properties of the BST thin film are characterized using a gold resonator incorporated with a planar capacitor. The dielectric loss tangent and capacitance of capacitors versus applied dc bias voltage are obtained under the influence of an applied dc voltage at about 2GHz. The tunability at low bias voltage is improved and a loss tangent of the BST thin film was decreased with a SRO electrode, it is thought that the SRO layer can provide a way to reduce the interface reaction, oxygen deficiency in the surface of the BST film.

REFERENCES

1. A. A. Golovkov, D. A. Kalinikos, A. B. Kozyrev, and T. B. Samoilova: Electronics letters, Vol. 34, No. 14 (1998) 1389.
2. A. T. Findikoglu, Q. X. Jia, X.D. Wu, G.J. Chen, T. Venkatesan, and D.Reagor: Appl. Phys. Lett., 68 (1996) 1651.
3. Karamanenko S.F., Dedyk A.L and Ter-Martirosyan L.T, 1998, Supercond. Sci. Technol., 11, 284~287
4. David Galt, John C. Price, James A. Beall, and Ronald H. Ono: Appl. Phys. Lett. 63, (1993) 3078.
5. Vendik O. G., Ter-Martirosyan L.T., Dedyk, A.L. Karamanenko S.F. and Chakalov R.A., 1993,
Ferroelectrics, 144, 33
6. H. C. Li, A. D. Si, A.D. West, and X.X. Xi: Appl. Phys. Lett., 73 (1998) 464.

Magnetic and Others

TUNABLE PHOTONIC CRYSTALS

A. FIGOTIN*, Yu.A. GODIN*, I. VITEBSKY* * Department of Mathematics, University of California,Irvine, CA 92697-3875,
afigotin@math.uci.edu

PHYSICAL MECHANISMS OF TUNABILITY

Photonic crystals are spatially periodic dielectric structures usually composed of two different components. We consider a photonic crystal formed by a periodic array of voids in an optically dense homogeneous substance as shown in Fig. 1. The electromagnetic (EM) spectrum of a photonic crystal with a given geometry is determined by the refractive indices of its constitutive components.

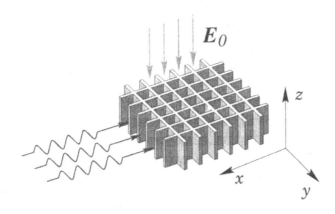

Fig. 1. A slab of a tunable tetragonal two-dimensional photonic crystal. Refractive index of the dielectric substance depends on the external field E_0 through a nonlinear mechanism. Changing E_0 one can control propagation of the EM signal.

Tunability of a photonic crystal assumes the possibility to alter controllably its electromagnetic properties by external magnetic or electric field, mechanical stress, strong electromagnetic radiation, or other external actions. To alter the performance characteristics of a photonic crystal, the controlling action must either modify the dielectric properties of the medium or alter its geometric characteristics. Since the alteration of the dielectric permittivity $\hat{\varepsilon}$ or magnetic permeability $\hat{\mu}$ seems to be more practical, it will be our primary focus. The other alternative is briefly discussed at the end of the paper.

It has long been known that sufficiently strong external electric and magnetic fields, as well as an intense electromagnetic radiation, can efficiently control electromagnetic characteristics of a variety of dielectric and magnetic materials (see, for example, [1-4] and references therein). In dielectric materials with pronounced nonlinearity a reasonably strong electric field $\vec{E}_0$ can substantially modify (usually reduce) the refractive index. Numerous examples of such nonlinear dielectrics can be found in [2] (see also [1] and [4]). Similarly, in magnetically ordered materials such as ferrites and antiferromagnets an external magnetic field $\vec{H}_0$ can dramatically alter the magnetic permeability tensor $\hat{\mu}$ [3]. In addition to that, external fields can change the macroscopic electrodynamics of a photonic crystal by introducing some qualitatively new effects, such as Faraday rotation, linear magnetoelectric response and more [5].

The controlling field $\vec{E}_0$ (or $\vec{H}_0$) must be strong enough to alter the property tensors $\hat{\varepsilon}$ (or $\hat{\mu}$) of the medium. On the other hand, the alternating electromagnetic fields $\vec{E}(t)$ and $\vec{H}(t)$ associated with the electromagnetic wave propagating through the photonic crystal should be weak enough to be considered as a linear signal. In most cases the above requires the amplitudes $\vec{E}(t)$ and $\vec{H}(t)$ to be relatively small compared to those of the controlling fields $\vec{E}_0$ (or $\vec{H}_0$)

$$E_0 \gg E(t) \text{ or } H_0 \gg H(t). \tag{1}$$

In general, the controlling field $\vec{E}_0$ (or $\vec{H}_0$) of a frequency Ω_0 must be treated as an inseparable part of the electrodynamic problem. However, a parametric dependence of the property tensor $\hat{\varepsilon}$ (or $\hat{\mu}$) on the external field $\vec{E}_0$ (or $\vec{H}_0$) can be materialized if the frequency of the controlled EM signal Ω is significantly higher than that of the controlling field

$$\Omega \gg \Omega_0. \tag{2}$$

For some dielectrics condition (2) can substantially relax the requirement (1) or even allow the controlling field to be smaller than the field of the signal. In this connection we consider two particular possibilities.

The first one is related to the ferroelectric or ferromagnetic media where relatively weak quasi-stationary field can dramatically affect the substance. Numerous examples of the kind can be found among so-called "soft" ferromagnetic or ferroelectric materials (see, for example, [2], [3], and [5]). At the same time, propagating EM signal having relatively high frequency Ω *may be treated as a small perturbation even if its amplitude does not meet the condition (1)*.

The second worth mentioning situation occurs when the external field of a frequency Ω_0 causes some kind of resonance response of the medium. The resonance behavior can result in a noticeable nonlinear effect even in the case of relatively small amplitude of the controlling field $\vec{E}_0$ (or $\vec{H}_0$). For the case of magnetic resonance, the detailed description of such phenomena can be found in [6]. At the same time, since the carrier frequency Ω of the signal is significantly different from Ω_0, it may not cause a resonance response of the medium and, therefore, can be treated as a linear EM wave.

In summary, the physical mechanisms responsible for the tunability effect in photonic crystals can be extremely diverse depending on the frequency range of interest and the materials used. Usually, the controlling field has to be relatively strong to ensure a noticeable alteration of the property tensor. But in some special cases the required controlling field can be even weaker than the EM field of the controlled signal.

WHY TUNABLE PHOTONIC CRYSTALS?

First of all, photonic crystal tunability is often a necessity rather than an additional feature. Indeed, the fabrication of perfectly periodic dielectric structures with strictly specified electromagnetic characteristics is more than a challenging and costly task. Usually it is simply impractical. For this reason, it is essential to have a possibility of tuning or adjusting the characteristics of a photonic crystal after it has been made instead of fabricating a new one to meet the specs. In addition to that, it can be used for different frequency ranges. Secondly, some of photonic crystal applications are based on the possibility to control their EM spectrum. For instance, a straightforward application of a tunable photonic crystal is a lossless EM switch [7]. For this device an external controlling field changes the EM spectrum so that the carrier frequency Ω falls in or out of a spectral gap resulting in a transition between transparent and opaque states. More sophisticated applications of tunable photonic crystals are considered in [8]. Thirdly, the tunability can do more than simply to control basic EM characteristics of a photonic crystal. An external field can also bring about some qualitatively new spectral features. For instance, an external magnetic field can cause a nonreciprocal behavior of the electromagnetic spectrum of a magnetic photonic crystal (some examples of nonreciprocal electromagnetic spectra in magnetic media can be found in [9]).

In this communication we focus on a few specific aspects of tunability related to the same physical origin. One of these aspects has been introduced in our resent publication [8]. It is related to a strong and very distinct spectral anomalies similar to those of well-known electronic topological phase transitions in metals [10].

Let us consider how continuous alteration of the spectrum affects the propagation of EM waves of a fixed frequency Ω. At first glance, the only noticeable effect of tunability is the possibility of switching between the transparent and the opaque states, depending on whether the carrier frequency Ω falls into a transmittance band or a photonic band gap. This transition is a remarkable effect by itself. In addition to this transition, another interesting phenomenon takes place. As we have shown in [8], virtually any photonic crystal can be "tuned" into a state with extreme anisotropy of EM properties. In this state the EM waves can propagate only within a narrow angle along one or a few special directions (see Fig. 2).

This effect allows to control efficiently the direction of EM wave propagation through the periodic dielectric structure. In terms of tunability, the mentioned extremely anisotropic states lie very close to those with virtually isotropic states. Hence, such a "photonic lens" works selectively for a very narrow frequency band.

The next important effect is a dramatic reduction of the group velocity of EM waves of certain frequencies. This phenomenon can be seen as an effective way to enhance nonlinear phenomena in a substance.

In addition to the mentioned above, there exists another distinctive anomaly. It occurs when a piece of the equifrequency surface loses its convexity. In this situation there will be *more than one plane EM wave of given frequency Ω and of the same polarization propagating in the same direction*. Their group velocities are essentially different.

In the course of the spectrum modification the conditions of EM wave propagation change dramatically. The total number of the critical spectral anomalies may vary for different photonic crystals and different spectral bands of the same crystal. On the other hand, the basic characteristic features of such transformations persist in any kind of tunable photonic crystals, because those changes are predetermined by the topology of the equifrequency surface of the spectrum (see [8] and [10]). For this reason, we believe, many of the discussed spectral features can be realized in virtually any tunable photonic crystal.

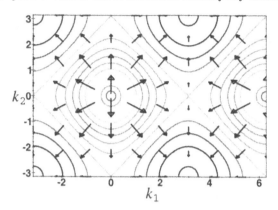

Fig. 2. Successive cross sections of the dispersion surfaces (equifrequency curves) arranged in ascending order of ε in the plane of the quasimomentum vector (k_1, k_2). The arrows point in the direction of the group velocity of electromagmetic waves. Near the saddle points the waves can propagate only in two directions. For higher values of ε this effect is even more pronounced [8].

It must be emphasised that there is a crucial difference between the electronic topological phase transitions in metals and the EM waves in tunable photonic crystals. In metals, the above remarkable effects are extremely rare and difficult to produce. By contrast, in tunable photonic crystals they are virtually guaranteed. Besides, in photonic crystals they might have some practical applications.

BASIC TECHNICAL PROBLEMS

Proper Choices of Tunable Dielectric Materials

In spite of the formal mathematical similarity, the cases when the tunability is achieved based on external electric field $\vec{E}_0$ or external magnetic field $\vec{H}_0$ may differ significantly because of the essential difference between electric and magnetic properties of the real dielectric materials. We consider here the materials with nonlinear dielectric response, which can be controlled by an external electric field. For this case our main requirements to the material are: (i) low-field electric nonlinearity, facilitating the control of the tensor $\hat{\varepsilon}$ without recourse to high voltage; (ii) sufficiently high electric permittivity and low dielectric losses at the frequency range selected. There exists a variety of dielectrics with pronounced nonlinearity and sufficiently high dielectric constant at the frequency range up to 10^{12} Hz (see, for instance, [1], [2] and references therein). Those materials are widely used in microwave applications as well as in nonlinear optics. Furthermore, the nonlinearity and high dielectric permittivity often come together and can be found in numerous ferroelectrics.

The Field Uniformity Problem

In the absence of external field, a photonic crystal is a spatially periodic composite structure. Either of the two constitutive components is usually homogeneous by itself, as is the case with the 2D photonic crystal in Fig. 1. Under applied electric field this situation can change even if the external field $\vec{E}_0$ is perfectly uniform. Indeed, an external electric field $\vec{E}_0$ induces electric polarization $\vec{P}$ in the medium which, in turn, produces an additional depolarization electric field $\vec{E}_i$. The resulting electric field $\vec{E}$ inside the system is the sum of the external and depolarization fields [5]. The depolarization field depends significantly on the geometry of the photonic crystal and the direction of applied field $\vec{E}_0$. More importantly, the depolarization field is not uniform $\vec{E}_i = \vec{E}_i(\vec{r})$ even if the field $\vec{E}_0$ is. Hence, the resulting electric field $\vec{E}$

$$\vec{E} = \vec{E}_0 + \vec{E}_i(\vec{r}) = \vec{E}(\vec{r}). \tag{3}$$

inside the photonic crystal is not uniform either.

Since the dielectric constant ε of the materials used in photonic crystals must be substantially greater than 1, the depolarization field $\vec{E}_i$ is always comparable in magnitude with the applied field $\vec{E}_0$ and the depolarization effects can not be ignored. Similar problem exists with the demagnetization field in a magnetic photonic crystal. Apart from a simple renormalization of the external field $\vec{E}_0$, the depolarization field $\vec{E}_i$ can cause two major problems. The first one is that the resulting electric field $\vec{E}$ can be nonuniform even within each of the two dielectric components of the photonic crystal. This may seriously complicate the EM spectrum calculation, because the nonuniformity makes the dielectric permittivity $\hat{\varepsilon}$ a complicated function of $\vec{r}$. In the case of 3D tunable photonic crystals the above problem is, in fact, unavoidable. In 1D and 2D periodic systems a solution can be usually found. For instance, the geometry chosen in Fig. 1 ensures the uniformity of the internal field $\vec{E}$ if $\vec{E}_0$ is uniform.

The second problem is even more serious. The internal field $\vec{E}(\vec{r})$ from (3) is a periodic function of $\vec{r}$ within the system if and only if the shape of the photonic crystal meets some strict requirements. These requirements are well known in electrostatics and magnetostatics [5], and if they are not met the resulting field $\vec{E}(\vec{r})$ will not be periodic in $\vec{r}$. This factor can make the photonic device inoperational. To ensure the periodicity of the internal electric field $\vec{E}(\vec{r})$ in the case of 2D photonic crystal shown in Fig. 1 one should choose the dimensions of the system in the xy-plane much greater than the thickness of the slab in the z-direction.

Photonic crystals with alterable geometry

The most likely candidates with this feature can be found among the polarized media such as ferroelectrics, ferromagnets and ferrites with naturally occurring periodic domain structures. Almost every single electrically or magnetically polarized substance supports some sort of periodic or nearly periodic domain structure. The problem usually lies in the inferior quality of periodicity. Still, there are some examples when the equilibrium domain structure displays nearly perfect geometry. One can mention so-called magnetic bubbles

[11]. In an external magnetic field $\vec{H}_0$ they can form perfectly periodic $2D$ hexagonal structures with the geometrical parameters, which can be easily and precisely controlled by $\vec{H}_0$.

ACKNOWLEDGMENT AND DISCLAIMER

Effort of A. Figotin, Yu. Godin and I. Vitebsky is sponsored by the Air Force Office of Scientific Research, Air Force Materials Command, USAF,under grant number F49620-99-1-0203. The US Government authorized to reproduce and distribute reprints for governmental purposes notwithstanding any copyright notation thereon. The views and conclusions contained herein are those of the authors and should not be interpreted as necessarily representing the official policies or endorsements, either expressed or implied, of the Air Force Office of Scientific Research or the US Government.

REFERENCES

1. B.E.A. Saleh and M.C. Teich. *Fundamentals of photonics*, (John Wiley & Sons, Inc., New York, 1991), chap. 18.

2. *Landolt-Bornstein Numerical data and functional relationships in science and technology*. vol. **16**. *Ferroelectrics and related substances*. New York, 1981.

3. *Landolt-Brnstein Numerical data and functional relationships in science and technology*. vol. **12**. *Magnetic and other properties of oxides and related compounds*. New York, 1978.

4. Y.Y. Zhu and N.B. Ming. *Dielectric superlattices for nonlinear Optical Effects*. Opt. and Quantum Electron. **31**, 1093 (1999).

5. L. Landau and E. Lifshits, *Electrodynamics of continuous media*, 2nd ed. (Butterworth-Heinemann, Oxford, 1995).

6. *Magnetism*, edited by G.T. Rado and H. Suhl (eds.), (vols. **1-5**, Academic Press, New York, 1963-1973).

7. M. Scalora, J. Dowling, Ch. Bowden, M. Bloemer, and M. Tocci, US Patent No. 5 740 287 (April, 1998).

8. A. Figotin, Yu.A. Godin, and I. Vitebsky, Phys. Rev. B **57**, 2841 (1998).

9. T.H. O'Dell. *The electrodynamics of magneto-electric media*, (North-Holland Publishers, Amsterdam, 1970).

10. Y.M. Blanter, M.I. Kaganov, A.V. Pantsulaya, A.A. Varlamov,Phys. Reports **245** (4), 159-257 (1994).

11. A. H. Eschenfelder. *Magnetic Bubble Technology*, 2nd ed. (Springer-Verlag, New York, 1981).

Strain Tuned Magnetic Properties of Epitaxial Cobalt Ferrite Thin Films

G. HU,[*] J.H. CHOI,[**] C.B. EOM[**] and Y. SUZUKI[*]
[*]Department of MS&E, Cornell University, Ithaca, NY 14853, gh47@cornell.edu
[**]Department of Mech. Engr. & Mater. Sci., Duke University, Durham, NC 22208

ABSTRACT

Epitaxial cobalt ferrite thin films have been fabricated on bare MgO substrates and compared to those grown on $CoCr_2O_4$ buffered $MgAl_2O_4$. Various structural characterizations, including x-ray diffraction, Rutherford backscattering spectroscopy and transmission electron microscopy, demonstrate excellent crystallinity of the films. Films grown under tension exhibit magnetic properties dominated by the stress anisotropy. In (110) oriented films grown on $CoCr_2O_4$ buffered $MgAl_2O_4$, post deposition annealing switched the in-plane easy and hard directions completely while no such behavior is observed in cobalt ferrite films grown on MgO. The anomalous behaviors observed in as-grown films and films after annealing can be explained in terms of lattice distortion and cation redistribution.

INTRODUCTION

Spinel structure thin film ferrites have been recognized to have significant technological potential in high frequency applications. The promising properties of ferrite thin films have led to many scientific studies[1-4] focusing on the materials issues, such as epitaxy, microstructure and strain, that are not present in the bulk, and their effects on electronic and magnetic behaviors. In the studies of epitaxial Fe_3O_4 grown on MgO substrates, Margulies et al.[1,2] have found that antiphase boundaries (APB), which are a result of the nucleation and growth mechanism, give rise to enhanced intrasublattice superexchange coupling and account for the anomalous magnetic properties of the magnetite films. Similar results were also observed in $CoFe_2O_4$ films grown on MgO substrates by Dorsey et al.[3] and $NiFe_2O_4$ films grown on $SrTiO_3$ by Venzke et al.[4]. In our studies of $CoFe_2O_4$ (CFO) thin films, we present our results on MgO substrates and compare them to those on a spinel structure substrate $MgAl_2O_4$ where we are able to eliminate the formation of antiphase boundaries. Structure characterization of the antiphase boundary free films by transmission electron microscopy demonstrates a high quality of the film, with a dislocation density less than $1.6 \times 10^{10}/cm^2$. Considering the large crystalline anisotropy and magnetostriction of cobalt ferrite, we expect microstructural effects such as cation distribution and stress on magnetic properties to be enhanced significantly.

Cobalt ferrite has an inverse spinel structure, where the O atoms make up an FCC lattice, an eighth of the tetrahedral (A) sites are occupied by Fe^{3+} ions and half of the octahedral (B) sites are randomly occupied by Co^{2+} and Fe^{3+} ions. Bulk work has shown that heat treatment can vary the cation distribution among the tetrahedral and octahedral sites of the spinel ferrites[5-9], thus modifying their magnetic properties. In particular, cobalt ferrite, with a degree of inversion ranging from 80% to 95%, exhibit more site sensitivity than others[9,10], given the large anisotropic effect of Co^{2+} ions. Studies of ferrite thin films upon post deposition annealing provide mixed results. While significant improvement in magnetic properties and crystallinity was found in $NiFe_2O_4$ films grown on $SrTiO_3$[4], the anomalous magnetic behaviors remain in the Fe_3O_4 films grown on MgO[1,2]. In our case, while post deposition annealing gave rise to an unexpected decrease of saturation magnetization in films grown on MgO substrates, dramatic change in magnetic anisotropy was observed in films grown on CCO buffered MAO substrates.

In our previous studies[11], the fabrication and characterization of epitaxial cobalt ferrite thin films were described and the magnetic behaviors were well explained by magnetoelastic theory. Here, we report our recent results on films grown under tension and films after annealing. The anomalous magnetic properties observed in these films will be explained in terms of lattice strain and cation disorder.

EXPERIMENT

Epitaxial CFO thin films have been fabricated on bare MgO substrates as well as $CoCr_2O_4$(CCO) buffered $MgAl_2O_4$(MAO) by pulsed laser deposition. Structural characterizations were carried out by X-ray diffraction (XRD), atomic force microscopy (AFM), transmission electron microscopy (TEM) and Rutherford backscattering spectroscopy (RBS). Magnetization measurements were performed on a Lakeshore vibrating sample magnetometer (VSM) and a Quantum Design SQUID magnetometer. The detailed deposition conditions of the films and the equipment related parameters are described elsewhere [11].

RESULTS

We have performed normal and grazing incidence XRD measurements of cobalt ferrite ($CoFe_2O_4$) films on MgO substrates on a four-circle diffractometer. For the samples with film thickness varying from 650Å to 1 μm, there is no evidence of the presence of film orientations other than those of the underlying substrate. Good alignment of the in-plane crystallographic orientations is indicated by twofold and fourfold symmetry of the φ-scans of the (111) reflections on (110) and (100) oriented samples, respectively. RBS channeling, with figure of merit of crystalline χ_{min} ~10% (figure 1), reveals excellent crystallinity. The transmission electron microscopy diffraction patterns also indicate very good epitaxial and crystalline quality of the films.

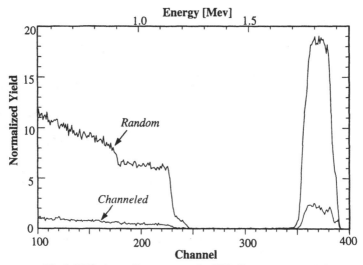

Fig.1. RBS channeling spectra of a CFO film grown on MgO

202

These CFO films grown under tension (on MgO), like those grown under compression (on CCO buffered MAO) exhibit magnetic behaviors quite different from that of the bulk. For (110) oriented films, the tensile strains make [1$\bar{1}$0] and [001] the in-plane easy and hard directions respectively, while the situation is opposite in films grown under compression[11]. CFO films grown on (100) MgO substrates have easy axes along film normal and hard film planes as shown in figure 2, despite the large demagnetization energy.

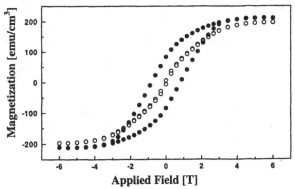

Fig.2. M-H loops for a film grown on (100) oriented MgO substrate. Solid and open circles indicate the magnetization along film normal and [010] direction in film plane, respectively.

All these observations can be explained well as a combination of magnetocrystalline, magnetoelastic and demagnetization energy. Compared to bulk materials, thin films are under biaxial strains, due to the lattice mismacth with the substrate. As cobalt ferrite has magnetostriction constants significantly larger than some other spinel materials, the stress anisotropy turns out to be an important part in the total free energy. To evaluate this magnetoelastic energy, we need to determine the stress states of the films. For (110) oriented films, the elastic moduli of the mutually perpendicular in-plane [001] and [1$\bar{1}$0] directions are different. After a transformation of coordinate systems, the elastic moduli along these two directions can be expressed as:

$$Y_{[001]} = C_{11} + C_{12} - C_{12}(C_{11}+3C_{12}-2C_{44})/(C_{11}+C_{12}+2C_{44}) \quad (1)$$

$$Y_{[1\bar{1}0]} = (2C_{11} + 6C_{12} + 4C_{44})/4 - (2C_{11} + 2C_{12} - 4C_{44})(C_{11}+ 3C_{12} - 2C_{44})/4(C_{11}+C_{12}+2C_{44}) \quad (2)$$

where, C_{ij} are elastic constants of cobalt ferrite.
In a cubic system, the magnetoelastic energy has the form

$$E_{me} = -(3/2)\lambda_{100}\sigma(\alpha_1^2\gamma_1^2 + \alpha_2^2\gamma_2^2 + \alpha_3^2\gamma_3^2) - 3\lambda_{111}\sigma(\alpha_1\alpha_2\gamma_1\gamma_2 + \alpha_1\alpha_3\gamma_1\gamma_3 + \alpha_2\alpha_3\gamma_2\gamma_3) \quad (3)$$

where, α_j are direction cosines of M_s, γ_j are direction cosines of stress σ, and λ_{hkl} are magnetostriction constants. Taking the literature value of the magnetostriction constants for bulk cobalt ferrite ($\lambda_{100} \sim -590 \times 10^{-6}$ and $\lambda_{111} \sim 120 \times 10^{-6}$) and assuming the following values for the elastic constants, $C_{11} = 2.73 \times 10^{12}$ dynes/cm^2, $C_{12} = 1.06 \times 10^{12}$ dynes/cm^2 and $C_{44} = 0.97 \times 10^{12}$ dynes/cm^2, then, for the (110) oriented films, the magnetoelastic energy associated with the different directions are $E_{me}^{1\bar{1}0} = 1158\varepsilon \times 10^6$ ergs/cm^3, $E_{me}^{1\bar{1}1} = 1673\varepsilon \times 10^6$ ergs/cm^3 and $E_{me}^{001} = 2704\varepsilon \times 106$ ergs/cm^3, where ε is strain. As the film normal is stress free, $E_{m.e}^{[110]} = 0$.

For films under tension and compression, stress anisotropy makes [1$\bar{1}$0] and [100] the magnetically easy in plane directions, respectively. As we are comparing in plane directions, demagnetization energy does not contribute to the difference of the total energy. However, there is a term due to the crystalline anisotropy energy, which renders the <100> and <111> directions to be the lowest and highest energy states, respectively. The observed easy axis will be determined by the minimization of the total energy, which shows that the stress effect dominates the magnetic anisotropy of the films. For films grown on (100) oriented MgO substrates, the similar arguments can be made. As the crystalline anisotropy energy is equivalent along the [100], [010] and [001] directions, magnetoelastic energy only competes with the shape anisotropy. It turns out that stress anisotropy overcomes the demagnetization energy and makes the film normal an easy axis.

Though $CoFe_2O_4$ films grown on (100) MgO substrates show easy axes along the film normal, the saturation along an easy axis can only be achieved at a field as high as 40kOe and the hard direction remains unsaturated at 70kOe. The measured saturation magnetization is only half of the bulk value at room temperature. Some possible explanations for the high saturation field and the reduced saturated magnetization, are enhanced intrasublattice superexchange interactions (as Margulies et al.[2]) or noncrystalline Co-Fe-O (as Venzke et al.[4]). In order to determine the importance of the two factors, we measured MH loops at low temperatures. Cooling down the sample from room temperature to 5K under 60 kOe, gave rise to shifted magnetization loops. The shift is evidence that there exists a non-negligible amount of second phase magnetic material that is exchange biased to the crystalline CFO material. We believe that this secondary phase material is amorphous since XRD reveals no other phase besides the spinel structure. Also, we found that the magnetization was strongly dependent on temperature as shown in figure 3. The presence of antiphase boundaries (APB) in Fe_3O_4 films grown on MgO substrates, was reported to be the origin of the high saturation field by Margulies et al.[1,2]. Considering that $CoFe_2O_4$ has a lattice parameter almost twice as big as MgO and the AFM images show that the growth mechanism of these films is not layer by layer, it is likely that antiphase boundaries exist in our CFO films too and give rise to the anomalous behaviors mentioned above. But, antiphase boundaries alone cannot explain the low saturation magnetization and the shifted loops. Also, in our study of films with a composition of $Co_{0.2}Fe_{2.8}O_4$, easily saturated M-H loops and reasonable saturation magnetization were obtained in films grown on MgO substrates. As antiphase boundaries did not produce any anomalous behaviors in those $Co_{0.2}Fe_{2.8}O_4$ films, we believe that cobalt concentration and cation distribution may be important factors here.

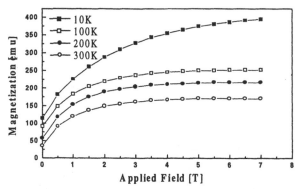

Fig.3. Magnetization measured along film normal as a function of field and temperature for (100) oriented CFO films grown on MgO.

It is well known that the magnetic properties of cobalt ferrite are sensitive to heat treatments[9,10]. After being annealed at 1000°C in air for 45mins, the film normal remains the easy direction in films grown on (100) oriented MgO substrates. In contrast to Venzke et al.'s observation, post annealing did not recover the saturation magnetization of the CFO films to its bulk value, but decreased it even more. Extended X-ray fine structure (EXAFS) measurements have been performed to examine the cobalt cation arrangement of the CFO films before and after annealing. Obvious migration of cobalt ions from tetrahedral sites to octahedral sites was observed by comparing the EXAFS data from as-grown and annealed films. Considering the nonequilibrium growth of the films, it is reasonable that the annealed films are closer to the thermal equilibrium state and have a higher inversion degree. Though the migration of cobalt ions is qualitatively consistent with the decrease of the saturation magnetization after annealing, to explain the low value of M_s and high saturated field, more detailed investigation is necessary.

On the other hand, post deposition annealing of films grown on CCO buffered (110) MAO, produced even more confusing magnetic behaviors. After being annealed in air at 1000°C for 45mins, the films exhibited a shift of the in-plane easy axis from [001] to [1$\bar{1}$0], in films with thickness below 8000Å. Annealing samples in N_2 atmosphere produced the same results, which convinced us that oxygen content should not be the reason. Since the easy and hard in-plane directions in films grown on the bare MgO and bare MAO substrates did not switch even after extensive annealing times and the magnetic anisotropy appeared to be time dependent, interdiffusion between the buffer layer and CFO film was considered to contribute to the switch of magnetic anisotropy in annealed films. Electron microprobe verifies the existence of an interdiffused layer of Co-Cr-Fe-O layer. Moreover electron energy loss spectroscopy (EELS) reveals that there is an increase in the number of Co^{2+} ions in the tetrahedral sites without the formation of Co^{3+} ions[15,16]. The shift of Co $2p_{3/2}$ peak towards a higher binding energy and the absence of a peak associated with the Co^{3+} ion after annealing in the EELS spectra supports this hypothesis (figure 4). This migration of Co^{2+} ions, in conjunction with stress relaxation as revealed by grazing incidence diffraction, can change the sign and magnitude of the magnetocrystalline anisotropy energy term relative to the stress anisotropy energy term, thus explaining the switch of the magnetically easy and hard directions in CFO films on CCO buffered MAO.

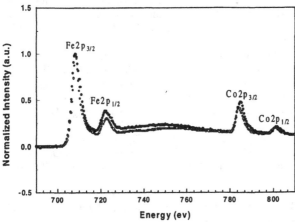

Fig.4. EELS spectra for CFO films grown on CCO buffered MAO before annealing (open circles) and after annealing (solid circles).

A similar shift of the easy direction, due to post deposition annealing of cobalt doped $Y_3Fe_5O_{12}$ (YIG) films has been reported by Dale et al.[17]. By comparing the magnetic behavior of cobalt doped YIG films to that of undoped ones, they concluded that the shift resulted from the migration of cobalt ions. This further convinced us that cobalt ions play a crucial role in the switch of the anisotropy. We believe that the first order magnetocrystalline anisotropy constant is negative in films after annealing, which accounts for the shift of easy direction from [001] to [1$\bar{1}$0].

CONCLUSIONS

In summary, we explain the anomalous magnetic behavior observed in epitaxial cobalt ferrite thin films on MgO in terms of lattice mismatch and cation distribution. The study shows that magnetic anisotropy can be modified significantly by stress effects as well as cation site occupancy.

ACKNOWLEDGEMENTS

We would like to thank Vince Harris for EXAFS measurements of CFO films on MgO substrates. We would also like to thank Maura Weathers, Mick Thomas and Peter Revesz for valuable discussions. This work was supported by an ONR Young Investigator grant and a David and Lucile Packard Foundation Fellowship (Y.S.). Structural characterization was carried out at the central facilities of Cornell Center for Materials Research.

REFERENCES

1. D.T. Margulies, F.T. Parker, F.E. Spada, R.S. Goldman, J. Li, R. Sinclair and A.E. Berkowitz, *Phys. Rev. B* **53**, 9175 (1996).
2. D.T. Margulies, F.T. Parker, M.L. Rudee, F.E. Spada, J.N. Chapman, P.R. Aitchson and A.E. Berkowitz, *Phys. Rev. lett.* **79**, 5162 (1997).
3. P.C. Dorsey, P. Lubitz, D.B. Chrisey and J.S. Horwitz, *J. Appl. Phys.* **79**, 6338 (1996).
4. S. Venzke, R.B. van Dover, Julia M. Philips, E.M. Gyorgy, T. Siegrist, C-H. Chen, D. Werder, R.M. Fleming, R.J. Felder, E. Coleman and R. Opila , *J. Mater. Res.* **Vol. 11** (1996).
5. G.A. Sawatzky, F. van Der Woude and A.H. Morrish, *Phys. Rev.* **187**, 747 (1969).
6. R.M. Persoons, E. De Grave, P.M.A. de Bakker and R.E. Vanderberghe, *Phys. Rev. B* **47**, 5894 (1992).
7. D.S. Erickson and T.O. Mason, *J. Solid State Chem.* **59**, 42 (1985).
8. G.A. Sawatzky, F. van Der Woude and A.H. Morrish, *J. Appl. Phys.* **39**, 1204 (1968).
9. Masatake Takahashi and Morris E. Fine, *J. Appl. Phys.* **43**, 4205 (1972).
10. J.G. Na, T.D. Lee and S.J. Park, *J. Mater. Sci. Lett.* **12**, 961 (1993).
11. Y. Suzuki, G. Hu, R.B. van Dover and R.J. Cava, *J. Magn. Magn. Mater.* **191**, 1 (1999).
12. G.Hu, T.K. Nath, C.B. Eom and Y. Suzuki (unpublished).
13. A. Paul and S. Basu, *Trans. J. Brit. Ceram. Soc.* **73**, 167 (1974).
14. A. Hauet, J. Teillet, B. Hannoyer and M. Lenglet, *Phys. Status Solidi A* **108** 257 (1987).
15. J. Grimblot, J.P. Bonnelle and J.P. Beaufils, *J. Electron Spectr. Related Phen.* **8**, 437 (1976).
16. T.J. Chuang, C.R. Brundle and D.W. Rice, *Surf. Sci.* **59**, 413 (1976).
17. Darren Dale, G. Hu, Vincent Balbarin and Y. suzuki, *Appl. Phys. Lett.* **74**, 3026 (1999).

ELECTRONIC STRUCTURE OF TUNABLE MATERIALS MnAl AND MnGa

A. N. CHANTIS*, D. O. DEMCHENKO*, A. G. PETUKHOV*, and W. R. L. LAMBRECHT**
*Department of Physics, South Dakota School óf Mines and Technology, Rapid City, SD 57701-3995
**Department of Physics, Case Western Reserve University, Cleveland, OH 44106-7079

ABSTRACT

We present first-principle calculations of equilibrium lattice constants, band structures, densities of states and magnetocrystaline anisotropy energy for bulk MnAl and MnGa. The linear-muffin-tin-orbital (LMTO) method has been used within the framework of the local spin density approximation (LSDA). Both the atomic sphere approximation (ASA) and the full-potential (FP) versions of the LMTO method were employed. Calculations of the equilibrium structures were performed both for paramagnetic and ferromagnetic phases of MnAl and MnGa. The results of these calculations indicate that the large tetragonal distortion of the crystal structure is caused by the spin polarization of the electronic subsystem. The magnetocrystalline anisotropy energy per unit cell for MnAl and MnGa is shown to be 0.244 meV and 0.422 meV respectively. This is in good agreement with previous calculations and some experimental data. Magnetic moments, density of states and dependence of magnetocrystalline anisotropy energy on the lattice constant ration c/a are also found to be in good agreement with previous results.

INTRODUCTION

MnAl and MnGa films have been grown epitaxially on semiconductor substrates and have received a lot of attention due to their intriguing magnetic and magneto-transport properties such as unusual magnetic anisotropy and extraordinary Hall effect (EHE). Being ferromagnetic, these compounds show a great promise for merging magnetism with semiconductor technology. For both MnAl and MnGa magnetization measurements by means of vibrating sample magnetometry and EHE indicate perpendicular magnetization with a remnant magnetization of about 200 emu/cm³ and EHE resistivity in the range of 4-8 $\mu\Omega$ and 0.5-4 $\mu\Omega$ respectively [1, 2, 3]. Both materials display a square-like hysteresis shape of the $M - H$ and EHE curves and reasonable coercivity. In addition to this, the measured value of magnetic anisotropy constant for MnAl is about $10^6 J/m^3$ at low temperatures. This means that MnAl and most likely MnGa exhibit rather strong magnetocrystalline anisotropy which is at least 10 to 100 times stronger than that of the other ferromagnetic materials such as Fe, Co or Ni [4].

The strong magnetic anisotropy and EHE in MnAl and MnGa create the possibility for future utilization of these materials in novel hybrid magnetoelectronic devices, such as non-volatile memory elements and spin-injection transistors [5, 6]. For the fabrication of these hybrid devices, it is important to perform predictive calculations of the magnetic and magneto-transport properties of MnAl/GaAs and MnGa/GaAs heterostructures. First-principle calculations of the electronic structures of bulk MnAl and MnGa materials is a necessary step in this direction. In particular, we would like to predict the easy magnetization axes and the values of magnetic anisotropy energy (MAE). Since the unusual magnetic anisotropy of these compounds is related to their structural anisotropy in the τ-phase, our primary goal is to find equilibrium atomic structures of τ-MnAl(Ga) which are characterized by certain tetragonal distortions. Moreover, we would like to explore the interplay between spin polarization, structural distortion, and magnetic anisotropy. Despite of a number of important previous works on the electronic structure of MnAl and MnGa [7, 8] the first-principle determination of their equilibrium atomic structure has not yet been done. Here we report the results of our calculations of basic electronic, magnetic and structural properties of these materials.

COMPUTATIONAL METHOD

MnGa and MnAl are both binary transition metal alloys, with CuAu-type tetragonal structure and experimentally measured lattice constants $\tilde{a}$=0.390 nm, $\tilde{c}$=0.362 nm and $\tilde{a}$=0.393 nm, $\tilde{c}$=0.356 nm respectively. This structure can be represented as two simple tetragonal sublattices of Mn and Al (or tetragonally distorted CsCl structure) with another set of lattice parameters $a = \tilde{a}/\sqrt{2}$ and $c = \tilde{c}$. In the discussion that follows we will mostly refer to this structure as to a *reduced* structure with lattice parameters a and c . All calculations were performed in the framework of the local spin density approximation (LSDA) and the linear-muffin-tin-orbital method (LMTO) in both its atomic sphere approximation (ASA) and full-potential (FP) versions. The latter was employed in the calculation of the equilibrium structure constants.

207

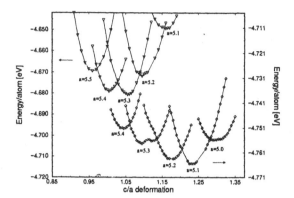

Figure 1: Total energies versus c/a for MnAl. Diamonds: spin-polarizad calculations; triangles: non-spin-polarized calculations.

In this case no spherical approximations can be made for the interstitial potential. Also the close packing of these particular structures allows the method to work with relatively high accuracy. Spin polarized and non-spin polarized equilibrium structure calculations were performed for both MnAl and MnGa.

Band structure, density of states and magnetic anisotropy energy calculations were performed with the LMTO-ASA method. In these calculations Ga $3d$ states in MnGa were treated as core orbitals. The density of states was calculated by means of the tetrahedron method on a regular mesh of 1000 points in the Brillouin zone. This was found to be sufficient to achieve self-consistency in the charge density with high accuracy. For the calculation of the magnetic anisotropy energy we used the force theorem, according to which, the MAE can be obtained as a difference of sums over occupied eigenvalues of one-electron Kohn-Sham equations corresponding to different directions of the magnetization [4]:

$$\Delta E(\hat{n}) = \sum_{i,\vec{k}}^{occ} \epsilon_i(\hat{n}, \vec{k}) - \sum_{i,\vec{k}}^{occ} \epsilon_i(\hat{n}_0, \vec{k}) \tag{1}$$

where $\hat{n}$ is the unit vector in the direction of magnetization and $\hat{n}_0$ is the unit vector corresponding to the magnetization in [001] direction. To rotate the direction of magnetization we introduced a rotation matrix which performed a unitary transformation on spin-orbit part of the LMTO Hamiltonian. In this way we calculated the MAE as the aforementioned difference between the band energies for the magnetizations in the [100] and [001] directions in the reduced structure. The Brillouin zone integration was performed with the tetrahedron method on a regular mesh of 13824 sample points.

Table I: Equilibrium lattice constants, magnetic anisotropy energy (MAE), and magnetic moments (in μ_B)

Compound	a (Å), SP.	a (Å), non-SP.	c/a SP	c/a non-SP	MAE (meV/unit cell)	μ_{Mn}	μ_{Al}, μ_{Ga}
MnAl	2.699	2.804	1.23	1.06	0.244	-2.192	0.041
MnGa	2.540	2.752	1.46	1.11	0.422	-2.274	0.065

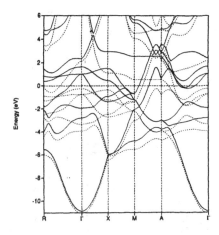

Figure 2: Spin-polarized band structure of MnAl

RESULTS

MnAl

Fig. 1 shows the results of non-spin polarized and spin-polarized equilibrium structure calculations for MnAl. In Table I we summarize all of the obtained results. The calculated lattice parameters are $a=2.699$ Å, $c=3.319$ Å and $a=2.804$ Å, $c = 2.972$Å from spin-polarized and non-spin-polarized calculations respectively. The energy difference between non-spin-polarized and spin polarized states is 85 meV/atom. The non-spin-polarized calculated value of lattice parameter a is closer to the experimental one. In fact, the difference between non-spin-polarized calculated and the experimental value is within the ordinary limits given by all other similar calculations. However the non-spin-polarized calculations fail to predict the correct value for the crystal deformation parameter, c/a. This value is more accurately predicted by spin-polarized calculations. This fact clearly indicates that the deformation of the crystal is caused mainly by its magnetization.

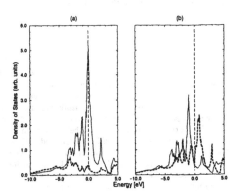

Figure 3: Density of states in MnAl: (a) non-spin-polarized calculations (solid line-Mn, dashed line-Al); (b) spin-polarized calculations (solid line-majority spins, dashed line-minority spins).

Fig. 2 shows the results of spin-polarized band structure calculations for MnAl. The symbols R, Γ,

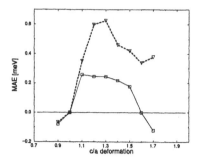

Figure 4: MAE versus deformation c/a, at equilibrium a, for MnGa(dashed line) and MnAl(solid line).

X, M, and A stand for the $(0,\frac{1}{2},\frac{1}{2})$, $(0,0,0)$, $(0,\frac{1}{2},0)$, $(\frac{1}{2},\frac{1}{2},0)$ and $(\frac{1}{2},\frac{1}{2},\frac{1}{2})$ symmetry points in the MnAl Brillouin zone respectively. It is found that there is no band gap at the Fermi level and that a large exchange splitting of about 2 eV exists between Mn 3d states with opposite spins. Fig. 3a shows the non-spin polarized density of states (DOS) for MnAl. The DOS of Mn has a sharp maximum at the Fermi level, which leads us to expect that MnAl is ferromagnetic.

The spin-polarized DOS for MnAl is shown in Fig. 3b. The long tail in the range between -10 and -5 eV is mainly due to Al s states and Mn s states. The Al p states give a small contribution to the density of states, in the region adjacent to -5 eV while Mn d states are spread all over the plotted range and have the largest contribution in the range between -2 and 2 eV, where the total density of states almost coincides with Mn d-DOS. The calculated magnetic moments for Mn and Al are -2.192 μ_B and 0.042 μ_B respectively. These values are slightly higher than the experimental ones [9], but the stoichiometry of the measured samples differs from the ideal one, for which all calculations are made.

The results of magnetocrystalline anisotropy energy calculations for both MnAl and MnGa are shown in Fig. 4 and Table I. At equilibrium, the MAE for MnAl is 0.244 meV per unit cell. This value is in good agreement with previous calculations and corresponds to $\hat{n}$ in the [100] direction of the tetragonal distorted reduced crystal structure of MnAl or MnGa (see Eq.(1)). This means that in bulk MnAl, the axis of easy magnetization lies in the c-plane rather than in x, y-plane. Within the accuracy of our calculations it is impossible to predict the exact direction of the easy magnetization axis in the c-plane. The calculated value of the MAE, leads to the value 1.4×10^6 J/m^3 for the magnetic anisotropy constant K_u which is close to the measured value as well. The dependence of MAE upon the deformation c/a, indicates that the easy magnetization axis turns from the c-plane to the x, y-plane whenever the deformation c/a is smaller than 1 or larger than 1.6.

MnGa

For MnGa we have almost the same picture with the exception that the obtained MAE, 0.422 meV per unit cell, is almost two times larger. This leads to a magnetic anisotropy constant K_u of $2.6 \times 10^6 J/m^3$. Unfortunately, we are not aware of any experimental data for the magnetic anisotropy constant of this material. The dependence of MAE on the c/a also differs considerably from that obtained for MnAl. In Fig. 4 it is seen that the easy magnetization axis turns from the c-plane to the x, y-plane only when c/a is below 1. Another interesting fact is that the deformation in the equilibrium structure is also much greater than that of MnAl. The difference between spin polarized calculated and non-spin polarized calculated equilibrium c/a for MnGa is about two times larger than that in MnAl. This fact and the fact that the MAE for MnGa is also about two times larger than that of MnAl is a clear indication that the tetragonal deformation and the magnetic anisotropy are closely related to each other.

The results of the equilibrium structure calculations are presented in Fig. 5. As in MnAl, it is seen that while non-spin-polarized calculations give accurate value of the lattice constant a they fail to predict the correct value of the deformation parameter c/a. Again, this value is more accurately predicted by spin-polarized calculations. The calculated DOS for the non-magnetic and ferromagnetic states, are shown in

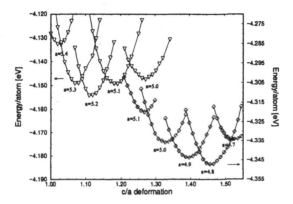

Figure 5: Total energies versus c/a for MnGa. Diamonds: spin-polarized calculations; triangles: non-spin-polarized calculations.

Fig. 6a and Fig. 6b respectively. Dispersion curves for the ferromagnetic phase are shown in Fig. 7. The calculated equilibrium lattice constants and magnetic moments are shown in Table I.

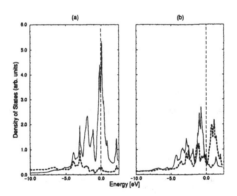

Figure 6: Density of states in MnGa: (a) non-spin-polarized calculations (solid line-Mn, dashed line-Ga); (b) spin-polarized calculations (solid line-majority spins, dashed line-minority spins).

CONCLUSIONS

We have performed several first-principle calculations of the magnetic and electronic properties of bulk MnAl and MnGa. Along with these, we have performed equilibrium structure calculations for paramagnetic and ferromagnetic phases of these materials. The obtained results for magnetic moments, density of states, band structure and magnetocrystalline anisotropy energy are in good agreement with experimental data and previous calculations. The MAE for MnGa is twice as large as that in MnAl. The lattice deformation of MnGa in the ferromagnetic state is also two times larger than that of MnAl. This shows how closely related are the structural deformation and the magnetic anisotropy.

211

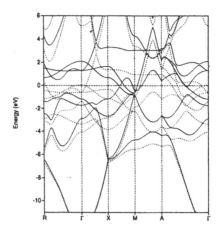

Figure 7: Spin polalirized band structure of MnGa

ACKNOWLEDGEMENTS

This work was supported by the AFOSR grant No. F49620-96-1-0383.

REFERENCES

1. M. Tanaka, J.P Harbison, T. Sands, B. Philips, T. L. Cheeks, V. G. Keramidas, *Appl. Phys. Lett.* **62**, 1565 (1993).

2. J. De Boeck, R. Oesterholt, A. Van Esch, H. Bender, C. Bruynseraede, C. Van Hoof and G. Borghs, *Appl. Phys. Lett.* **68**, 2744 (1996).

3. W. Van Roy, H. Bender, C. Bruynseraede, J. De Boeck, G. Borghs, *J. Magn. Magn. Mater.* **148**, 97 (1995).

4. G. H. Daalderop et al., *Phys. Rev. B* **41**, 11919 (1990).

5. G. A. Prinz, *Science* **250**, 1092 (1990).

6. J. De Boeck, T. Sands, J. P. Harbison, A. Scherer, H. Gilchrist, T. L. Cheeks, M. Tanaka and V. G. Keramidas, *Electronics Lett.* **29**, 421 (1993).

7. A. Sakuma, *J. Phys. Soc. Japan* **63**, 1422 (1994).

8. A. Sakuma, *J. Magn. Magn. Mater.* **187**, 105 (1998).

9. N. I. Vlasova, G. S. Kandaurova, YA. S. Shur and N. N. Bykhanova, *Phys. Met. Metall.* **51**, 1 (1981).

NANOCRYSTALLINE THIN FILMS FOR TUNABLE HIGH Q UHF/VHF DEVICES

WINSTON N. WIN, JIN-YOUNG PARK and RODGER M. WALSER
ECE Department, The University of Texas at Austin, TX 78712, wwin@mail.utexas.edu

ABSTRACT

Recent studies indicate that reactively sputtered FeCrTaN nanocrystalline thin films have many of the properties [large magnetization ($4\pi M_s$), anisotropy (H_{an}), and resistivity (ρ)] required for application in high Q (>1000), magnetically tunable devices operating in the VHF/UHF (~50 MHz to 500 MHz) frequency range. These films had thickness (d) of ~0.1-μm, but film thickness of ~1-μm may be required for many devices. Although most previous research has shown that the magnetic properties of sputtered films are significantly deteriorated when d~1-μm, there are recent reports that those of reactively sputtered nanocrystalline films of similar alloys (FeTaN, CoAl/SiO) are independent of thickness to d≈1-2-μm. Accordingly, this work investigated the possibility of reactively sputtering FeCrTa alloys in O/N gas mixtures to obtain films with device quality properties, that are independent of thickness to 1-2μm. The correlations between their magnetic and electrical properties, and their nano-heterogeneous microstructure, were studied to determine optimum reactive gas mixtures and sputtering parameters. Cross-section TEM studies were conducted to investigate the origins of the thickness independent properties, and the unexpected increases in anisotropy.

INTRODUCTION

For low loss, magnetically tunable device applications in the VHF/UHF (~50 MHz to 500 MHz) frequency range, ferromagnetic films must have large permeances (Λ), and low spin resonance and eddy current losses. These conditions can be met by maximizing the film magnetization ($4\pi M_s$) and thickness (d), minimizing spin damping, increasing the magnetic anisotropy (H_{an}) so that ferromagnetic resonance (FMR) occurs well above the device operating frequency, and simultaneously maximizing the resistivity (ρ).

For the ferromagnetic material to have high Q (low tan δ), requires a FMR well above its VHF/UHF operating frequency, and a large M_s and an in-plane anisotropy H_{an} of 100-1000 Oe. To achieve a large device fill factor, and a high permeance (product of thickness and permeability), films are required with thickness > than 1 μm and ρ > 100 $\mu\Omega$-cm.

The main objective of this research was to develop magnetic thin films for tunable high Q VHF/UHF devices [1]. As discussed in [1], the materials issues to be considered include the ferromagnetic spin resonance, eddy current loss, device fill factor, magnetic bias and tunability. For a large figure of merit, the material must have a high Q ($=\mu'/\mu''$) and a high device fill-factor or permeance. Hence, the main focus of the work reported here was to investigate processes that could increase the film permeance, ρ, and H_{an}, without reducing M_s or deteriorating the quality of their single-domain state.

Metallic ferromagnetic films, such as permalloy, have a maximum thickness above which an out of plane component of magnetization is developed. One approach to increase the thickness employs grain-growth interruption, in which multilayers with alternating magnetic and non-magnetic thin layers are deposited. However, this approach has several disadvantages, such as a reduced resistivity, and the requirement for sequential deposition [2].

Recently, ferromagnetic ternary alloy films, reactively sputtered in oxygen and nitrogen, have received interest as possible magnetic materials for applications in UHF/VHF devices. Their microstructure consists of nanometer size magnetic grains surrounded by amorphous oxide or nitride grain boundaries. The latter are responsible for increasing both their M_s and ρ [3]. Previous reports [4] indicated that the complex permeability of several micron ($\sim 2\mu m$) thick FeTaN films prepared by rf-diode sputtering was nearly that of an ideal single domain state. With the addition of Cr, alloy films with larger M_s (1.95 T) and H_{an} (50-60 Oe) were obtained by Jin et al [5]. However, the thickness of the films in this study were ≤ 0.1 μm.

The present study, investigated the possibility of increasing the thickness of this type of alloy film to ≥ 2 μm. FeCrTa:N/O films with thickness 1 μm to 2 μm were deposited by rf-diode reactive sputtering. The objective was to obtain films in which selective reactions with O/N gases could produce Fe and FeN grains with an increased M_s, and an increased ρ produced by surrounding the nanocrystal grains by amorphous Cr and Ta oxynitride boundaries. By adjusting the deposition parameters, an attempt was made to increase the H_{an} by generating a morphology with a large internal shape anisotropy. The investigation focused on a study of the changes in the sputtered film morphology, and in the resultant magnetic properties, as a function of variations in the deposition/process parameters.

EXPERIMENT

Films were RF-diode sputtered in Ar+N or Ar+(O+N) reactive gas mixtures. The amount of reactive gas incorporated was determined by the relative flow ratios of N and (N+O) to Ar, and as independently controlled by individual digital mass flow controllers. The chamber base pressure was always in the 10^{-7} Torr range and the total working gas pressure during deposition was a few mTorr. Films were deposited on 22 mm cover glass in the stationary mode. The typical deposition time for a 1 μm thick film was about 45 minutes, but inversely proportional to the RF power.

Thin film thickness was determined using a stylus profilometer. The DC resistivity was obtained by a conventional four-point probe method and the DC magnetic properties were obtained by either a vibrating sample magnetometer, or hysteresis loop tracer. Complex RF permeability spectra for 20 – 200 MHz were obtained with a permeameter. Higher frequency studies will be reported elsewhere. The morphology / microstructure investigations were made by XRD and TEM.

During all depositions, the film thickness was $\approx$1-2 μm, the magnetic field bias $\approx$ 50 Oe, and the substrate temperature was near ambient. Film were made under the following conditions: (i) fixed RF power and total flow rate, and variable N flow ratio, (ii) fixed RF power and N/(N+Ar) flow ratio, and variable total flow rate, and (iii) fixed total and N_2 flow rate, and variable RF power, and (iv) fixed RF power, total flow, and N+O flow ratio, and variable O/(N+O) flow ratios.

RESULTS

FeCrTaN Films: Microstructure

Figures (1a), (1b) and (1c) show the XRD data obtained for the FeCrTa:N films as functions of variations in the ratio of the N_2 to total flow rate, pressure and RF power respectively. The crystalline bcc Fe (110) peak was most prominent in all the data. Since no peaks related to Cr and Ta were observed, they were presumed to be in amorphous states. With increasing N_2 content, the the Fe (110) line width broadened, signifying a decrease in grain size. The grain size also decreased with total flow rate as shown in Figure (1b). With higher RF power, the grain size increased [Figure (1c)]. The cross section TEM results are shown in Figure

(2). Fe-rich grains with dimensions ranging from 50 to 150 nm, and amorphous grain boundaries were observed.

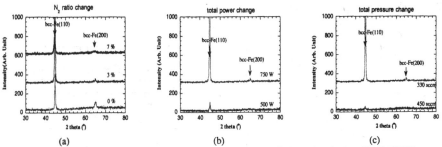

(a)   (b)   (c)

Figure 1. X-ray diffraction scans showing the dependence of the crystallinity and grain size of FeCrTaN films on: (a) nitrogen flow ratio, (b) RF power, and (c) total flow rate.

Figure 2. Cross-section TEM micrographs of FeCrTaN film with 7% N in the total flow. These films had an average grain size ~100 nm (left), with amorphous intergranular regions (right).

FeCrTaN Films: Magnetic Properties and Resistivity

The magnetic properties and resistivity of the sputtered FeCrTaN film are shown in Figures (3) and (4) as functions of the relative N_2 flow rate.

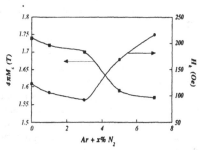

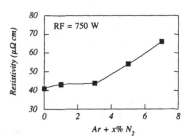

Figure 3. Variation of M_s & H_{an} (H_k)with N_2 flow ratio.   **Figure 4.** Variation of ρ with N_2 flow ratio.

H_{an} and ρ varied little with the initial increase in the % N_2 in the flow, but increased sharply for a N_2 flow ratio >3%. M_s, decreased slowly and then more rapidly for a N_2 flow ratio >3%. The increases in H_{an} and ρ observed for N_2 flow ratios >3% are likely associated with the thickening of a non-magnetic, amorphous grain boundary as seen in Figure (2). This is evident in the film with ≈ 5% N_2 flow ratio. The hysteresis loop and VHF complex permeability spectrum for this film are shown in Figures (5) and (6).

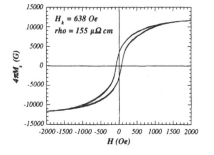

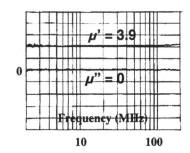

Figure 5. Hysteresis loop for isolated grains in FeCrTa:N film with high nitrogen content. ($H_k=H_{an}$)

Figure 6. μ' = 4 expected for spherical grains matched experimental observation (see text).

The hysteresis loop was isotropic with a remanent magnetization $M_r \sim M_s / 3$, indicating that the grains in this film were exchange isolated. The increase of the resistivity of these films to 155 $\mu\Omega$-cm.was consistent with this view, as was the observation from Figure (6) that the intial permeability was 3.9. This value is close to the theoretical value of 4 expected for a random, magnetically decoupled, array of ferromagnetic spheroids.

The estimated FMR of this film is over 9 GHz. However, their permeance was only ~ 4 μm. Oxygen was added to the Ar+N reactive gas in an attempt to increase the film permeance. The additional oxygen was intended to selectively react with Cr and Ta to give the high ρ, while a large M_s was maintained by the formation of Fe and FeN rich phases.

FeCrTaN/O Films: Magnetic Properties and Resistivity

Figures (7) and (8) show the variations in M_s, H_{an} and ρ observed when N_2 was partially replaced by O_2 in a fixed total flow with the total (N_2+O_2) flow fixed at a constant 5% of the total gas flow. With increasing O_2, M_s initially increased from 1.55 T (Tesla) to 1.8 T, then decreased for O_2/N_2 flow ratios >20%. H_{an} decreased mononotonically from 400 to 20 Oe with increasing O_2. For O_2/N_2 flow ratios < 20%, the resistivity decreased to nearly 1/2 of its initial value, then increased to within ~80% of its initial value. A comparison of Figures (3) and (7) shows that the replacement of N_2 by O_2 in the reactive gas flow, resulted in an increased $4\pi M_s$ and H_{an}. A comparison of Figures (4) and (8) shows the replacement also resulted in an increased ρ.

Based on these results, the optimum sputtering conditions (corresponding to the peak in Ms in Figure 7) were used to investigate the deposition of a thick film. The hysteresis loop and VHF μ spectra for the 2 μm thick FeCrTa:O/N film deposited under these conditions are shown in Figures (9) and (10). From this data it was determined that the film had a M_s=1.9 T, a coercive force H_c=45 Oe, remanent magnetization M_r=0.57 T, and resistivity ρ=71 $\mu\Omega$-cm. In the range 20-200 MHz, μ'=89 and $\mu''\approx$ 0, with a Q≈450. The permeance of this film ~180 μm was >10 times that previously reported for other high Q films in this frequency range.

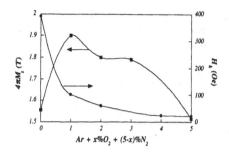

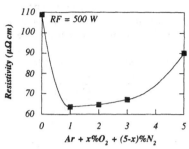

Figure 7. Variation in M_s and H_{an} (H_k) of FeCrTaN/O films as a function of O_2/N_2 flow ratio. ($O_2 + N_2$) flow fixed at 5% of fixed total flow.

Figure 8. Variation in resistivity of FeCrTaN/O films as a function of O_2/N_2 flow ratio. ($O_2 + N_2$) flow fixed at 5% of fixed total flow.

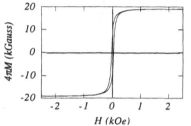

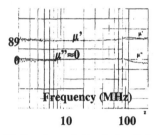

Figure 9. Hysteresis loop of 2 μm thick FeCrTa:N/O film deposited under optimum conditions (see text).

Figure 10. Complex μ of a 2μm thick optimized FeCrTa:N/O film. $\mu''\sim0$ indicates that FMR>200 MHz.

DISCUSSION

The variation in the magnetic properties of most sputtered films with increasing thickness, is determined by the influence of the structure of the initially nucleated film, on the microstructure of the film growth far from the interface. The microstructure of the thick 2 μm nanocrystalline was investigated by comparing the TEM of cross-sections near and far from the interface. The results showed that that these cross sections were identical to those in Figure (2). This result indicates that nanostructural growth of the reactively sputtered nanocrystalline films is due to a process limited growth, featuring a random renucleation of individual grains. The independence of the renucleation on the initial film growth is responsible for observed independence of the film properties on film thickness.

The large H_{an} (>100 Oe) observed in these films increases their FMR, and is desired for obtaining a high Q at VHF/UHF frequencies. The increase in H_{an} is attributed to the increase in internal shape anisotropy associated with the thickening of the grain boundaries with the increase in the reactive gas flow rate. This was confirmed by the limiting case in which films containing randomly oriented, magnetically isolated, spheroidal grains were observed to have a $\mu'=3.9$ [Figure (6)] which is agrees well with the value of 4 predicted by theory [6].

Table I compares the properties of two films with and without additional O. The M_s and ρ increased, and H_{an} decreased with the addition of 2%O_2. These observations suggest the possbility that oxygen preferentially bonded with Cr and Ta to increase the ρ, and that the available N bonded with Fe to increase the magnetization. The decrease in H_{an} may be due to the an increase in the intergranular magnetic coupling of Fe/FeN grains through the weakly magnetic CrO, and/or thinner Cr/Ta:O grain boundaries. The changes observed in the magnetic properties and the resistivity are consistent with the expected behavior.

Table I. Variation of film properties with the addition of oxygen gas.

Reactive gas	3%N_2*	3%N2+2%O2 **
$4\pi M_s$ (T)	1.7	1.8
H_{anh} (Oe)	92	59.9
ρ ($\mu\Omega$ cm)	43	64.7
μ' at 20 MHz	174	496

* RF power=750W   ** RF power=500W

CONCLUSIONS

FeCrTaN and FeCrTaO/N films with thickness > 1 μm were sputtered in reactive N and (N+O) gases. Their magnetic properties were dominated by the film microstructure and were independent of film thickness. Significant changes were observed in M_s, H_{an}, and ρ as the deposition parameters were varied. Large increases in H_{an} were obtained from the internal shape demagnetization associated with the film morphology produced by the O/N reacting with the sputtered FeCrTa alloys. The granular morphology of the nanocrystalline film varied with the reactive gas flow rates and deposition parameters. With increasing thickness, the resistivity of the film was reduced. The addition of O_2 to the N_2 reactive gas flow increased the film M_s, H_{an}, and ρ. For FeCrTa:O/N films with thickness $\approx$ 2μm, a magnetic Q $\geq$ 450 was obtained at VHF frequencies. FMR was calculated to occur in the thick films at 6 GHz, indicating that they may find application in low loss, high Q magnetic devices operating at UHF/VHF frequencies.

ACKNOWLEDGMENTS

The authors gratefully acknowledge the support of DARPA-Frequency Agile Materials for Electronics (FAME) Program under the direction of Dr. Stuart Wolf.

REFERENCES

[1] R.M. Walser, A.P. Valanju, and W. Win, "Materials issues in the application of ferromagnetic films in tunable high Q VHF/UHF devices", these proceedings.

[2] H. Fujimori, Scripta Met. **33**, 1625 (1995); M. Senda and Y. Nagai, J. Appl. Phys. **65**, 3151 (1989).

[3] A. Makino and Y. Hayakawa, Matls. Sci. Eng. A, **A181/A182**, 1020, (1994); S. Ohnuma, H. Fujimori, S. Mitani, and T. Matsumoto, J. Appl. Phys. **79**, 5130, (1996).

[4] E. van de Reit and R. Roozeboom, J. Appl. Phys. **81**, 350, (1997).

[5] S. Jin, W. Zhu, R.B. van Dover, T.H. Tiefel, V. Korenivski, and L.H. Chen, Appl. Phys. Lett. **70**, 3162 (1997).

[6] R.M. Walser, W. Win and P.M. Valanju, IEEE Trans. Mag. **34**, 1391 (1998).

Fundamentals

MECHANISMS OF DIELECTRIC LOSS IN MICROWAVE MATERIALS

ALEXANDER TAGANTSEV
Ceramics Laboratory, EPFL, Swiss Federal Institute of Technology, 1015 Lausanne, Switzerland

ABSTRACT

The present knowledge on mechanisms for dielectric loss at microwave and higher frequencies is reviewed. A special attention is paid to the intrinsic loss mechanisms. Relation between the loss and the dielectric permittivity as well as the effect of dc electric field on the loss are discussed. The theoretical knowledge is compared to the existing experimental data.

INTRODUCTION

The absorption of the energy of electromagnetic field, the frequency ω of which is much lower than that of the typical far-infrared absorption lines is called dielectric loss. For solid crystalline materials this frequency range corresponds to the frequencies which are much lower than that of the lowest optical phonon branch, the latter being typically 10^3-10^4 GHz. The dielectric loss is the factor that mainly limits the use of dielectric materials in microwave devices. An efficient use of a material in these devices requires understanding the nature of dielectric loss. Specifically, it is desirable to be able to distinguish between the contributions of the intrinsic (related to the ideal crystalline lattice) and extrinsic (related to crystal imperfections and ferroelectric domains) loss mechanisms in a given material. This knowledge can provide guidelines for the modification of the dielectric properties of materials.

At present, the existing concepts of dielectric loss in solids can be roughly divided into six categories: (i) interaction with local dipoles (Debye loss), (ii) fundamental phonon mechanisms (interaction with thermal phonons and multi-phonon processes) (see e.g.,[1]), (iii) excitation of acoustic phonons at the frequency of the ac electric field (see e.g.,[1-3]), (iv) interaction with the ferroelectric domain walls (see e.g.,[4]) (v) loss related to the "universal relaxation law" (see e.g.,[5]), (vi) contribution of the Ohmic conduction to the loss. In principle, any of these mechanisms can substantially contribute to the dielectric loss, however, the better the quality of a low-loss material and the higher the frequencies of the ac field used, the more important the role of the fundamental phonon mechanisms, (ii). The present paper is devoted to a discussion of these mechanisms and their role in the absorption of ac electric field at microwave and higher frequencies.

FUNDAMENTAL PHONON LOSS MECHANISMS

The origin of the fundamental loss is the interaction of the ac field with the phonons of the material. The theory of the loss stemming from this interaction has been developed for crystalline materials with a well defined phonon spectrum (see for a review[1]), i.e., for materials where the damping of phonons [average frequency of the inter-phonon collisions], Γ, is much smaller than their frequencies. According to this theory, in terms of quantum mechanics, the fundamental loss mainly corresponds to the absorption of the energy quantum of the electromagnetic field $\hbar\omega$ [ω is the ac field frequency] in collision processes with the thermal

221

phonons, which have much higher energies. This large difference in the energies makes it difficult to satisfy the conservation laws in these processes. In such a complicated situation, there exist three efficient schemes of absorption of the $\hbar\omega$-quanta, which correspond to the three main fundamental loss mechanisms: 1) three-quantum, 2) four-quantum, and 3) quasi-Debye.

Three-quantum mechanism

The three-quantum mechanism corresponds to a process involving a $\hbar\omega$-quantum and two phonons. The theory of this contribution has been developed by different authors[1,6-9]. Owing to a very big difference between the energy quantum of the electromagnetic field $\hbar\omega$ and those of the thermal phonons, three-quantum processes can take place only in the regions of the wave-vector space ($\bar{k}$-space) where the difference between the frequencies of two different phonon branches is small, specifically, of the order of ω and/or of the phonon damping Γ. These regions are usually located in the vicinity of the degeneracy lines of the spectrum, i.e. the lines in the $\bar{k}$-space where the frequencies of different branches are equal. In crystals exhibiting a phonon spectrum with low-lying (soft) optical modes, the degeneracy lines formed with a participation of these modes are of primary importance for the loss. Since the degeneracy of the spectrum is mainly controlled by the symmetry of the crystal, the explicit temperature dependence [which does not take into account the temperature dependence of the dielectric permittivity] and frequency dependence of the three-quantum loss are very sensitive to the symmetry of the crystal. Depending on crystalline symmetry, temperature interval, frequency range and some parameters of the phonon spectrum, the temperature and frequency dependence of the imaginary part of the dielectric permittivity can be described by power laws[1], i.e.

$$\varepsilon'' \propto \omega^n T^m \tag{1}$$

where $n = 1 - 5$; $m = 1 - 9$.

Consider an example of a cubic centrosymmetric crystal at $T \geq \hbar\Omega_{TO}/k_B$ where $\hbar$ and k_B are the Planck and Boltzmann constants and Ω_{TO} is the frequency of the lowest transverse optical phonon branch at $\bar{k} = 0$. In this case, the contribution of the transverse optical phonons to the imaginary part of the dielectric permittivity via the three-quantum mechanism is given by the expression[8,10]:

$$\varepsilon'' \propto \omega\Gamma^2 \left[\ln\left(\frac{\Omega_{LO}^2}{\omega^2 + 4\Gamma^2}\right) + \frac{\omega}{\Gamma}\tan^{-1}\frac{\omega}{2\Gamma} \right] \tag{2}$$

Here Γ is the damping of these phonons ($\Gamma \propto T$) and Ω_{LO} is the typical frequency of longitudinal optical phonons. A characteristic feature of this expression is a correlated crossover in the frequency and temperature dependences of the loss when one passes from the frequency range where $\omega \ll \Gamma$ to that where $\omega \gg \Gamma$. Indeed, from this expression one finds (within the logarithmic accuracy): $\varepsilon'' \propto \omega T^2$ for $\omega \ll \Gamma$ and $\varepsilon'' \propto \omega^2 T$ for $\omega \gg \Gamma$. This crossover can be distinguished in the data on the submillimeter frequency range loss reported by Stolen and Dransfeld[11] for alkali halide crystals (Fig.1a). One concludes from this figure that, in accordance with the theory, as the frequency increases, the "low frequency" behavior $\varepsilon''/T \propto T$ evolves towards $\varepsilon''/T = \text{const}$.

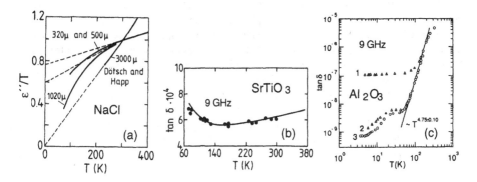

Fig.1 Temperature dependence of the dielectric loss in NaCl (a - imaginary part of the relative permittivity divided by T; the wavelength of the ac electric field in free space is indicated), $SrTiO_3$ (b – loss tangent; the frequency of the ac electric field is 9 GHz; solid line shows the prediction of the phonon theory[10]) and Al_2O_3 (c – loss tangent; the frequency of the ac electric field is 9 GHz, the numbers 1, 2, and 3 indicate the data for crystals of different quality). After Refs.[11-13].

Four-quantum mechanism

The four-quantum mechanism corresponds to the field-quantum absorption processes involving three phonons[11,14,15]. The conservation (energy and quasi-momentum) laws do not impose strong restrictions on the type and energy of the phonons participating in these processes. For this reason, in contrast to the three-quantum processes, not only the degeneracy lines but also the whole thermally excited part of the $\bar{k}$-space nearly homogeneously contributes to the loss. Due to this fact, the explicit temperature and frequency dependence of the contribution of this mechanism to the imaginary part of the dielectric permittivity appears to be insensitive to the symmetry of the crystal. For $T \geq \hbar\Omega_{TO}/k_B$, this mechanism provides $\varepsilon'' \propto \omega T^2$; whereas at lower temperatures its contribution to the loss drastically decreases with decreasing temperature, e.g., for $T << \hbar\Omega_{TO}/k_B$, this mechanism provides[1] $\varepsilon'' \propto \omega T^9$.

Fundamental loss mechanism in centrosymmetric crystal

The three- and four-quantum mechanisms are the only two mechanisms which control the intrinsic loss in centrosymmetric crystals[1]. In the case of practical interest of the low-microwave-loss materials, the theory of phonon transport, which has been used to obtain the results reviewed in the two previous sections, is applicable with quite a good accuracy. Specifically, that means that, possessing the complete information on the phonon spectrum and the constants of inter-phonon and ac field-phonon couplings, one can calculate the intrinsic dielectric loss with an accuracy equal to the relative damping of the typical phonons participating in the absorption of the ac field[1]. Thus, the result that the intrinsic loss in centrosymmetric crystals is given by the sum of the three- and four-quantum contribution can be considered as perfectly justified. However, in practice, when discussing the intrinsic microwave loss, one often uses two models differing from that discussed above. These two models are the Vendik model[15,16] and the so-called damped oscillator model[17]. Let us discuss the correspondence between the results of the phonon transport theory and of these models.

The Vendik model, as was mentioned in the original paper[15], is a version of the theory of the four-quantum mechanism. It gives correct dependences of the four-quantum-contribution to the loss tangent on frequency, temperature and dielectric permittivity for the temperature range of practical importance (the same as for Eq.(2)).

In the damped oscillator model, the loss is calculated by using the expression for the shape of the far-infrared absorption lines (associated with transverse optical branches) at the ac field frequency, which is much below the resonance frequencies. In this framework, the contribution to the complex dielectric permittivity from one branch (with frequency Ω_{TO}) is presented in the form[17]

$$\varepsilon^* \propto \frac{1}{\Omega_{TO}^2 - \omega^2 + 2i\omega\Gamma} \qquad (3)$$

and used at $\omega \ll \Omega_{TO}$ for the description of the dielectric loss. The analysis in terms of the phonon transport theory shows that the use of this formula for calculating the dielectric loss at $\omega \ll \Omega_{TO}$ is not justified[1], except for crystals of symmetry $\frac{4}{m}$ and $\frac{6}{m}$. Thus, the damped oscillator model, in general, is not supported by the comprehensive theory. Table 1 compares the predictions of the two aforementioned models with these of the phonon transport theory. The comparison is performed for the case of a cubic incipient ferroelectric at $T \geq \hbar\Omega_0 / k_B$ where Ω_0 is the soft-mode frequency (like SrTiO$_3$ or KTaO$_3$ at $T \geq 20 - 30$ K). One concludes from this table that, for the frequency range $\omega \leq \Gamma$ (for SrTiO$_3$ or KTaO$_3$ that means $\omega \leq 100$ GHz), to within the logarithmic factors, the Vendik model gives correct functional dependences of the loss tangent on the frequency, temperature and the dielectric permittivity, whereas the prediction of damped oscillator model substantially differs from that of the comprehensive phonon theory.

Table 1 Functional dependence of the loss tangent $\tan\delta = \varepsilon''/\varepsilon$ of a cubic incipient ferroelectric according to different models used for the description of the intrinsic loss.

Frequency range	Phonon transport theory	Vendik model	Damped oscillator
$\omega \leq \Gamma$	$\omega T^2 \varepsilon^{3/2}[1 + b\ln(\Omega_{TO}/\Gamma)]$	$\omega T^2 \varepsilon^{3/2}$	$\omega T\varepsilon$
$\omega \gg \Gamma$	$\omega^2 T\varepsilon^{3/2}$	———	$\omega T\varepsilon$

Quasi-Debye loss mechanism

The origin of this mechanism is the relaxation of the phonon distribution function of the crystal[1,7,17,18]. In non-centrosymmetric crystals, the phonon frequencies are linear functions of a small electric field applied to the crystal[7]. Thus, the oscillations of the ac field result in time modulation of the phonon frequencies; the latter in turn induces a deviation of the phonon distribution function from its equilibrium value. A relaxation of the phonon distribution functions gives rise to dielectric loss in a similar way as a relaxation of the distribution function of the dipoles gives rise to the loss in the Debye theory[19]. This analogy is expressed by the name "quasi-Debye". The frequency dependence of the quasi-Debye contribution to the loss factor is of the Debye type, the average relaxation time of the phonon distribution function playing the role of the Debye relaxation time. The contribution of the quasi-Debye mechanism to the tensor of the imaginary part of the relative dielectric permittivity of a crystal, $\varepsilon''_{\alpha\beta}$, can be expressed in terms of phonon dispersion curves $\Omega_j(\vec{k})$, phonon damping $\Gamma_j(\vec{k})$ and the so-called vector of electrophonon potential $\vec{\Lambda}_j(\vec{k})$. Here the index j numerates the phonon branches and $\vec{k}$ is the phonon wave-vector. Calculated in the relaxation-time approximation $\varepsilon''_{\alpha\beta}$ can be presented in quadratures[1]:

$$\varepsilon''_{\alpha\beta} = \varepsilon_o^{-1} \sum_j \int \frac{d^3k}{(2\pi)^3} N_j(N_j+1) \frac{\hbar^2 \Omega_j^2 \Lambda_j^{(\alpha)} \Lambda_j^{(\beta)}}{k_B T} \frac{2\Gamma_j \omega}{4\Gamma_j^2 + \omega^2} \tag{4}$$

where ε_o and $N_j = [\exp(\hbar\Omega_j/k_B T) - 1]^{-1}$ are the permittivity of free space and the Planck distribution function; the integration volume is the first Brillouin zone. The electrophonon potential $\vec{\Lambda}_j(\vec{k})$ is defined as the relative change of the phonon frequencies per unit of the applied electric field[1]:

$$\Delta\Omega_j(\vec{k}) = \Omega_j(\vec{k})\Lambda_j^{(\alpha)}(\vec{k})E^{(\alpha)} \quad \text{or} \quad \Lambda_j^{(\alpha)}(\vec{k}) = \frac{1}{\Omega_j(\vec{k})} \frac{\partial\Omega_j(\vec{k})}{\partial E_\alpha}. \tag{5}$$

One can notice an analogy between Eq.(4) and the expression for the Debye loss contribution[19]

$$\varepsilon'' \propto \frac{p^2}{k_B T} \frac{\tau\omega}{1+(\omega\tau)^2} \tag{6}$$

[p is the dipole moment of the Debye dipoles] as well as between Eq.(5) and the expression for a change of the energy of a dipole in an electrical field:

$$\Delta U = -p^{(\alpha)}E^{(\alpha)}. \tag{7}$$

with a correspondence $p \Leftrightarrow -\hbar\Lambda\Omega$ and $\tau \Leftrightarrow 1/2\Gamma$. Apparently, in non-centrosymmetric crystal, the phonons behave similarly to the Debye dipoles, as far as the dielectric loss is concerned[7].

An important feature of the quasi-Debye mechanism is that, once allowed by the symmetry, it can be easily a few orders of magnitude greater than that of three- and four-quantum mechanisms. To give an idea about the strength of this mechanism, rough estimates for the three-, four-quantum and quasi-Debye contributions to the loss tangent $\tan \delta = \varepsilon''/\varepsilon$ are presented in Fig.2. Specifically, this figure shows the contributions arising from the interaction of the ac field with the soft-mode phonons of a perovskite-type ferroelectric in the ferroelectric phase. It is seen that in this material the quasi-Debye contribution dominates the intrinsic loss in the frequency range $\omega \leq \Gamma$ (for SrTiO3 and KTaO3 at $T \geq 20 - 30$ K that means $\omega \leq 100$ GHz).

To the best knowledge of the author the quasi-Debye contribution has not been experimentally identified in non-centrosymmetric crystals.

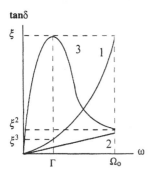

tanδ

Fig.2 Results of phonon transport theory. Schematic off-scale plot of the frequency dependence of the three-quantum (1), four-quantum (2), and quasi-Debye (3) contributions to the loss tangent due to the interaction of the ac electric field with the phonons of the soft mode of a displacive type ferroelectric. Ω_0 and Γ are the soft-mode frequency and damping, respectively. $\xi = \Gamma/\Omega_0$. The range of applicability of the theory is $\Gamma \ll \Omega_0$ and $\omega \ll \Omega_0$ (at ω tending to Ω_0, the curves are only shown to guide the eye).

INTRINSIC LOSS VS. DIELECTRIC PERMITTIVITY RELATION

It is well known that materials with a higher dielectric permittivity usually exhibit a higher loss tangent. This trend is unfavorable for application of high-permittivity materials. In the present section we will discuss this trend for intrinsic loss in centrosymmetric materials.

Loss-permittivity relation for three-quantum contribution

The phonon transport theory enables an evaluation of the relation between the dielectric loss and electric permittivity of a crystal. This relation has been theoretically studied in the case of practical interest – the three-quantum contribution in cubic centrosymmetric crystals[20]. The calculations have been made for a model, in which the dielectric response is assumed to be controlled only by one transverse optical (TO) mode. This is a good approximation for crystals with a ferroelectric soft mode or for cubic diatomic ionic crystals. For more complicated non-ferroelectric materials, this approximation is expected to give a qualitatively correct result. It been found that the three-quantum contribution to the imaginary part of the dielectric permittivity of a cubic centrosymmetric crystal can be evaluated (for $T \geq \hbar \Omega_{TO}/k_B$) as

$$\varepsilon'' \cong A\omega\left[\ln\left(\frac{\Omega_{LO}^2}{\omega^2 + 4\Gamma^2}\right) + \frac{\omega}{\Gamma}\tan^{-1}\frac{\omega}{2\Gamma}\right]; \qquad A = \frac{\mu\Gamma\varepsilon^2}{12\Omega_{LO}^2}\left(\frac{\varepsilon}{\varepsilon_\infty}\right)^{x-2}, \qquad (8)$$

where ε_∞ and Ω_{LO} are the optical permittivity and the typical frequency of longitudinal optical phonons; μ is the small anharmonicity parameter[1,21], which can be evaluated as $k_B T / M v^2$ (M and v are the average mass of the atoms and the average sound velocity in the crystal). The parameter x has been calculated for two limiting cases: (i) for the case of a strong dispersion of the TO mode (the typical case for ferroelectric soft modes) and (ii) for the case of a negligible TO mode dispersion (the case close to the situation in non-ferroelectric crystals). In these cases, $x = 2.5$ and $x = 5$ have been found, predicting a fast growth of dielectric loss with increasing permittivity.

The theoretical result has been compared to the data of submillimeter frequency studies of dielectric loss in a system of complex perovskites $Ba(M^{(l)}_{1/2} M^{(ll)}_{1/2})O_3$ ($M^{(l)}$ = Mg, In, Y, Gd, Nd and $M^{(ll)}$ = W, Ta, Nb)[20,22]. These are cubic non-ferroelectric (pseudo-cubic) centrosymmetric materials with the relative dielectric permittivity ranging from 20 to 40. In this frequency range the imaginary part of dielectric permittivity is expected to be controlled by the intrinsic contribution given by Eq.(8). A value of x close to 5 is expected due to the non-ferroelectric nature of these materials. The comparison of the theory to the experiment is shown in Fig 3. This figure shows that the experimental data on ε'' measured at 300 GHz do reveal a strong correlation between ε'' and ε, which can be approximated as a power law with the exponent close to 4. This value is close to the upper theoretical bound for this exponent. The difference between the upper bound for x and its measured value is attributed to a small but appreciable dispersion of the optical modes in the studied materials.

The case of the ferroelectric soft mode also has been tested[1,10] by using the experimental data on microwave dielectric loss in $SrTiO_3$. In this case, Eq.(8) with $x = 2.5$ and $\omega \ll \Gamma$ leads to $\tan\delta \equiv \varepsilon'' / \varepsilon \propto \omega T^2 \varepsilon^{1.5}$. As shown in Fig.1b, this expression is in good agreement with experimental data reported for crystalline $SrTiO_3$ by Buzin[12].

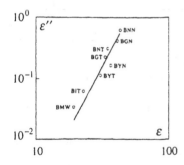

Fig.3 Imaginary part of the dielectric permittivity plotted vs. its real part for $Ba(Mg_{1/2}W_{1/2})O_3$ (BMW), $Ba(In_{1/2}Ta_{1/2})O_3$ (BIT), $Ba(Y_{1/2}Ta_{1/2})O_3$ (BYT), $Ba(Y_{1/2}Nb_{1/2})O_3$ (BYN), $Ba(Gd_{1/2}Ta_{1/2})O_3$ (BGT), $Ba(Nd_{1/2}Ta_{1/2})O_3$ (BNT), $Ba(Gd_{1/2}Nb_{1/2})O_3$ (BGN), and $Ba(Nd_{1/2}Nb_{1/2})O_3$ (BNN) measured at ω =300 GHz . The best fit $\varepsilon'' = B\varepsilon^x$; x = 4.1 ± 0.5. After Ref. [22].

Loss-permittivity relation and relevance of intrinsic loss at microwave frequencies

The strong dependence of the intrinsic dielectric loss on the electrical permittivity of crystals discussed above enables an evaluation of the relative importance of intrinsic loss in the total balance of the microwave loss.

For "hard" dielectric materials with typical values of the relative permittivity below 10, a very low level of the intrinsic loss at microwave frequencies is expected. Actually, a delicate experimental technique and high quality materials are needed to distinguish the intrinsic contribution to the loss in this case. This kind of study has been performed on high quality

Al_2O_3 crystals by Braginsky et al.[13] (Fig.1c). The measured frequency and temperature dependence of the loss tangent, $\tan\delta \propto \omega^{1.7}T^{4.75}$, is close to the theoretical prediction for thee-quantum mechanism in the limit $\omega \gg \Gamma^1$, $\tan\delta \propto \omega^2 T^4$. This attests to the intrinsic nature of the measure dielectric loss[1]. However, the level of measured loss has been found to be very low. Thus, one concludes that in materials with a very low-dielectric constant, the intrinsic loss is too low to be the limiting factor for at least standard microwave applications.

For material with slightly elevated electric permittivity like complex perovskites discussed in the previous section ($\varepsilon = 20 - 40$), the intrinsic loss becomes more important due to a strong ε'' vs. ε dependence. This is illustrated by Table 2, in which ε'' of different complex perovskite ceramics measured at 10 GHz is compared with the values of ε'' corresponding to the expected intrinsic contribution. The evaluation of the intrinsic contribution at 10 GHz has been performed by extrapolation of ε'', measured at 300 GHz, down to 10 GHz, using Eq.(8). One concludes from this table that the expected intrinsic and measured losses are comparable, though the measured loss is always greater than the expected intrinsic contribution. However, in some cases, the intrinsic loss controls more than 50% of the total loss at 10 GHz. Thus, we see that, in the materials with elevated electric permittivity, the intrinsic loss at microwave frequencies does not seem to make the main contribution, though in some materials it can be of a considerable importance.

Table 2 Imaginary part of the electric permittivity of complex perovskites (acronyms are explained in Fig.3) measured at 10 GHz and extrapolated from 300 to 10 GHz by using Eq.(8). After Ref. [22].

$10^3\varepsilon''$	BNT	BGT	BYT	BIT	BNN	BGN	BYN	BMW
Measured At 10 GHz	8.2	40.3	6.8	2.0	21	62	12	1.6
Extrapolated From 300 to 10 GHz	6.0	2.9	1.7	0.7	16	11	3.9	0.5

When passing from non-ferroelectric materials to materials with ferroelectric soft modes, one encounters a strong increase of the dielectric permittivity. Even though, in the latter case the dependence $\tan\delta \propto \varepsilon^x$ ($x = 3 - 4$) gives way to a weaker dependence, $\tan\delta \propto \varepsilon^{1.5}$, the increase of ε results in a substantial increase of intrinsic contribution to dielectric loss. Finally, the role of the intrinsic loss becomes more important so that, at least in good quality ferroelectric crystals (in the paraelectric phase or incipient ferroelectrics), the intrinsic loss is expected to control the dielectric loss at microwave frequencies. A good example is the microwave absorption in $SrTiO_3$: an excellent agreement between the theoretical predictions for the intrinsic loss and the experiment is illustrated in Fig.1b.

EFFECT OF DC FIELD IN INTRINSIC LOSS IN CENTROSYMMETRIC CRYSTALS

One of the promising applications of microwave materials is their use for tunable microwave devices like phase shifters, tunable resonators, tunable filters etc[23,24]. These applications are based on a strong dc electric field dependence of the dielectric permittivity observed in paraelectric perovskites (e.g. $SrTiO_3$ and $KTaO_3$). However, the practical use of this effect is hindered by a recently found phenomenon: the application of the dc field to high quality tunable dielectric materials gives rise to a dramatic increase of the loss factor[25,26]. Figure 4 illustrates this effect, which consists of the fact that a relatively small decrease (30%) of the permittivity is accompanied by an increase of loss tangent of several hundred percent. Though seemingly unexpected, this phenomenon was theoretically predicted by the present author many years ago[18] and a quantitative theory of this effect for $KTaO_3$ has been recently developed[27]. The mechanism of this phenomenon, the so-called field-induced quasi-Debye mechanism, is presented below.

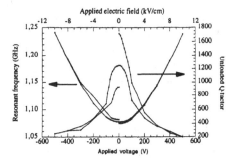

Fig.4 The resonant frequency and the unloaded Q-factor of a $KTaO_3$ resonator with super-conductive electrodes as a function of the dc bias. The resonant frequency is inverse proportional to the square root of the dielectric permittivity of the material. $1/Q$ gives an estimate for the loss tangent of the material. After Gevorgian et al.[26]

Field-induced quasi-Debye mechanism

As was discussed above, the intrinsic dielectric loss in crystals is controlled by three mechanisms: three-quantum, four-quantum, and quasi-Debye mechanisms; in centrosymmetric crystals, only three-quantum and four-quantum mechanisms being active. In non-centrosymmetric crystals all three mechanisms are active, including the quasi-Debye mechanism. The contribution of the latter is substantial: it can easily be a few orders of magnitude greater than that of the three-quantum and four-quantum mechanisms. Clearly, the application of a dc field to a centrosymmetric crystal breaks its central symmetry and therefore gives rise to the quasi-Debye mechanism. In the case where the quasi-Debye mechanism is induced in a centrosymmetric crystal by application of a dc electric field, we call it "field-induced quasi-Debye mechanism". Though the contribution of the field-induced quasi-Debye mechanism tends to zero as far as the dc field does, this mechanism is so efficient that the induced relative variation of the intrinsic loss is usually much larger than the induced relative variation of the permittivity[18]. One can elucidate this feature of the field-induced quasi-Debye mechanism as follows.

Let us compare the dielectric loss in three situations. (i) A cubic centrosymmetric material in the paraelectric phase, the electric permittivity being equal ε and the dc polarization $P = 0$. (ii) The same material under a dc electric field E, which is weak enough to induce only a small relative change of the permittivity $\Delta\varepsilon / \varepsilon$, the dc polarization $P = P_{dc} \approx \varepsilon\varepsilon_0 E$. (iii) The same material in the ferroelectric phase (or hypothetical ferroelectric phase), the electric permittivity having *the same value of* ε and the dc polarization having its spontaneous value P_s. In the first case, the three- and four-quantum mechanisms control the loss. Let us denote their contribution to the imaginary part of the dielectric permittivity as ε''_{cs}. In the third case, three-, four-quantum and quasi-Debye mechanisms contribute to the loss, however, in the frequency range of interest, $\omega \leq \Gamma$, the loss is dominated by the latter. Let us denote its contribution to the imaginary part of the dielectric permittivity as ε''_{qD}, $\varepsilon''_{qD} \gg \varepsilon''_{cs}$. One can easily conclude that, in the ferroelectric phase, the quasi-Debye contribution is proportional to P_s^2. Indeed, the electrophonon potential Λ [defined by Eq.(5)] is non-zero due to the lack of the center of symmetry in the ferroelectric phase, and thus should be proportional to the spontaneous polarization P_s. In turn, the quasi-Debye contribution, being proportional to Λ^2 (see Eq.(4)), should be proportional to P_s^2. The situation with the centrosymmetric material under a small dc bias is intermediate between the two others. Now the three-, four-quantum and quasi-Debye mechanisms contribute to the loss, however, being induced by a relatively weak field, the latter does not work in its full power. On the lines of the above consideration, one finds that the contribution of the field-induced quasi-Debye mechanism should by proportional to the square of the dc field-induced polarization, P_{dc}^2. Thus, compared to the case of the ferroelectric phase, in the case of the material under the bias, the quasi-Debye contribution is scaled down by a factor of $(P_{dc}/P_s)^2$. This factor can be roughly evaluated as $\Delta\varepsilon/\varepsilon$. Finally, the field-induced variation of the imaginary part of the dielectric permittivity can be evaluated as $\Delta\varepsilon'' \cong \varepsilon''_{qD}(P_{dc}/P_s)^2 \cong \varepsilon''_{qD}\left(\dfrac{\Delta\varepsilon}{\varepsilon}\right)$. Since $\varepsilon''_{qD} \gg \varepsilon''_{cs}$, one finds that the field-induced relative variation of ε'' should be much greater than the field-induced variation of the permittivity, indeed: $\dfrac{\Delta\varepsilon''}{\varepsilon''} = \dfrac{\Delta\varepsilon''}{\varepsilon''_{cs}} \cong \dfrac{\varepsilon''_{qD}}{\varepsilon''_{cs}}\dfrac{\Delta\varepsilon}{\varepsilon} \gg \dfrac{\Delta\varepsilon}{\varepsilon}$. According to Ref. [18] and the field-induced quasi-Debye mechanism is not the only one controlling the field dependence of the loss in centrosymmetric crystals; however, it is the only mechanism which can provide a strong field-induced increase of the intrinsic loss.

To calculate the quasi-Debye contribution to the loss, one should calculate $\bar{\Lambda}_j(\bar{k})$ by using Eq.(5) and then integrate and summate according to Eq.(4). These calculations can be performed, if one knows the dispersion curves and damping of the phonon spectrum as well as the dependence of the spectrum on the electric field. In ferroelectrics, the most obvious contribution to the quasi-Debye loss appears due to the well-known sensitivity of the soft-mode frequency to the electric field. Using the classical expression for the soft-mode dispersion curve[28]

$$\Omega_{TO}^2(k) = \frac{\lambda}{\varepsilon(E)} + sk^2 \tag{9}$$

where λ and s are constants, the equation of the dielectric non-linearity[28] $\varepsilon^{-1}(E) = \varepsilon^{-1} + 3B\varepsilon_o P^2$, and Eq.(5) one evaluates the corresponding component of the electrophonon potential as

$$\Lambda_{TO} = \frac{\lambda}{\Omega_{TO}^2(k)} 3B\varepsilon_o^2 \varepsilon P. \tag{10}$$

The contribution to the field-induced dielectric loss related to this component of the electrophonon potential, which is actually due to the relaxation of the distribution function of the soft-mode phonons, was calculated in Ref. [18]. Another contribution is brought about by the relaxation of the distribution function of the acoustic phonons, whose spectrum is also rather sensitive to the electric field due to their interaction with the field-dependent soft mode. This contribution has been found to be very important in paraelectric perovskites[27], where the interaction between the soft mode and acoustic modes is known to be strong[29,30].

The field-induced quasi-Debye mechanism has been recently evaluated for KTaO$_3$ at relatively small dc fields[27], i.e. the fields inducing $\Delta\varepsilon/\varepsilon \ll 1$. It has been found that the main contribution to the field-induced loss comes from the relaxation of the distribution function of transverse acoustic phonons. The calculations have been performed for $T = 80$ K, in the approximation of the isotropic phonon spectrum, the direction of the dc and microwave fields having been set parallel to the 4-fold axis of the crystal. The dc-field-dependent contribution to the loss tangent has been evaluated as

$$\tan\delta \cong \omega \frac{\Delta\varepsilon}{\varepsilon} C \qquad (11)$$

where ω should be taken in GHz and the constant $C = (1.2 - 4.3) \times 10^{-2}$. Comparison of Eq.(11) with the experimental data shows reasonable agreement. According to Ref. [26], in KTaO$_3$ at the dc field inducing $\Delta\varepsilon/\varepsilon = 0.35$, $\omega \approx 1$ GHz, and temperature below but not much lower that 90 K, the loss tangent is about 0.5×10^{-2} whereas Eq.(11) gives $\tan\delta \approx (0.4 - 1.5) \times 10^{-2}$.

Thus, the fact that the theory provides a correct qualitative interpretation and a reasonable numerical estimate of the effect strongly suggests that the field-induced quasi-Debye mechanism governs the dramatic field-induced increase of the loss factor in incipient ferroelectrics.

CONCLUSIONS

(i) The intrinsic dielectric loss in crystals is controlled by three mechanisms: three-quantum, four-quantum, and quasi-Debye mechanisms. In centrosymmetric materials only the two weakest - three-quantum and four-quantum mechanisms - are allowed by the crystalline symmetry. In non-centrosymmetric materials the quasi-Debye mechanism is also allowed. In this case, this mechanism is expected to dominate the intrinsic loss.

(ii) The trend – "the higher the electric permittivity, the higher the loss" – is strongly pronounced for the intrinsic loss. For intrinsic loss mechanisms in cubic centrosymmetric crystals, the relation between the real ε and imaginary ε'' parts of the electric permittivity is evaluated as $\varepsilon'' \propto \varepsilon^x$ ($x = 2.5 - 5$).

(iii) Due to the aforementioned trend, the role of the intrinsic loss at microwave frequencies increases with increasing electric permittivity of the material. This role is usually negligible for materials with a dielectric constant below 10; it might be important in non-ferroelectric materials with a dielectric constant of a few tens. In ferroelectric materials at microwave frequencies, however, the intrinsic dielectric loss should generally be of primary importance.

(iv) The application of dc field induces the quasi-Debye mechanisms in centrosymmetric crystals. This effect explains a drastic dc-electric-field-induced increase of the dielectric loss recently reported for SrTiO$_3$ and KTaO$_3$ incipient ferroelectrics.

ACKNOWLEDGMENTS

This work was supported by the Swiss National Foundation. The author thanks Stephan Stuecklin for reading the manuscript and useful comments.

REFERENCES

1   V. L. Gurevich and A. K. Tagantsev, Adv. Phys. **40**, 719 (1991).

2   E. Schlöman, Phys.Rev. **135**, A413 (1964).

3   O. G. Vendik and L. M. Platonova, Sov. Phys. Solid State **13**, 1353 (1971).

4   J. O. Gentner, P. Gerthsen, N. A. Schmidt, and R. E. Send, J. Appl. Phys. **49**, 4585 (1978).

5   A. K. Jonscher, *Universal relaxation law* (Chelsea Dielectrics Press, London, 1996).

6   V. S. Vinogradov, Fiz.Trerd. Tela **4**, 712 (1962).

7   V. L. Gurevich, Fiz.Tverd.Tela **21**, 3453 (1979).

8   A. K. Tagantsev, Sov.Phys. JETP **53**, 555 (1981).

9   K. A. Subbaswamy and D. L. Mills, Phys. Rev. B **33**, 4213 (1986).

10 A. K. Tagantsev, Sov.Phys.JETP **59**, 1290 (1984).

11 R. Stolen and K. Dransfeld, Phys. Rev. **139**, 1295 (1965).

12 I. M. Buzin, Vestn. Mosk. Univ. Fiz. Astron., **18**, 70 (1977).

13 V. B. Braginsky, V. S. Ilchenko, and K. S. Bagdassarov, Phys. Lett. A **120**, 300 (1987).

14 B. Y. Balagurov, V. G. Vaks, and B. I. Shklovskii, Fiz. Tverd. Tela **12**, 89 (1970).

15 O. G. Vendik, Sov.Phys. Solid State **17**, 1096 (1975).

16 O. G. Vendik, L. T. Ter-Martirosyan, and S. P. Zubko, J.Appl.Phys. **84**, 993 (1998).

17 G. J. Cooms and R. A. Cowley, J. Phys. C **6**, 121 (1973).

18 A. K. Tagantsev, Sov.Phys.JETP **50**, 948 (1979).

19 C. Kittel, *Introduction to Solid State Physics*, 4th Edition ed. (John Wiley & Sons,Inc., New York, London, 1971).

20 A. K. Tagantsev, J. Petzelt, and N.Setter, Solid State Commun **87**, 1117 (1993).

21 V. L. Gurevich, *Transport in Phonon Systems* (North-Holland, Amsterdam, 1986).

22 R. Zurmulen, J. Petzelt, S. Kamba, G. Kozlov, A. Volkov, B. Gorshunov, D. Dube, A. Tagantsev, and N. Setter, J. Appl. Phys. **77**, 5351 (1995).

23 M. J. Lancaster, J. Powell, and A. Porch, Supcund. Sci. Technol. **11**, 1323 (1998).

24 O. G. Vendik, E. K. Hollmann, A. B. Kozyrev, and A. M. Prudan, Journal of Superconductivity **12**, 325 (1999).

25 O. G. Vendik, E. Kollberg, S. S. Gevorgian, A. B. Kozyrev, and O. I. Soldatenkov, Electr. Lett. **31**, 654 (1995).

26 S. Gevorgian, E. Carlsson, E. Wikborg, and E. Kollberg, Intergrated Ferroelectrics **22**, 245 (1998).

27 A. K. Tagantsev (unpublished).

28 M. E. Lines and A. M. Glass, *Principles and applications of ferroelectric and related materials* (Clarendon Press, Oxford, 1977).

29 J. D. Axe, J. Harada, and G. Shirane, Phys.Rev. **1**, 1227 (1970).

30 R. Comès and G. Shirane, Phys. Rev. **5**, 1886 (1972).

PERMITTIVITY AND MICROSTRUCTURE OF (Ba,Sr)TiO₃ FILMS: TEMPERATURE AND ELECTRIC FIELD RESPONSE

YU.A. BOIKOV*· **, T. CLAESON*, Z. IVANOV*, E. OLSSON***
*Physics and Engineering Physics, Chalmers University of Technology & Göteborg University, S-41296 Göteborg, Sweden, boikov@fy.chalmers.se
**Ioffe Physico-Technical Institute Russian Academy of Sciences, 194021 St.Petersburg, Russia
***Analytical Materials Physics, The Ångström Laboratory, Uppsala University, S-75121 Uppsala, Sweden

ABSTRACT

Epitaxial heterostructures $(001)(Y,Nd)Ba_2Cu_3O_{7-\delta} \| (100)SrTiO_3 \| (001)(Nd,Y)Ba_2Cu_3O_{7-\delta}$, $(100)SrRuO_3 \| (100)Ba_{0.8}Sr_{0.2}TiO_3 \| (100)SrRuO_3$, $(100)SrRuO_3 \| (100)SrTiO_3 \| (100)SrRuO_3$ and $(100)SrTiO_3 \| (001)YBa_2Cu_3O_{7-\delta}$ have been grown by laser ablation. There was only a small difference of the dielectric permittivity, in the temperature range 180-300K, between a bulk single crystal and an epitaxial $(100)SrTiO_3$ layer inserted between either high-T_C superconducting or $SrRuO_3$ electrodes. At T<150K, on the other hand, the response of the dielectric permittivity of the $SrTiO_3$ layer on temperature or electric field depended to a large extent upon the materials used as bottom and top electrodes in the heterostructures. The temperature dependence of the dielectric permittivity for the $SrTiO_3$ layer in $(100)SrRuO_3 \| (100)SrTiO_3 \| (100)SrRuO_3$ was well extrapolated by a Curie-Weiss relation in the range of T=80-300K, with about the same Curie constant ($C_0=7.5 \times 10^4$ K) and Curie temperature ($T_{Curie}=21$K) as in a bulk single crystal. At temperatures higher the phase transition point (65 K), the electric field response of the permittivity of the $SrTiO_3$ layer between high-T_C superconducting or metallic oxide electrodes was well extrapolated by the same relation used for a bulk single crystal. The smallest loss factor, $\tan\delta$, was measured for the capacitance $(100)SrRuO_3 \| (100)SrTiO_3 \| (100)SrRuO_3$ (T≈50-300K, f=100 kHz). The measured conductance G for the $SrTiO_3$ layer in the $(001)(Y,Nd)Ba_2Cu_3O_{7-\delta} \| (100)SrTiO_3 \| (001)(Y,Nd)Ba_2Cu_3O_{7-\delta}$ heterostructure fitted well the relation $\ln G \sim -(E_D/kT)$, with E_D=0.08-0.09 eV in a temperature range close to 300K. Pronounced hysteresis was observed in the temperature dependence of the dielectric permittivity for the $(100)Ba_{0.8}Sr_{0.2}TiO_3$ layer at temperatures close to the phase transition point, like in the case of a bulk single crystal. The permittivity of the $(100)Ba_{0.8}Sr_{0.2}TiO_3$ layer decreased more than 50% when an electric field of 2.5×10^6 V/m (T≈300K, f=100 kHz) was applied.

INTRODUCTION

High quality epitaxial thin layers of $Ba_{1-x}Sr_xTiO_3$ are attractive in a wide variety of applications due to their non-linear dielectric behaviour:
a) (x>0.8,T≈77K) in tunable microwave components based on low loss high-T_C superconducting films [1],
b) (x=0.3-0.5,T≈300K) in varactor type capacitance structures [2], and
c) (x<0.3,T≈300K) as elements of electrooptic modulators and switches [3].

A relatively large dielectric permittivity, ε, a considerable remnant polarisation and a pyroelectric coefficient imply an application of $Ba_{1-x}Sr_xTiO_3$ layers in dynamic random excess and ferroelectric memories and low cost IR detectors [4].

233

To be used in microelectronic circuits or in an integrated microwave element, the non-linear ferroelectric material should be manufactured as a thin film grown on a desirable substrate and combined with conducting electrodes. Available data in the literature [5] show, however, that dielectric properties of thin (Ba,Sr)TiO$_3$ films inserted in multilayer heterostructures differ dramatically from those of a corresponding bulk single crystal or ceramic sample. The ε, and the loss factor, tanδ, for (Ba,Sr)TiO$_3$ depend crucially on defects in the microstructure and deviations from stoichiometry.

Because of the low mobility of nucleated particles at the surface of the growing film, selective reevaporation of components, chemical interactions with the substrate, and mechanical stress, the microstructure of thin films of multi-component perovskite like oxides grown by laser ablation is considerably distorted as compared with that for a corresponding single crystal. The dielectric non-linearity of the (Ba,Sr)TiO$_3$ layer may be dramatically masked by the parasitic influence of the interface zone at a ferroelectric/electrode contact. Dielectric saturation and oxygen deficiency [6] in the region of the (Ba,Sr)TiO$_3$ layer which is in contact with the metallic electrode are the main reasons of the low dielectric permittivity in this region.

As there are fundamental similarities in structure and lattice matching between (Ba,Sr)TiO$_3$ and high-T$_C$ superconducting (Nd,Y)Ba$_2$Cu$_3$O$_{7-\delta}$ (N,Y)BCO and oxide metallic SrRuO$_3$ (SRO) films [7,8] the latter may be good electrode materials to decrease the negative influence of the interface zones to the ferroelectric layer.

A comparative study of the structure and dielectric properties of epitaxial SrTiO$_3$ and Ba$_{0.8}$Sr$_{0.2}$TiO$_3$ layers, combined with high-T$_C$ superconducting, oxide metallic as well as ordinary metallic electrodes, have been undertaken in order to understand the mechanisms responsible for the degradation of the ferroelectric thin layer in multilayer heterostructures.

EXPERIMENTAL TECHNIQUES

Laser ablation (KrF, λ=248 nm, τ=30 ns) was used to grow (200 nm)SrRuO$_3$/(800 nm)SrTiO$_3$/(200 nm)SrRuO$_3$ (SRO/STO/SRO) and (200 nm)SrRuO$_3$/(800 nm)Ba$_{0.8}$Sr$_{0.2}$TiO$_3$/(200 nm)SrRuO$_3$ (SRO/BSTO/SRO) heterostructures on the surface of (100) polished (LaAlO$_3$)$_{0.3}$+(Sr$_2$AlTaO$_6$)$_{0.7}$ (LSATO) plates (5x5x1 mm^3). Structure and surface morphology of the epitaxial layers in YBCO/STO/YBCO and NBCO/STO/NBCO grown by laser ablation were investigated in refs. [7,9,10]. Polycrystalline SRO, STO and BSTO targets, prepared by standard ceramic technology, were ablated with an energy density of 1.8 J/cm^2. The substrate temperature T$_S$ and oxygen pressure P$_O$ during ablation were 790°C and 0.4 mbar, respectively. Top silver contacts were thermally evaporated through a mask on the surface of the STO layer in the (100)STO $\|$ (001)YBCO heterostructure.

Structure and phase composition of the layers were determined by x-ray Philips X´pert MRD ($\omega/2\theta$-, ϕ-scans and rocking curves). The out-of-plane lattice parameter for the STO and BSTO layers was determined from x-ray $\omega/2\theta$ scans with incident and scattered x-ray beams in the plane normal to the substrate surface. To determine the in-plane lattice parameter of a ferroelectric layer, the substrate with its deposited trilayer was fixed in such a way that (110)BSTO was normal to the plane containing the incident and scattered x-ray beams. An Atomic Force Microscope, AFM (NanoScope-IIIa), was used to study the surface morphology of the layers in the heterostructures.

The superconducting transition temperature T$_C$ for the YBCO and NBCO films, combined with the ferroelectric layer in a multilayer, was determined from temperature dependencies of the resistivity, ρ, and the magnetic susceptibility, χ. The critical current density jc was calculated from I-V curves measured on microbridges (50 μm length, 8 μm width).

Photolithography and Ar-ion milling (0.2 mA, 500V) were used to form the top electrodes, (100x100 μm²), and openings in the ferroelectric layer to contact the common bottom electrode.

The capacitance, C, loss factor, tanδ, and conductance, G, for the parallel plate capacitors were measured by an hp 4263A LCR meter (f=1-100 kHz). C and tanδ were measured with and without DC bias voltage V_b applied to the electrodes. V_b was marked as positive when "+" was applied to the top electrode. Electric field, E=V_b/d, was marked in the same way as V_b; d is the thickness of the ferroelectric layer.

RESULTS

The following epitaxial orientation relationships were determined by x-ray (ω/2θ- and φ-scans) for the STO and BSTO layers grown on SRO electrodes - (100)[010]STO∥(100)[010]SRO∥(100)[010]LSATO and (100)[010]BSTO∥(100)[010]SRO∥(100)[010]LSATO, see Fig.1a and 1b, (a cubic unit cell was used for the BSTO during identification of the peaks in the x-ray scans). The top NBCO film in the NBCO/STO/NBCO heterostructure was well c-axis oriented (c-axis normal to substrate plane), but a-axis oriented (c-axis parallel to substrate plane) particles were detected by x-ray and AFM for the top YBCO one. Because of the tetragonal distortion of the BSTO unit cell, the x-ray peaks in the ω/2θ x-ray scans (measured at 300K) for the SRO/BSTO/SRO trilayer were non symmetric and essentially broader than those of STO in SRO/STO/SRO. The full width at half maximum

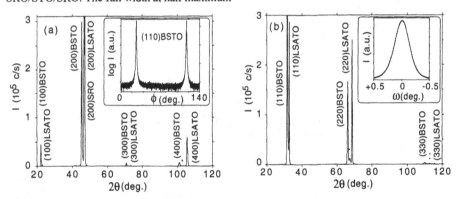

Fig.1. a) An x-ray (ω/2θ, CuK$_\alpha$) scan for a (100)SRO∥(100)BSTO∥(100)SRO heterostructure grown on (100)LSATO. The substrate plane was normal to the plane formed by incident and scattering x-ray beams during the measurement. * - denotes the (400)SRO x-ray peak. The insert shows a (φ, CuK$_\alpha$) scan for the (110)BSTO x-ray reflection. b) The x-ray (ω/2θ, CuK$_\alpha$) scan measured for the same heterostructure as in (a) when the plane formed by incident and scattering x-ray beams was normal to the (110)BSTO. •-and : - (220) and (330)SRO x-ray peaks. The insert plots the rocking curve for the (200)BSTO x-ray reflection.

(FWHM=0.4 deg.) of a rocking curve measured for the (200)BSTO x-ray reflection for the SRO/BSTO/SRO heterostructure was several (3-8) times larger than that for the (200)STO reflection of the STO/(N,Y)BCO and STO/SRO layers. This is, to a large extent, an indication of the tetragonal distortion of the BSTO unit cell at 300 K. The out-of plane lattice parameters measured by x-ray for the STO and BSTO layers were 3.909 Å and 3.993 Å, respectively. The measured in-plane lattice parameters for the BSTO (3.994 Å) and STO (3.909 Å) were roughly the same as the out-of-plane ones. The small difference between out-of-plane and in-plane lattice parameters for the BSTO layer implies that there are roughly equal fractions of grains oriented with the tetragonal c-axis normal to the substrate surface as along the latter.

The STO films grown on the surface of (100)SRO or (001)(N,Y)BCO films were pin-hole free, see Fig.2a. Grooves decorate grain boundaries in the ferroelectric layer grown on a (N,Y)BCO bottom electrode [11]. Well developed a-axis particles and spirals of growth were detected at the surface of the top YBCO in the YBCO/STO/YBCO trilayer [8,10]. Because of hillock like structures and pinholes, the surface of the top NBCO film in the NBCO/STO/NBCO heterostructure was rough, see Fig.2b. A high density of pinholes, detected by AFM, at the surface of the NBCO film, in the NBCO/STO/NBCO heterostructure, is an indication of nucleation of microinclusions of a second phase at the surface of the growing superconducting film.

a)   b)

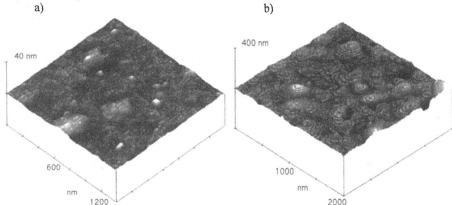

Fig.2. a) The AFM image of the surface of a (800 nm) thick (100)STO film grown on (001)NBCO. b) The AFM image of the surface of a (240 nm) thick (001)NBCO film grown on (100)STO∥(001)NBCO. Pinholes and growth steps (1.2 nm height) are well detectable at the surface of the NBCO film.

About the same $\varepsilon/\varepsilon_0$ was measured, at 300K, for STO layers sandwiched between YBCO, NBCO, or SRO electrodes, see Fig.3a (ε_0 dielectric constant of vacuum). At T<150K, ε for the STO layer between SRO electrodes increases much faster than that for STO inserted between high-T_C superconducting electrodes, see Fig.3a. The $\varepsilon(T)$ dependence for the STO layer became weaker when the top YBCO film in the YBCO/STO/YBCO heterostructure was replaced by Ag one.

A pronounced maximum was observed at T≈335 K in the measured $\varepsilon(T)$ curve for the BSTO layer during cooling (410-10 K) of the SRO/BSTO/SRO heterostructure, Fig.3b. The dielectric permittivity for the BSTO layer was suppressed (up to 10%) and maximum was shifted about 10 K towards higher temperature if the curve was measured during heating. The maximum in $\varepsilon(T)$ became broader and was shifted to higher temperature if a (-2 V) bias voltage was applied to the SRO electrodes, see Fig.3b.

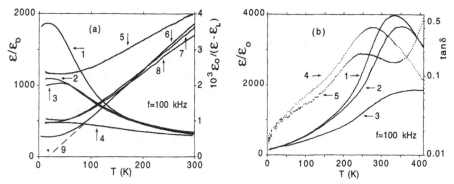

Fig.3. a) The temperature dependencies of $\varepsilon/\varepsilon_0$ (1-4) and $\varepsilon_0/(\varepsilon-\varepsilon_L)$ (5-8) for the STO layer in the (100)SRO∥(100)STO∥(100)SRO (1,6), (001)NBCO∥(100)STO∥(001)NBCO (2,7), (001)YBCO∥(100)STO∥(001)YBCO (3,8) and Ag/(100)STO∥(001)YBCO (4,5) heterostructures. 9 - tangent of curve (6) in the range 80-300 K (dashed line). T_{Curie}=21 K was determined as the temperature corresponding to the point where the tangent of curve (6) crosses the T-axis, shown by ♥. f=100 kHz. b) The temperature dependencies of $\varepsilon/\varepsilon_0$ (1,2,3) and tanδ (4,5) for the BSTO layer in the (100)SRO∥(100)BSTO∥(100)SRO heterostructure. The curves (1,3,4,5) were measured during cooling of the sample, but (2) during heating. Curves (3,5) were measured with V_b= -2 volts applied between the SRO electrodes.

The measured $\varepsilon(E)$ curves for the STO layer in the SRO/STO/SRO heterostructure were roughly symmetric relative to the point E=0, see Fig.4a. The maxima in the $\varepsilon(E)$ dependencies for the STO layer combined with YBCO or NBCO electrodes were shifted by 4×10^5 V/m and 6×10^5 V/m, respectively, towards the negative side of the origin.

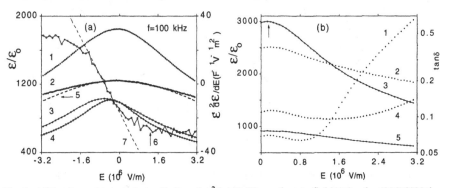

Fig.4. a) The dependence of $\varepsilon/\varepsilon_0$ (1-5) and $\varepsilon^{-2}d\varepsilon/dE$ (6) on electric field E for the (100)STO layer in (100)SRO∥(100)STO∥(100)SRO (1,2,5), (001)NBCO∥(100)STO∥(001)NBCO (3,6) and (001)YBCO∥(100)STO∥(001)YBCO (4). (7) denotes the tangent of curve (6) at low E. 1- measured at T=20K, 2-at 77K, 3 and 4 - at 63 K, f=100 kHz. The non-linearity parameter ξ=10x10^9 $F^{-3}m^5V^{-2}$ was used to extrapolate the $\varepsilon/\varepsilon_0$(E,T=77K) curve (5). b) The dependence of tanδ (1,2,4) and $\varepsilon/\varepsilon_0$ (3,5) on electric field E for the BSTO layer in the SRO/BSTO/SRO heterostructure. 3 -measured at T=290K and f=100 kHz, 1,2,4,5- at 175 K, 2,5-100 kHz, 4-10 kHz, 1-1 kHz. The arrow shows the position of the maximum in curve (3).

237

The shift of the maximum increased further (≈2.4x10⁶ V/m) if the top YBCO film in YBCO/STO/YBCO was replaced by Ag layer. The shift of the maximum is an indication of an internal electric field induced in the STO layer because of the difference in electronic properties of the top and bottom ferroelectric/electrode interfaces in the heterostructure. The strongest electric field response of ε for the BSTO layer in the SRO/BSTO/SRO heterostructure was observed at a temperature near the maximum in the $\varepsilon(T)$ dependence, see Fig.4a.

The smallest tanδ was measured for the STO layers in SRO/STO/SRO heterostructures and the largest ones for STO combined with YBCO or NBCO electrodes, see Fig.5. The tanδ measured for the STO layer inserted between the bottom YBCO and top Ag electrodes was hardly temperature dependent. There was a maximum at T≈270 K in the tan$\delta(T)$ curve for the BSTO layer when the loss factor was measured from low to high temperature, see Fig.3b. Like in the case of the $\varepsilon(T)$ curve, the maximum in the tan$\delta(T)$ dependence was shifted towards the low temperature side when tanδ was measured from 410 K down to 10 K. The maximum in the tan$\delta(T)$ curve of BSTO was shifted downwards if -2 V bias voltage was applied to the SRO electrodes, see Fig.3b. At T≥175 K, tanδ for the BSTO layer was very dependent on frequency, see Fig.4b.

At T<200K, the conductance of the STO layer in a (N,Y)BCO/STO/(N,Y)BCO heterostructure was hardly temperature dependent but increased rapidly with temperature at T≈ 250K), see Fig.5b. The T_C's = 86-88K and j_C's = (0.5-1.0)x10⁶ A/cm² measured for the top high-T_C superconducting film in the (N,Y)BCO/STO/(N,Y)BCO heterostructures were suppressed, as compared with those for (N,Y)BCO films grown on a single crystal STO substrate. The T_C's determined from resistive measurements agreed well with those determined from the temperature dependence of the magnetic susceptibility χ, see Fig.5b for the latter.

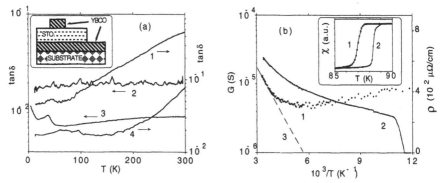

Fig.5. a) The temperature dependence of tanδ for the STO layer in (001)YBCO∥(100)STO∥ (001)YBCO (1), Ag/(100)STO∥(001)YBCO (2), (100)SRO∥(100)STO∥(100)SRO (3) and (001)NBCO∥(100)STO∥(001)NBCO (4) heterostructures. A sketch of the trilayer capacitance heterostructure used for measurement of ε, tanδ, and G for the ferroelectric layer is shown in the insert. f=100 kHz. b) The temperature dependencies of G (1) for the STO layer and resistivity ρ (2) for the top NBCO film in the NBCO/STO/NBCO heterostructure. (3)-tangent for curve 1 at T close to 300K. The insert plots the temperature dependence of the magnetic susceptibility χ for NBCO/STO/NBCO (1) and YBCO/STO/YBCO (2) heterostructures.

DISCUSSION

The investigated STO layers induced in the multilayers were grown epitaxially at the surface of the (N,Y)BCO and SRO electrodes. Because of tetragonal distortion of the unit cell, BSTO layer combined with SRO electrodes posses polycrystalline structure with (001) or (100) oriented grains.

The STO layer grown on the surface of (001)YBCO or (001)NBCO posses, unfortunately, granular structure [10,11]. The surface of the STO and BSTO layers grown on YBCO or NBCO is rough because of grooves which decorate the low angle grain boundaries [17]. Possible reasons for distortion of structure of the STO layer grown on YBCO have been discussed in [8].

STO is an incipient ferroelectric and remnant polarisation is zero up to a low temperatures [12]. $Ba_{1-x}Sr_xTiO_3$ solid solutions posses remnant polarisation at $T<T_{Curie}$. Curie temperature T_{Curie} for the bulk samples of the BSTO decrease about liner when x increase in the range 0-1.0. The T_{Curie} for a bulk ceramic sample of the BSTO is close to 340K [12].

Temperature Dependence

As follow from data presented in the Fig.3a, the strongest dependence of ε upon temperature, $\varepsilon(15K)/\varepsilon(300K)\approx6$, was measured for the STO layer sandwiched by two SRO electrodes. The ratio for the STO layer in the YBCO/STO/YBCO and NBCO/STO/NBCO heterostructures was roughly two times smaller. The ε and $d\varepsilon/dT$ decreased further if the top superconducting electrode in the YBCO/STO/YBCO was replaced by metallic one (Ag). The maximum in the $\varepsilon(T)$ curve for the STO layer in the SRO/STO/SRO heterostructure was observed at $T\approx15K$. The maximum was shifted on 10-30K to a side of high temperatures in the case of (N,Y)BCO/STO/(N,Y)BCO heterostructures. The maximum in the $\varepsilon(T)$ dependence for Ag/STO/YBCO heterostructure was broad and the largest ε was measured at T=60K.

The measured lower ε for the STO layer in the Ag/STO/YBCO capacitance heterostructure, as compared with that for the STO in the YBCO/STO/YBCO one, is due to a small parasitic capacitance of the interface zone Ag/STO connected in series with that induced by ferroelectric layer volume. Dielectric saturation in a strong electric field, which arise because of difference in work function for Ag and electron affinity for STO, and high density of oxygen vacancies near the interface zone are responsible for decrease of ε in the region of the STO layer adjacent to contact ferroelectric/metal. When both electrodes in the trilayer heterostructure are made from the cuprate superconductors or oxide metal, effect of interface zones on the measured at $T\approx300K$ dielectric permittivity of the STO layer is small. As result of that, the measured at room temperature ε for the STO layer combined with YBCO, NBCO or SRO electrodes are about the same like $\varepsilon[100]$ for the bulk single domain single crystal [16].

At T>150K the measured $\varepsilon(T)$ curves for the STO layer in the YBCO/STO/YBCO and NBCO/STO/NBCO heterostructures fit well that for the SRO/STO/SRO one. At 15K<T<120K $d\varepsilon/dT$ for the STO layer combined with SRO electrodes is essentially larger than that for high-T_C superconducting electrodes. SRO and STO are chemically compatible up to high temperature (800°C [18]). Better microstructure of the STO layer [8] and sharper STO/SRO interface zones in the SRO/STO/SRO heterostructure, as compared with those for the STO in (N,Y)BCO/STO/(N,Y)BCO, are responsible for the observed enhancement of the temperature dependence of the ε for the ferroelectric layer combined with strontium rutenate electrodes.

An experimental $\varepsilon(T)$ curve for the STO single crystal was well extrapolated in a wide temperature range by modified Curie-Weiss relation [19]

$$\varepsilon = \varepsilon_L + \varepsilon_0 C_0/(T - T_{Curie}) \qquad (1)$$

where $C_0 = 7.1 \times 10^4$ K and $\varepsilon_L/\varepsilon_0 = 44$ - dielectric constant extrapolated to infinite temperature. The T_{Curie} for the bulk single domain STO single crystal is in the range 30-40K [16,19].

In the temperature range 80-300K, the measured $\varepsilon(T)$ for the STO layer in the SRO/STO/SRO heterostructure was well extrapolated by relation (1) with $C_0 = 7.5 \times 10^4$ K and $T_{Curie} = 21$K, see Fig.3a. As follow from data presented at Fig.3a, the temperature interval where ε^{-1} for the STO layer in the YBCO/STO/YBCO, NBCO/STO/NBCO and Ag/STO/YBCO heterostructures increase about linearly with a temperature is smaller and shifted to a higher temperatures, as compared with that for the SRO/STO/SRO heterostructure. The T_{Curie}'s determined from the slope of $\varepsilon^{-1}(T)$ dependence for the STO layer in the YBCO/STO/YBCO, NBCO/STO/NBCO or Ag/STO/YBCO heterostructures are negative. The C_0's for the heterostructures with both high-T_C superconducting electrodes were in the range $(8.6-9.0) \times 10^4$ K. The C_0 increased up to 12.5×10^4 K when the top YBCO electrode in trilayer heterostructure YBCO/STO/YBCO was replaced by Ag one. The calculated larger C_0's for the STO layer in the heterostructures with high-T_C superconducting electrodes, as compared with that for SRO/STO/SRO one, is an other indication of the degradation of the microstructure of the STO layer and distortion of the stoichiometry of the interface zones.

Position of the maximum (T≈335 K) in the $\varepsilon(T)$ dependence for a BSTO layer is well matched with that for a bulk samples [12]. Histeresis was observed in the $\varepsilon(T)$ dependence for a bulk single crystal of $BaTiO_3$ [21] at a temperatures near phase transition point.

Electric Field Response

The strongest response of the ε for the STO layer on electric field was observed at the temperatures close to the maximum in the $\varepsilon(T)$ curve, see Fig.4a. The $\varepsilon(E)$ curve for the STO layer sandwiched between SRO electrodes was about symmetric relative to the point E=0. Observed shift of the maximum in the $\varepsilon(E)$ curves measured for the STO sandwiched between high-T_C superconducting electrodes, see Fig.4a, is an indication of internal electric field induced by difference in electronic parameters of the top and bottom superconductor/ferroelectric interfaces. The nucleation process for the STO layer on (N,Y)BCO film is strongly influenced by distortion of the stoichiometry of the adsorbed phase as result of interdiffusion exchange of the components with superconducting film. An observed decrease of the T_c's and j_c's for the top (N,Y)BCO in the (N,Y)BCO/STO/(N,Y)BCO heterostructures as well as AFM data on the surface morphology indicate on degradation of the microstructure of the superconducting film grown on the STO intermediate layer. Because of strong internal electric field in the region of Ag/STO interface, response of ε for the STO layer in the Ag/STO/YBCO heterostructure on external electric field was weak and essentially dependent upon polarity of an applied V_b [10]. In distinguish from the case of the YBCO/STO/YBCO and NBCO/STO/NBCO heterostructures, ε for the STO sandwiched between SRO electrodes, at T>125 K, was independent on electric field ($E < 3.2 \times 10^6$ V/m).

Devonshire [13] has shown that free energy, F, of ferroelectric material may be expressed in powers of the electrical polarisation, P, and strain, ζ. When ζ is zero, the F(P) dependence follow relation [14]

$$F\,(T,P) = F(T,0) + (1/2)\eta(T)P^2 + (1/4)\xi P^4 + (1/6)\,\gamma P^6 \ldots \quad (2)$$

where $\eta(T) = \varepsilon_0^{-1} C_1^{-1}(T-T_0)$, C_1 and T_0 - Curie constant and Weiss temperature respectively, F(T,0)- free energy of unpolarised state, terms of the order P^8 and higher are neglected. Since

$$dF/dP\;=\;E \qquad\qquad (3)$$

and

$$d^2F/d^2P = \varepsilon^{-1} \qquad\qquad (4)$$

equation (2) may be transformed [15] as

$$\varepsilon^{-1} = \eta(T) + 3\eta^{-2}(T)\xi E^2 + \ldots \quad (5)$$

At $T > T_{Curie}$ the substance is in paraelectric state and from (5), at E=0, follow Curie-Weiss relation

$$\varepsilon/\varepsilon_0 = C_1/(T-T_0) \qquad\qquad (6)$$

The difference between T_0 and T_{Curie} for $Ba_{1-X}Sr_XTiO_3$ alloys is small [12].

The measured $\varepsilon(E)$ dependencies for the STO layers fitted well those extrapolated by relation (5), see curves 2 and 5 in the Fig.4a. The parameter $\eta(T)$ for the STO layer may be determined from $\varepsilon(T)$ at E = 0 and the non-linearity parameter , ξ, from the $\varepsilon(E)$ dependence at small E. In accordance with (5)

$$\varepsilon^{-2}d\varepsilon/dE \;=\; -\,6\eta^{-2}(T)\xi E \qquad\qquad (7)$$

The $\varepsilon^{-2}d\varepsilon/dE(E)$ dependence for the STO layer in the NBCO/STO/NBCO heterostructure and tangent for the curve at low E are shown on Fig.4a. The ξ's determined for the STO [9] and BSTO [20] layers from the slope of the $\varepsilon^{-2}d\varepsilon/dE(T)$ curve at low E were in the range $(6-10) \times 10^9$ $F^{-3}V^5m^{-2}$. The ξ for the STO and BSTO layers was hardly dependent on temperature. The estimated ξ for the STO layer in the trilayers agree well with that obtained from Raman spectroscopy data on shift of the soft mode frequency in electric field for the bulk single crystal of the strontium titanate [15].

Devonshire [13] has shown, that, when $\zeta \neq 0$, right side of the relation (2) have to include electromechanical terms which lead to piezoelectric effect and electrostriction. Because of mechanical effect, free energy of ferroelectric is decreased by a term proportional to the square of the elastic strain [22]. This strain is proportional to the square of polarisation, so electromechanical term in the expression for free energy of the ferroelectrics have to be proportional to P^4. The electromechanical coupling does not alter Curie-Weiss relation, but influence T_{Curie} [13,22]. The shift of Curie temperature ΔT_{Curie} in electric field may be expressed as [14]

$$\Delta T_{Curie} = C_2\,E \qquad\qquad (8)$$

C_2 - is dependent on coefficients η, ξ,γ for the powers of the polarisation in expansion of the free energy. The $C_2 = 1.9\times10^{-3}$ [14], for E(V/cm), was determined from plots of electrical polarisation P versus applied electric field E, measured at different temperatures.

An observed essential shift ($\Delta T_{Max}\approx45$ K) of the maximum in the ϵ(T) curve for the BSTO layer when -2V bias voltage was applied to the SRO electrodes relative to that in the case of $V_b=0$ agrees well with an estimation made on the base of the relation (8). External electric field during measurements of the curve 3 at Fig.3b was $V_b/d=25\times10^3$ (V/cm). According to (8), the shift of the Curie point for the BSTO layer in electric field E = 25 kV/cm have to be 48 K. From the relative shift of maximum in the curve 1 and 2 in the Fig.5 it is possible to estimate an effective electric field ($E_{ef}\approx8$ kV/cm, T$\approx$300 K) induced in the BSTO layer during cooling at T<T_{Curie}. It is internal electric field which responsible for difference in $\epsilon_{Max}/\epsilon_0$ for the curve 1 and 2 at the Fig.3b.

Losses

The measured tanδ and conductance G (f =1-100 kHz) for the STO layers in the trilayer capacitance heterostructures obey the relation G=Cωtanδ, where the angular frequency $\omega=2\pi$f. Because of relatively small difference in ϵ, measured at T>150 K, see Fig.3a, for the STO layers in the (Y,N)BCO/STO/(N,Y)BCO and SRO/STO/SRO heterostructures, an observed much smaller loss factor for the capacitors with SRO electrodes, see Fig.5a, have to be attributed in-part to a higher resistivity of the corresponding ferroelectric layer. That is well matched with an available data on [8]crystallinity of the STO layer in the heterostructures. The measured rocking curves for the (200)STO x-ray reflection for the YBCO/STO/YBCO heterostructure was several times broader than that for SRO/STO/SRO one. An effective conductance for the Ag/STO/YBCO trilayer capacitor is determined, to a large extent, by the height of Shottky barrier formed at the Ag/STO interface.

At T=300 K, the measured tanδ for the STO layer combined with high-T_C superconducting electrode about liner decreased when frequency increased in the range 1-100 kHz. That is clear indication of an essential contribution of the DC conductance of the ferroelectric layer to loss factor [23]. At T>200K, temperature and electric field dependence of G for the STO layer in the (Y,N)BCO/STO/(Y,N)BCO heterostructures may be well explained in the frameworks of the Poole-Frenkel emission model [24]. An experimental G(T,E) curves may be well extrapolated by the relation [20]

$$\ln G \sim -(E_D-\beta E^{1/2})/rkT \qquad (9)$$

where E_D - the energy difference between impurity (emission) levels and the conduction (valence) band, $\beta = (e^3/\pi\epsilon_r{}^{hf}\epsilon_0)^{1/2}$, $\epsilon_r{}^{hf}$ - relative dielectric constant at high frequency ($\epsilon_r{}^{hf} =n^2$, n- refractive index), e- charge of electron, r- depends on the charge carrier density and the density of donor and acceptor sites and may be in the range 1-2 [23]. As follow from data presented at Fig.5b, lnG for the STO layer in the NBCO/STO/NBCO heterostructure depends linearly upon 1/T within a temperature range close to 300 K. The E_D determined from the slope of the lnG(1/T) dependence is 0.09 eV (T$\approx$300 K, r=1). The measured values of E_D for the epitaxial STO layer agree well with the ionisation energy of donor centers (oxygen vacancies) in reduced STO bulk samples.

The observed at T$\approx$270K maximum in the tanδ(T) curve for the BSTO layer, see curve 4 in the Fig.3b, is due to non elastic interaction of electromagnetic wave with ferroelectric domain walls [12]. Alignment of the ferroelectric domains in external electric field promote decrease of the tanδ, see curve 5 at the same figure, for the BSTO layer at T<300 K. Beside electric field, loss factor for the BSTO layer in the SRO/BSTO/SRO heterostructure was essentially dependent on frequency f, see Fig.4b. Observed decrease of the tanδ (at E=0) when f increased in the range 1-100 kHz is due to decrease of ferroelectric losses. The largest contribution of the ferroelectric losses to tanδ have to be observed when $\omega\tau_D \approx 1$ [25], where τ_D - characteristic time of relaxation of the ferroelectric domain wall. When frequency decrease in the range 100-1 kHz, difference 1- $\omega\tau_D$ for the BSTO layer increase. An observed pronounced increase of the loss factor for the BSTO layer with electric field at f=1 kHz, see Fig.4b, indicate essential contribution of the DC conductance to the tanδ. Because of distortion of stoichiometry of the intergrain zones, DC conductance along grain boundaries in the BSTO layer as well as in STO one combined with (N,Y)BCO electrodes have to be enhanced, as compared with that for the grain volume.

CONCLUSIONS

The dielectric permittivity of the STO layer between electrodes, in a heterostructure, depends to a large extent on the microstructure and parameters of the interface zones. Largest values of ε were measured for the STO and BSTO layers sandwiched between SRO electrodes. The strongest electric field response of the ε for epitaxial STO in a multilayer heterostructure was observed at temperatures close to the maximum in the ε(T) dependence. The measured sharp maximum and pronounced hysteresis in the ε(T) dependence, at a temperatures close to the Curie point, indicate a high quality of the microstructure of the BSTO layer and sharp interfaces ferroelectric/electrodes. The conductance of the STO layer is enhanced by a high density of oxygen vacancies. Because of a large DC conductance at T>200K, an STO layer between superconducting electrodes has a much higher loss factor, tanδ, than corresponding layer in the SRO/STO/SRO heterostructure.

ACKNOWLEDGMENTS

The research was performed within the framework of the scientific co-operation between the Royal Swedish and the Russian Academies of Sciences. It was supported, in part by the Swedish Institute, TFR, the Materials Consortium on Superconductivity and the project 98-02-18222 of the Russian Foundation of Basic Research.

REFERENCES

1. I.S. Gergis, J.T. Cheung, T.H. Trinh, E.A. Sovero and P.H. Kobrin, Appl.Phys.Lett. **60**, p. 2026, (1992).

2. K.M. Johnson, J.Appl.Phys. **33**, p. 2826, (1962).

3. D.M. Gill, B.A. Block, C.W. Conrad, B.W. Wessels, S.T. Ho, Appl.Phys.Lett. **69**, p. 2968, (1996).

4. C.S. Hwang, Mater.Sci.Eng. B **56**, p. 178, (1998).

5. C. Basceri, S.K. Streiffer, A.I. Kingon, and R. Waser, J.Appl.Phys. **82**, p. 2497, (1997).

6. J.J. Lee, C.L. Thio, and S.B. Desu, J.Appl.Phys. **78**, p. 7073, (1995).

7. Yu.A. Boikov and T. Claeson, J.Appl.Phys. **81**, p. 3232, (1997).

8. Yu.A. Boikov and T. Claeson, Physica C, Submitted for publication.

9. Yu.A. Boikov, V.A. Danilov , T. Claeson, and D.Erts, Phys.Solid State **41**, p. 355, (1999).

10. Yu.A. Boikov, Z.G. Ivanov, A.N. Kiselev, E. Olsson and T. Claeson, J.Appl.Phys.**78**, p. 4591, (1995).

11. M.E. Tidjani, R. Gronsky, J.J. Kingston, F.C. Wellstood and J. Clarke, Appl.Phys.Lett. **58**, p.765, (1991).

12. A.D. Hilton and B.W. Ricketts, J.Phys.D: Appl.Phys. **29**, p.1321, (1996).

13. A.F. Devonshire, Phil.Mag. **40**, p. 1040, (1949).

14. W.J. Merz, Phys.Rev. **91**, p. 513, (1953).

15. J.M. Worlock and P.A. Fleury, Phys.Rev.Lett. **19**, p. 1176, (1967).

16. R.C. Neville, B. Hoeneisen and C.A. Mead, J.Appl.Phys. **43**, p. 2124, (1972).

17. Yu.A. Boikov, V.A. Danilov, E. Carlsson, D. Erts and T. Claeson, Physica B **262**, p. 104, (1999).

18. S.Y. Hou, J. Kwo, R.K.Watts, J.-Y. Cheng, and D.K. Fork, Appl.Phys.Lett. **67**, p. 1387, (1995).

19. G. Rupprecht and R.O. Bell, Phys.Rev. **135**, p. A748, (1964).

20. Yu.A. Boikov and T. Claeson, Supercond.Sci.Technol. **12**, p. 654, (1999).

21. W.J. Merz, Phys.Rew.**76**, p.1221, (1949).

22. J.C. Slater, Phys.Rew.**78**, p. 748, (1950).

23. G. Rupprecht and R.O. Bell, Phys.Rev. **125**, p. 1915,(1962).

24. J.R. Yeargan and H.L. Taylor, J.Appl.Phys. **39**, p. 5600, (1968).

25. K. Fossheim and B. Berre, Phys.Rev. B **5** , p. 3292, (1972).

SOFT-MODE PHONONS in SrTiO₃ THIN FILMS STUDIED by FAR-INFRARED ELLIPSOMETRY and RAMAN SCATTERING

A. A. SIRENKO *, C. BERNHARD **, A. GOLNIK **, I. A. AKIMOV *†,
A. M. CLARK *, J.-H. HAO *, and X. X. XI *
* Department of Physics, the Pennsylvania State University, University Park, PA 16802,
sirenko@phys.psu.edu
** Max-Planck-Institut für Festkörperforschung, D-70569 Stuttgart, Germany
† A. F. Ioffe Physico-Technical Institute, St. Petersburg 194021, Russia

ABSTRACT

We report the experimental studies of the vibrational spectra of SrTiO₃ films with the thickness of 1 μm grown by pulsed laser deposition. Fourier-transform infrared ellipsometry between 30 and 700 cm^{-1} and electric field-induced Raman scattering have been utilized for investigation of the phonon behavior. These results can be used for comparison with the low-frequency measurements of the static dielectric constant. In the films, the soft mode reveals hardening compared to that in bulk crystals. This observation is in agreement with the Lyddane-Sachs-Teller formalism.

INTRODUCTION

Ferroelectric materials possess an electric polarization and the direction of this polarization can be reversed by an external electric field. At the temperature above the paraelectric-to-ferroelectric phase transition, these materials are characterized by extremely high values of the dielectric constant (of the order of 30000). Application of a moderate external electric field, which is usually much lower than the breakdown value, results in a dramatic change, or tuning, of the dielectric constant. These properties of ferroelectrics became very promising for practical application after development of modern thin film deposition techniques, which satisfy the requirements of the compact size and compatibility with the planar technology. Recently, ferroelectric materials have been widely utilized in microelectronics as a base for capacitor structures, high density random access memories (DRAMs), nonvolatile ferroelectric memories (FRAMs), hydrogen sensors, optical switches, filters, and waveguides (see Ref. [1] for more detail).

One of the incipient ferroelectrics, which have recently renewed great interest in their fundamental properties, is SrTiO₃ (STO). Several growth techniques, such as pulsed laser deposition (PLD) [2,3], rf magnetron sputtering [4], and molecular beam epitaxy [5] have been used to produce high quality STO thin films on different substrates with the thickness ranging from a few monolayers to a few microns. Although the crystalline quality of available epitaxial STO films is high, which is usually reported based on the result of x-ray diffraction, it has been demonstrated that at low temperatures STO films have much lower static dielectric constant ε_0 and higher losses compared to that of STO single crystals. The typical low temperature values of ε_0 vary from about 10^3 [2,3] to 5×10^3 [4] depending on the thin film growth technique. At present, the cause of this difference, i.e. the so-called "size effect", is not well understood. One of the models which have been offered to explain this effect, proposes the existence of a "dead layer" with a low dielectric constant located at the interfaces of the films [6,7]. In another approach, the size effect is explained as a consequence of the structural defects in the films induced by impurities and oxygen vacancies or strain due to the lattice mismatch

245

with the substrate [8]. Whether the poorer film properties are related to the bulk film material or are mostly determined by the interface quality and local-field effects has remained an open question. The answer is considered to be the key issue for optimization of the dielectric properties of STO and a number of other ferroelectric thin films. As it will be shown further, the understanding of this problem is impossible without detailed experimental studies of the vibrational spectra in these materials.

SOFT MODES in SrTiO₃

The phonon behavior is of the central importance for the theory of ferroelectricity. When the temperature of the ferroelectric phase transition T_c is approached, the zone-center frequency of the so called "soft mode", which is the lowest frequency transverse optical phonon, falls to zero:

$$\omega_{TO1}(T) = const \times |T - T_c|^{1/2}. \tag{1}$$

For ferroelectrics this instability in the spectrum of the optical phonons was predicted by Fröhlich [9] and later was theoretically studied using lattice dynamic calculations [10] and a microscopic approach [11]. The Lyddane-Sachs-Teller (LST) relationship for a crystal with N infrared-active optical modes ($N = 3$ for STO) connects the frequency of the soft mode to the static ε_0 and high frequency ε_∞ dielectric constants:

$$\varepsilon_0 = \varepsilon_\infty \prod_{j=1}^{N} \frac{\omega_{LOj}^2}{\omega_{TOj}^2}, \tag{2}$$

where ω_{LOj} and ω_{TOj} are eigenfrequencies of the longitudinal and transverse optical phonons, respectively. It is generally found that the eigenfrequencies of the phonons other than the soft mode exhibit no sizable variation with temperature and, hence, the temperature variation of the soft mode frequency and the static dielectric constant are connected as follows:

$$\varepsilon_0(T) \propto \omega^{-2}_{TO1}(T), \tag{3}$$

which is consistent with the Curie-Weiss law for the temperature dependence of the static dielectric constant:

$$\varepsilon_0(T) = const \times |T - T_c|^{-1}. \tag{4}$$

In bulk STO crystals, the experimental data for the soft modes and static dielectric constant obey Eqs.(1) and (4) above 50 K (with T_c=32 K) but deviate from the Curie-Weiss law at lower temperatures [12]. This deviation was theoretically understood in terms of zero-point lattice fluctuations of the Ti atom with respect to the oxygen octahedral [13] and, consequently, hardening of the soft modes, which never reach zero frequency, was attributed to their strong coupling with transverse acoustic phonons. As a result, at normal conditions STO does not reveal a ferroelectric phase transition even at very low temperatures. However, a strong variation of the soft mode frequency from 90 to 13 cm^{-1} with decreasing temperature from room to the liquid helium value, respectively, is consistent with dramatic increase of the static dielectric function at low temperature ($\varepsilon_0 \cong 24000$ at T= 4 K). Before proceeding to the experimental methods of the soft mode studies in STO thin films we will briefly describe basic properties of the optical phonons.

In cubic bulk STO crystals with a perovskite structure every atom in the primitive cell is at the center of inversion and, consequently, the optical phonons are of the odd symmetry. Three phonon branches of the F_{1u} symmetry are infrared active and one branch of the F_{2u} symmetry (so called "silent mode") is neither infrared nor Raman active. The crystal field splits phonons into longitudinal and doubly degenerated transverse modes, which are usually labeled in literature according to an increase of their frequency: ω_{LOj} and ω_{TOj} with $j = 1,2,3,4$. The soft mode is the TO_1 optical phonon. At low temperatures, below the structural cubic-to-tetragonal phase transition, which takes place in the bulk at about T_a=105 K, the infrared-active optical phonons split into modes of the A_{2u} and E_u symmetry remaining infrared active. The A_{2u} and the doubly degenerate E_u phonons are polarized along and perpendicular to the tetragonal axis, respectively. The splitting of the optical phonons in the tetragonal phase imply an anisotropy of the dielectric function tensor. In contrast to the cubic phase, where the dielectric function tensor is degenerate into a scalar, in the tetragonal phase it has two different components, $\varepsilon_{0\parallel}$ and $\varepsilon_{0\perp}$, for electric field direction parallel and perpendicular to the tetragonal axis, respectively. The frequencies of the splitted components of the soft mode are still connected to the corresponding components of the dielectric function tensor by the LST relationship. The second group of the phonons of interest is the structural, or R modes. The softening of these phonons at T_a=105 K, when their frequency drops to zero, results in the cubic-to-tetragonal structural phase transition, which in contrast to that in $BaTiO_3$ is not associated with the appearance of the ferroelectric phase. The doubling of the primitive cell in the tetragonal phase leads to the folding of the R phonons into the zone center and splitting into two Raman active mode of the A_{1g} and E_g symmetry.

EXPERIMENTAL TECHNIQUES for the SOFT MODE STUDIES in SrTiO₃

The aforementioned specifics of the optical phonons in STO determine the experimental techniques, which can be utilized for their studies. Phonon spectra in STO bulk crystals have been investigated for many decades using far infrared (IR) spectroscopy [12], neutron scattering [14], Raman scattering [15], and hyper-Raman scattering [16]. Further we will consider the applicability of these experimental methods to the phonon studies in STO thin films.

Neutron scattering: The geometrical factor restricts this very powerful method since it usually requires several grams [17] of the solid state sample. Thus, the conventional neutron scattering cannot be directly applied to the phonon studies in STO films with the typical thickness less than 1 μm and at present this method is not of the practical interest.

Hyper-Raman scattering: This technique can, in principle, provide useful information about the infrared-active optical phonons [16]. However, these experiments utilize very high excitation laser power density and only transparent in the infrared spectral range samples can be studied by conventional hyper-Raman scattering with a pulsed Nd:YAG laser excitation. Special efforts should be made to choose the right transparent substrate with no hyper-Raman response in the frequency range of the STO optical phonons. Our preliminary results on hyper-Raman scattering demonstrated that the defect-induced absorption of the exciting laser usually results in a damage of the studied STO films. Moreover, the STO films grown on conducting substrates, which are of the practical interest, can by no means be studied with hyper-Raman scattering because of the strong absorption in the substrate layer and a consequent damage of the whole sample.

Infrared spectroscopy: In bulk STO this method allowed measurements of the dielectric function in the far-IR spectral range and determination of the frequencies of the IR-active optical phonons. To obtain the real and imaginary parts of the dielectric function in reflection or transmission experiments, it requires application of the Kramers-

Kronig transformation. The uncertainties of this approach limit the accuracy of the measurements. Far-infrared ellipsometry provides the same kind of information but without relaying on the Kramers-Kronig transformation, since both the real and imaginary parts of the dielectric function are measured at the same time. Recently we have applied the Fourier transform infrared (FTIR) ellipsometry to thin STO films using synchrotron radiation [18]. The high brightness of the synchrotron light source allows us to probe thin films with a small in-plane area, thus minimizing the influence of the film thickness fluctuation.

Raman scattering: The conventional Raman scattering is limited in bulk STO by the odd parity of the optical phonons since at the normal conditions they are not Raman-but IR-active. First order Raman scattering has been measured in bulk STO only in the case in which ferroelectricity was induced by application of hydrostatic or uniaxial stress, or the central symmetry of the system was broken by impurities. Application of external electric field also removes the center of inversion and makes all the optical phonons Raman-active [15]. Our previous studies of thin STO films [19] demonstrated the observation of the zone center optical phonons in the Raman scattering experiments. The possible reason for that is the existence of the local polarization due to the presence of oxygen vacancies [8]. In the same manner as external electric field, this polarization removes the center of inversion in STO films. Raman scattering in STO thin films under application of external electric field (see Ref. [8] for more detail) is of the special interest since it allows measurements of the soft mode frequency tuning. These experiments will be described below along with the results of FTIR ellipsometry.

EXPERIMENT

STO film with the thickness of $1\,\mu$m was epitaxially grown using PLD on a single crystal [001]-oriented STO substrate. A conductive $SrRuO_3$ (SRO) buffer layer with the thickness of 0.35 μm was inserted between the film and the substrate to screen the optical signal from bulk STO. The SRO buffer layer also serves as the bottom electrode. To apply an external electric field normal to the STO film plane, a transparent indium-doped tin oxide (ITO) layer was deposited on top of the STO film at 200 C° in 16 mTorr of oxygen pressure, resulting in high conductivity of ITO. The lattice constant mismatch between SRO and STO (~0.64%) results in slight tensile in-plane stress in the STO films. X-ray diffraction [20] and Raman scattering by R phonons [8] demonstrated that STO films grown in the same conditions have a cubic structure at high temperatures with a transition to a tetragonal structure at about 125 K, which is higher than that in the bulk (T_a^{bulk} =105 K). Narrow full width at half maximum (FWHM) is found for both the Bragg peaks (~0.1°) and the rocking curves (~0.16°).

The sample geometry [see insets to Fig. 1(a,b)] allows the ellipsometry, Raman scattering, and low-frequency dielectric measurements to be performed on the same structure. The sample was attached to a cold finger inside an optical He-flow cryostat with a temperature range between 5 and 300 K. Raman scattering measurements were taken in the conventional backscattering and in close-to-90° configurations. The z axis is perpendicular to the film plane, while x and y axes are parallel to the [100] and [010] directions of the cubic STO structure, respectively. The 514.5 nm line of an Ar^+-ion laser was used for excitation. Raman spectra were recorded with a SPEX Triplemate spectrometer equipped with a charge-coupled device (CCD) detector. During the Raman scattering measurements, a dc voltage was applied to the sample providing an electric field inside the STO film up to 30×10^4 V/cm. FTIR ellipsometry experiments have been carried out at the National Synchrotron Light Source in Brookhaven National Laboratory using synchrotron radiation [21]. No electric field has been applied to the sample during the ellipsometry measurements.

RESULTS AND DISCUSSION

Fig. 1(a) presents an imaginary part of the effective dielectric function, $\varepsilon''(\omega)$, for the 1 μm-thick STO film at 10 K. The soft TO_1 phonon mode is clearly observed along with other TO modes, and their positions are marked in the figure with arrows. The phonon frequencies were determined by the best fit of the experimental spectra to the factorized form of the dielectric function [12]:

$$\varepsilon(\omega) = \varepsilon_\infty \prod_{j=1}^{N=3} \frac{\omega_{LOj}^2 - \omega^2 + i\gamma_{LOj}}{\omega_{TOj}^2 - \omega^2 + i\gamma_{TOj}}, \qquad (5)$$

where $\gamma_{TO(LO)j}$ represents damping of the TO_j (LO_j) modes. Note that the LST relation is the $\omega \to 0$ limit of Eq. (5). The peak at 480 cm^{-1} is close to the frequency of the LO_3 phonon and corresponds to the Berreman mode originated from the energy losses at the interface between STO and SRO [22].

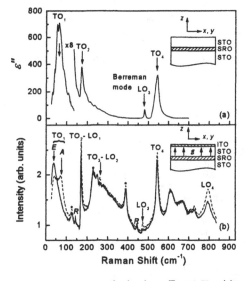

Figure 1.
Ellipsometry and Raman spectra for 1 μm-thick STO film at T=10 K. The insets show the schematics of the investigated structure grown on an STO substrate.

(a) The imaginary part of the effective dielectric function. Positions of the optical phonons are marked with arrows.

(b) Raman spectra without (solid line) and in the presence of external electric field of 22×10^4 V/cm directed normal to the film (doted line). The soft mode components are labeled A and E. Structural modes are denoted with R. Optical phonons from the SRO buffer layer are marked with stars.

Raman spectra obtained at $T = 10$ K with and without application of external electric field $\mathcal{E}$ are shown in Fig.1(b). The phonon frequencies determined by Raman scattering are similar to the results of FTIR ellipsometry. Raman spectrum at $\mathcal{E} = 0$ (solid line) contains a strong peak at about 40 cm^{-1} originated from the E_u component of the soft mode (labeled E). Application of the external electric field [see dotted line in Fig. 1(b)] leads to appearance of the A_{2u} component of the soft mode (labeled A). The selection rules for the A and E peaks change with an increase of the electric field and in the strong field-limit the nonzero components of the Raman tensor are α_{zz} and $\alpha_{xz,yz}$ for the A and E phonon modes, respectively. The splitting between them at low temperatures and in the absence of electric field is induced by the tetragonal distortion. The soft mode frequency is higher at low temperature than that in bulk STO crystals. This result is in a quantitative agreement with the LST formalism.

CONCLUSIONS and ACKNOWLEDGMENTS

We have demonstrated that FTIR ellipsometry and electric field-induced Raman scattering in STO thin film provide experimental determination of the optical phonon frequencies including that of the soft mode. It is possible to measure the temperature and electric field dependencies of the soft mode frequency and the splitting of the soft mode induced by the tetragonal distortion. The details of these studies will be published elsewhere [8,18]. Based on the LST relation, our measurements provide a non-contact determination of the static dielectric constant, which can be compared with the experimental values measured in the same sample with, for example, an LCR-meter. Systematic experimental studies of the soft mode behavior will help to find an explanation of the difference between the dielectric properties of the STO films and bulk crystals.

ACKNOWLEDGMENTS

The authors are grateful to L. Carr and G. Williams at the beam-line U4IR and U12IR at the NSLS. This work was partially supported by DOE under Grant No. DFFG02-84ER45095 and by NSF under Grant No. 9702632. The experiments have been carried out in part at the National Synchrotron Light Source (NSLS), Brookhaven National Laboratory, which is supported by DOE, Division of Materials Sciences and Division of Chemical Sciences under contract number DE-AC02-98CH10886.

REFERENCES

1. O. Auciello, J. F. Scott, and R. Ramesh, Physics Today **51**, 22 (1998).
2. H.-Ch. Li, W. Si, A. D. West, and X. X. Xi, Appl. Phys. Lett. **73**, 190 (1998).
3. H.-Ch. Li, W. Si, A. D. West, and X. X. Xi, Appl. Phys. Lett. **73**, 464 (1998).
4. D. Fuchs, C. W. Schneider, R. Schneider, and H. Rietschel, J. Appl. Phys., **85**, 7362 (1999).
5. R. A. McKee, F. J. Walker, and M. F. Chisholm, Phys. Rev. Lett., **81**, 3014 (1998).
6. C. Zhou and D. M. Newns, J. Appl. Phys. **82**, 3081 (1997).
7. C. Basceri, S. K. Streiffer, A. I. Kingon, and R. Waser, J. Appl. Phys. **82**, 2497 (1997).
8. I. A. Akimov, A. A. Sirenko, A. M. Clark, J.-H. Hao, and X. X. Xi, unpublished.
9. H. Fröhlich, Theory of Dielectrics (Oxford: Clarendon Press, 1949).
10. W. Cochran, Adv. Phys. **9**, 387 (1960).
11. R. A. Cowley, Adv. Phys. **12**, 421 (1963).
12. J. L. Servoin, Y. Luspin, and F. Gervais, Phys. Rev. B **22**, 5501 (1980).
13. A. S. Chaves, F.C.S. Barreto, and L.A.A. Ribeiro, Phys. Rev. Lett. **37**, 618 (1976).
14. J. D. Axe, J. Harada, and G. Shirane, Phys. Rev. B **1**, 1227 (1970).
15. J. M. Worlock and P. A. Fleury, Phys. Rev. Lett. **19**, 1176 (1967).
16. H. Vogt, Phys. Rev. B **51**, 8046 (1995).
17. P. Lindner and G. Wignall, MRS Bulletin **24** (No. 12), 34 (1999).
18. A. A. Sirenko, C. Bernhard, A. Golnik, A. M. Clark, J.-H. Hao, W. Si, and X. X. Xi, accepted for publication in Nature (2000).
19. A. A. Sirenko, I.A. Akimov, J.R. Fox, A.M. Clark, H.-Ch. Li, W. Si, and X.X. Xi, Phys. Rev. Lett. **82**, 4500 (1999).
20. B. O. Wells *et al.*, unpublished.
21. R. Henn, C. Bernhard, A. Wittlin, M. Cardona, and S. Uchida, Thin Solid Films **314-315**, 643, (1998).
22. D. W. Berreman, Phys. Rev. **130**, 2193 (1963).

Materials Characterization

Broadband Determination of Microwave Permittivity and Loss in Tunable Dielectric Thin Film Materials

James C. Booth, Leila R. Vale, and Ronald H. Ono*
*National Institute of Standards and Technology, 325 Broadway, Boulder, CO, 80303

ABSTRACT

We demonstrate a new method for determining the frequency-dependent dielectric properties of thin-film materials at microwave frequencies using coplanar waveguide (CPW) transmission line measurements. The technique makes use of the complex propagation constant determined from multiline thru-reflect-line (TRL) calibrations for CPW transmission lines to determine the distributed capacitance and conductance per unit length. By analyzing data from CPW transmission lines of different geometries, we are able to determine the complex permittivity of the dielectric thin film under study as a function of frequency from 1 to 40 GHz. By performing these measurements under an applied bias voltage, we are able in addition to determine the tuning and figure of merit that are of interest for voltage-tunable dielectric materials over the frequency range 1 to 26.5 GHz. We demonstrate this technique with measurements of the permittivity, loss tangent, tuning, and figure of merit for a 0.4 μm film of $Ba_{0.5}Sr_{0.5}TiO_3$ at room temperature.

INTRODUCTION

Voltage-tunable materials are technologically important for frequency-agile microwave applications, such as tunable filters[1], delay lines, and phase shifters[2]. The ability to electronically tune the frequency of operation of such devices, by changing the permittivity of a constituent material under an applied electric field, will enable many new high-performance microwave applications. In order to fabricate integrated devices, tunable materials in thin-film form are highly desirable. To date, however, thin-film materials that display the requisite tuning are hampered by very high microwave losses. In an effort to reduce the losses of tunable thin film materials, new candidate materials are being developed in ever increasing numbers. However, these materials are commonly evaluated for permittivity and loss at frequencies much lower than those required by the intended applications. Techniques at microwave frequencies that utilize well-characterized resonant cavity measurements are made difficult for these materials by the small volume occupied by films that are typically less than 1 μm in thickness. Characterization of these tunable materials at microwave frequencies is therefore made, if at all, using patterned thin-film devices, where the results often depend upon the specific device geometry used. While such measurements are adequate for optimizing a particular material system, they can be unsatisfactory for comparing different materials or for the subsequent design of devices that differ significantly from the test geometry used.

To address the need for measurements to characterize tunable thin-film materials at microwave frequencies, we have developed an experimental technique based on measurements of coplanar waveguide (CPW) transmission lines over the frequency range 0.1 to 40 GHz. We determine the loss and permittivity of dielectric thin films over this broad frequency range by analyzing the measured propagation constant for CPW transmission lines with and without the dielectric thin film under test. We make use of a distributed-circuit model (see Fig. 1) in order to determine the capacitance per unit length $C(\omega)$ and conductance per unit length $G(\omega)$ of the transmission lines incorporating the dielectric thin film from the measured propagation constants. We then analyze the distributed capacitance per unit length $C(\omega)$ and conductance per unit length $G(\omega)$ obtained by the use of different transmission-line geometries to extract geometry-independent quantities for the loss and permittivity of the dielectric film under test over the frequency range 0.1 to 40 GHz. Because these measurements can easily be performed in the CPW geometry under a bias voltage (which is applied using standard bias tees), the change in permittivity with applied voltage, which defines the film tuning, can also be determined. The loss and tuning can in turn be used to obtain a single frequency-dependent figure of merit for the performance of the tunable dielectric thin film; the figure of merit is independent of specific device geometry.

These experimental determinations of the thin film loss and permittivity can be used to compare the performance of different materials, or to optimize the performance of a specific material. The coplanar waveguide geometry is particularly well suited for the planar devices that can be fabricated using epitaxial thin films on single-crystal substrates. The measurements described here are performed in a cryogenic microwave probe station at arbitrary temperatures from 350 K down to 20 K. Because the properties of interest of these tunable thin-film materials typically depend strongly on temperature, the cryogenic probe station is an ideal configuration for optimizing the microwave response of these materials. Knowledge of both the frequency dependence and the temperature dependence of the relevant dielectric properties can provide important physical clues to the origin of losses in these materials. Since the material properties extracted using this technique are independent of measurement geometry and span a broad frequency range, the results of these experiments can also be used by designers to predict the performance of devices incorporating these thin-film materials in any planar geometry at arbitrary frequency.

In what follows, we apply this new measurement technique to determine the dielectric properties of a 0.4 μm film of $Ba_{0.5}Sr_{0.5}TiO_3$ grown by pulsed laser deposition on a $LaAlO_3$ substrate. $Ba_{0.5}Sr_{0.5}TiO_3$ in thin-film form has been studied by a number of researchers for voltage-tunable applications[3-8]. We discuss in detail our determination of the propagation constant of our CPW transmission lines which are fabricated both on the dielectric thin film sample and on the bare substrate for a number of different CPW geometries. We also describe how we obtain the distributed circuit parameters ($C(\omega)$, $G(\omega)$) of the transmission lines, using two different analytical techniques. We then extract geometry-independent measures of loss, permittivity, tuning and figure-of-merit, as a function of frequency from 0.1 to 26.5 GHz. We conclude with a discussion of the errors and limitations of this technique for determining loss, permittivity, and tuning of dielectric thin-film materials.

EXPERIMENT

Capacitance and Conductance in Coplanar Waveguide Transmission Lines

We obtain the complex propagation constant $\gamma = \alpha + i\beta$, where α is the attenuation constant and β is the phase constant, of our CPW transmission lines as a function of frequency by performing a multiline thru-reflect-line (TRL) calibration[9]. The TRL calibration uses planar CPW calibration sets consisting of individual elements that are all fabricated using the same geometry (center conductor linewidth, gap spacing, metalization thickness, etc., see Fig. 1(b)). An individual calibration set consists of a thru (defined as a transmission line of zero length), a short-circuit reflect, and transmission lines of different lengths. For the measurements described here, we fabricate the CPW devices using silver films 0.5 to 1 μm thick for the metallic conductors, with a center conductor linewidth (2s in Fig. 1(b)) of 55 μm. We fabricate three distinct calibration sets which have gap spacings (g in Fig. 1(b)) of 100 μm, 50 μm, and 25 μm. By

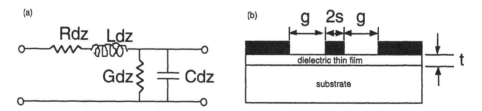

Fig. 1. Schematic diagram showing (a) distributed-circuit model for a transmission line, and (b) CPW cross-section. R, L, C, G are the resistance, inductance, capacitance, and conductance per unit length.

measuring the S-parameters of the individual devices in each calibration set using movable air coplanar probes, we are able to obtain the complex propagation constant of the CPW transmission lines in that particular calibration set using the Multical algorithm developed at NIST [9]. Figure 2 shows the measured attenuation and relative phase constant obtained in this manner for the 50 μm gap calibration set fabricated on a bare LaAlO$_3$ substrate, and on a substrate onto which a 0.4 μm film of Ba$_{0.5}$Sr$_{0.5}$TiO$_3$ had been deposited. The increase in both the attenuation and phase constant due to the Ba$_{0.5}$Sr$_{0.5}$TiO$_3$ thin film is substantial. Also shown in Fig. 2 is the effect of a bias voltage of 50 V on the propagation constant. For these measurements, the bias tees that we use limit the frequency range to 0.1 to 26.5 GHz.

Once we have obtained the propagation constant for both the test (dielectric film and substrate) and the reference (bare substrate only) transmission lines, we use two different techniques to extract the capacitance and conductance per unit length of the transmission lines on the dielectric thin film under study. The first technique, called the calibration comparison technique[10], uses a comparison of the TRL calibrations of the test and reference calibration sets

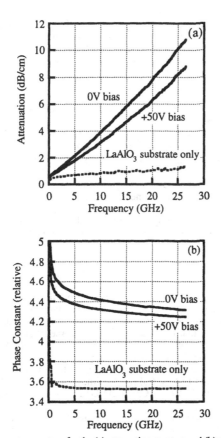

Fig. 2. Measured values at room temperature for the (a) attenuation constant and (b) relative phase constant for CPW transmission lines on a bare LaAlO$_3$ substrate on a LaAlO$_3$ substrate onto which a 0.4 μm film of Ba$_{0.5}$Sr$_{0.5}$TiO$_3$ has been deposited. The CPW metalization is silver 0.5 μm thick with a center conductor linewidth of 55 μm and a gap spacing of 50 μm. Also shown are values for the attenuation and relative phase constant under the application of a dc bias voltage of 50V.

with a TRL calibration on a third sample that has been well characterized. The third sample consists of gold lines on a quartz substrate. These lines are fabricated with embedded lumped resistors that are used to determine the capacitance per unit length[11] and hence the characteristic impedance[12] of the quartz calibration sets. We use the calibration comparison technique to obtain the (complex) characteristic impedance Z_0 of the lines on our test and reference samples. We then combine the characteristic impedance with our measured values of the propagation constant γ to obtain the distributed circuit parameters R, L, C, and G vs. frequency[13] for the both the test and reference samples:

$$\frac{\gamma}{Z_0} = G + i\omega C$$

$$\gamma Z_0 = R + i\omega L \quad ,$$

(1)

where R and L are the distributed resistance and inductance per unit length, while C and G are the distributed capacitance and conductance per unit length. We then assume that the contribution to the capacitance and conductance due to the dielectric film is given by the difference between the capacitance and conductance of the test and reference samples.

We also determine the capacitance and conductance per unit length $C(\omega)$ and $G(\omega)$ of the CPW transmission lines fabricated on the dielectric thin film under study using the equivalent-impedance method described by Janezic and Williams[14]. This method makes use of the fact that the ratio of the propagation constant of the lines on the test sample (incorporating the dielectric thin film) to the propagation constant of the lines on the reference sample (bare substrate) is given by the following:

$$\frac{\gamma_t}{\gamma_r} = \sqrt{\frac{(R_t + i\omega L_t)(G_t + i\omega C_t)}{(R_r + i\omega L_r)(G_r + i\omega C_r)}} \quad ,$$

(2)

where R, L, C, and G are the distributed circuit parameters of the test lines (film and substrate, subscript t) and the reference lines (substrate only, subscript r) respectively. Since we expect the loss tangent of $LaAlO_3$ to be approximately 10^{-4} or smaller, we assume that the conductance per unit length of the bare substrate is negligible, so that we can ignore G_r in Eq. (2). We assume that the inductance and resistance per unit length are identical for the test and reference samples ($R_t = R_r$, $L_t = L_r$). In order to realize this experimentally, we fabricate the metallic lines on the bare substrate and dielectric sample simultaneously in the same deposition system. We will present below experimental evidence that this assumption ($R_t = R_r$, $L_t = L_r$) is valid for the calibration sets described here. With these assumptions, we can obtain the conductance and capacitance per unit length of the test sample as long as the capacitance per unit length of the reference lines is known:

$$G_t + i\omega C_t = i\omega C_r \left(\frac{\gamma_t}{\gamma_r}\right)^2 .$$

(3)

To obtain $C_r(\omega)$ in Eq. (3) above, we use results from the calibration comparison technique described previously. Figure 3 shows the capacitance per unit length for CPW lines on a bare $LaAlO_3$ substrate determined by the calibration comparison method. Also shown in this figure is the capacitance per unit length that is calculated for the geometry under consideration by use of a finite element analysis with a relative permittivity of 23.2 for the $LaAlO_3$ substrate. The finite element model uses three dimensional conductors that are perfect conductors, and the results for $C(\omega)$ show a frequency dependence that is similar to that obtained using the calibration comparison technique. We use a quadratic fit to $C_r(\omega)$ for the lines on the bare substrate in the equivalent-impedance method, Eq. (3), to determine $C_t(\omega)$ and $G_t(\omega)$ for the lines incorporating the dielectric thin film.

We also use the calibration-comparison technique to obtain the resistance and inductance per unit length for the lines on the bare substrate (reference sample) as well as for the lines incorporating the dielectric thin film (test sample), to determine if they are indeed equivalent, as required by the equivalent-impedance method. Such a comparison is shown in Fig. 4. From this figure it is clear that the assumptions are in fact valid for the sample considered here.

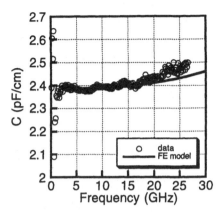

Fig. 3. The measured capacitance per unit length vs. frequency for CPW transmission lines on a bare LaAlO₃ substrate at room temperature. The CPW metalization is silver 0.5 μm thick with a center conductor linewidth of 55 μm and a gap spacing of 100 μm. The solid line is the result of a finite element calculation for the same geometry using a relative dielectric constant of 23.2 for LaAlO₃.

We can now use the determination of $C_t(\omega)$ in Fig. 3 to calculate $C_t(\omega)$ and $G_t(\omega)$ for the lines on the dielectric film under test using the equivalent-impedance method Eq. (3). The results are illustrated in Fig. 5, which shows the capacitance per unit length $C_t(\omega)$ and the device loss tangent $G_t(\omega)/\omega C_t(\omega)$ determined using the equivalent-impedance technique as well as the calibration comparison technique. This figure shows that both techniques yield similar values for $C_t(\omega)$ and $G_t(\omega)/\omega C_t(\omega)$, although the two different techniques make use of very different assumptions. The equivalent impedance method gives less scatter in the determination of $C_t(\omega)$ and $G_t(\omega)/\omega C_t(\omega)$.

Once we have determined the device capacitance per unit length, we obtain the contribution that is due to the dielectric film by simply subtracting the substrate capacitance per unit length from the total device capacitance per unit length. This is equivalent to assuming that the capacitance due to the dielectric film adds in parallel to that of the substrate and air to give the total capacitance. We also assume that the device conductance is due entirely to the dielectric film. We determine $C_{film}(\omega)$ and $G_{film}(\omega)$ for calibration sets having different gap spacings, which are shown in Fig. 6. As expected, as the gap spacing decreases both the capacitance and conductance per unit length increase. Note that the $G_t(\omega)/\omega C_t(\omega)$ in Fig. 6 is the loss tangent of the entire device, which can be related to the loss tangent of the dielectric thin film approximately by

$$\tan\delta_{film} \approx \frac{G_t(\omega)}{\omega C_t(\omega)} \frac{C_t(\omega)}{C_{film}(\omega)} \tag{4}$$

Permittivity and Loss Tangent Determination

In order to obtain the complex permittivity as a function of frequency, we need an expression that relates ε' and ε'' to C_{film} and G_{film}, respectively. One can use a conformal mapping approach[15] to obtain an analytical expression for the contribution to the capacitance due to a thin film in the CPW geometry:

$$C_{film} = \varepsilon_0 \left(\varepsilon_r - \varepsilon_{sub}\right) \frac{K(k)}{K(k')}$$

$$k = \frac{\sinh(\pi s/2t)}{\sinh(\pi(s+g)/2t)} \tag{5}$$

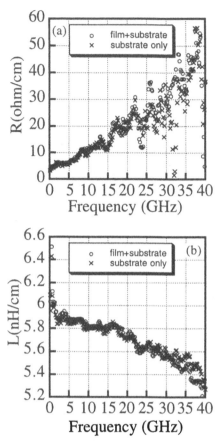

Fig. 4. Measured values for the (a) resistance per unit length and (b) inductance per unit length vs. frequency for a CPW transmission line with and without a 0.4 μm film of Ba$_{0.5}$Sr$_{0.5}$TiO$_3$. The CPW metalization is silver 0.5 μm thick with a center conductor linewidth of 55 μm and a gap spacing of 100 μm. The substrate is LaAlO$_3$.

In Eq. (5) ε_r is the relative permittivity of the film, ε_{sub} is the relative permittivity of the substrate, 2s is the center conductor linewidth, g is the gap spacing, and t is the film thickness (see Fig. 1(b)). Here K is the complete elliptic integral of the first kind, and $k' = (1-k^2)^{1/2}$. For the case of thin films (t << s, g), this expressions simplifies to

$$C_{film} \approx \varepsilon_0 \left(\varepsilon_r - \varepsilon_{sub} \right) \frac{2t}{g} \qquad (t << s, g) \quad . \qquad (6)$$

If we now plot the measured film capacitance (per unit length) at a fixed frequency as a function of t/g, we can extract the permittivity from the slope of the resulting straight line. This is illustrated in Fig. 7, which shows that the film capacitance at 10 GHz is indeed linear in t/g, for dimensions such that t/g < 0.01. We obtain the real part of the permittivity at this frequency from the slope of C_{film} vs. t/g, and obtain the imaginary part of the permittivity from the slope of G_{film}/ω vs. t/g. Note from Fig. 7 that both quantities C_{film} and G_{film}/ω have finite intercepts as t/g approaches zero. This feature is not predicted by the conformal mapping expression Eq.(6), which predicts that C_{film} and G_{film}/ω approach zero as t/g approaches zero.

In order to gain further insight into the behavior of C_{film} and G_{film}/ω as a function of CPW geometry, we have performed a number of finite element simulations for the geometries discussed

258

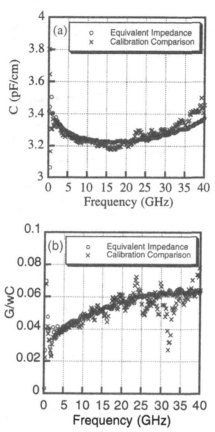

Fig. 5. Measured values using two different techniques of (a) the capacitance per unit length and (b) device loss tangent vs. frequency for CPW transmission lines fabricated on a 0.4 μm film of $Ba_{0.5}Sr_{0.5}TiO_3$. The CPW metalization is silver 0.5 μm thick with a center conductor linewidth of 55 μm and a gap spacing of 100 μm. The substrate is $LaAlO_3$.

here using a commercially available software package. We extract the film capacitance and conductance per unit length using the equivalent impedance analysis of the numerically calculated propagation constants. Analysis of such data yields plots very similar to Fig. 7, which have a finite intercept in plots of C_{film} and G_{film}/ω vs. t/g. This analysis of the simulated data gives values for the permittivity correct to within about 1 %, suggesting that the analytical technique presented here has reasonable accuracy.

We can now extract values for the real and imaginary parts of the permittivity as a function of frequency by performing the linear fits described above at each frequency point for the data shown in Fig. 6. We plot the results as $e_r'(\omega)$ and as loss tangent $\tan\delta(\omega)$ in Fig. 8 for the $Ba_{0.5}Sr_{0.5}TiO_3$ film at room temperature. Note from Fig. 8 the considerable frequency dependence observed in both $e_r'(\omega)$ and $\tan\delta(\omega)$ over the range 1-26.5 GHz.

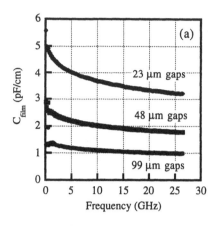

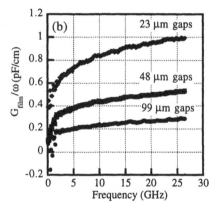

Fig. 6. Measured values for three different CPW geometries at room temperature of (a) the film capacitance per unit length and (b) film conductance per unit length vs. frequency for lines fabricated on a 0.4 μm film of $Ba_{0.5}Sr_{0.5}TiO_3$. The CPW metalization is silver 0.5 μm thick with a center conductor linewidth of 55 μm. The substrate is $LaAlO_3$.

Tuning And Figure of Merit Determination

For materials such as $Ba_{0.5}Sr_{0.5}TiO_3$ the quantities of technological interest are the loss tangent and the amount of tuning that can be achieved for a given bias voltage. We can determine the tuning by simply measuring the change in the capacitance per unit length for a given applied bias voltage. We then define the relative tuning T as:

$$T(V) = \frac{\varepsilon_r(0) - \varepsilon_r(V)}{\varepsilon_r(0)} = \frac{\left[C_{film}(0) - C_{film}(V)\right]}{\varepsilon_r(0) \cdot 2\varepsilon_0 t/g} \quad , \tag{7}$$

where we have used Eq. (6) to relate the change in ε_r to the change in C_{film} so that Eq. (7) is valid only in the thin film limit ($t \ll s, g$). Note that the tuning for a particular film depends on the applied bias voltage, so that the value specified for the tuning is meaningful only if a corresponding bias voltage (or field) is specified. An example of the tuning measured for a $Ba_{0.5}Sr_{0.5}TiO_3$ thin film at room temperature for a bias field of 2 V/μm is shown in Fig. 9(a) (we assume that the average bias field E_{bias} in the film is related to the applied voltage V and gap spacing as $E_{bias} = V/g$).

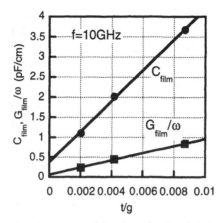

Fig. 7. Measured capacitance and conductance per unit length as a function of t/g at a measurement frequency of 10 GHz. The CPW metalization is silver 0.5 μm thick with a center conductor linewidth of 55 μm. The substrate is LaAlO₃.

Defined in this manner, the tuning divided by 2 gives the relative change in frequency or phase achievable for a tunable filter or phase shifter. Note that this definition is independent of the film thickness used. The film thickness and specific geometry must be used, however, to calculate the tuning of any actual device.

In order to compare the performance of different tunable dielectric materials, measured perhaps at different temperatures, it is convenient to define a figure of merit that takes into account the tuning and loss of a thin film material, since it is quite common for materials that have a large tuning factor to also have high losses. We use the following definition for a figure of merit K(V):

$$K(V) = \frac{T(V)}{\tan \delta_{max}} \qquad (8)$$

The higher the value for K(V) for a given applied voltage, the better the performance of the material. We show in Fig. 9(b) the figure of merit at room temperature for a $Ba_{0.5}Sr_{0.5}TiO_3$ thin film. Although this material exhibits considerable tuning (see Fig. 9(a)), the high loss tangent results in a K(V) value at $E_{bias} = 2$ V/μm of order unity over most of the frequency range. Also note from Fig. 9(b) that the figure of merit drops by a factor of 2 from 5 GHz to 25 GHz.

Measurement Errors and Resolution

The sensitivity of this measurement technique can be defined as the smallest value for the loss tangent, or the smallest change in capacitance that can be extracted from the measured quantities. Examination of Eq. (3) shows that errors in G_t and C_t result directly from errors in the experimental determination of C_t and the relevant propagation constants. Additional errors come from the extraction of C_{film} from C_t. In the determination of the permittivity, additional errors come from uncertainties in the CPW dimensions t and g. Note, however, the loss tangent and tuning determinations (and hence also the figure of merit determination), are independent of the film thickness and CPW dimensions (and associated errors).

In order to estimate some of these uncertainties, it is necessary to determine the reproducibility errors associated with the propagation constant measurements. This is easily accomplished by using the calibration comparison technique[10] which was introduced previously for determining the characteristic impedance of a given calibration set. By using this technique to compare two calibrations performed on the same calibration set, it is possible to determine the upper bound for errors in measured S-parameters for a given calibration. By relating the errors in S-parameters to errors in the propagation constant, we are able to place a limit on the smallest measurable loss tangent and tunability for a given calibration set. We find that the resolution of the

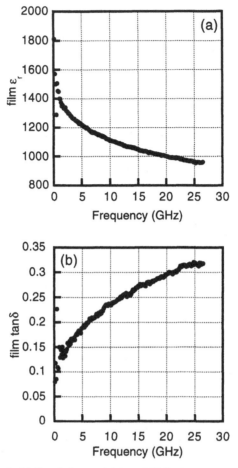

Fig. 8. Measured values for the (a) film relative permittivity and (b) film loss tangent as a function of frequency for a 0.4 μm film of $Ba_{0.5}Sr_{0.5}TiO_3$ at room temperature.

measurement for $tan\delta_{film}$ and $\Delta C/C_{film}$ is proportional to the repeatability errors discussed above, and inversely proportional to the filling factor of the dielectric as well as the absolute phase of the transmission line. This means that we obtain better resolution for higher dielectric constants, higher frequencies, and longer transmission lines. For the geometries considered here, the smallest resolution we can achieve is on the order of 0.01 for both $\Delta C/C$ and $tan\delta$, for frequencies greater than about 5 GHz.

The resolution quoted above is the best one can achieve using this technique if all other quantities are known exactly. In practice the determination of the errors involved in the measurement of $C_r(\omega)$ are much more difficult to quantify than errors in the propagation constant. We do note that our determination of C_r agrees reasonably well with simulation results for a reasonable value for the substrate relative permittivity ($\varepsilon_{sub} \sim 23.3$), see Fig. 3. For the permittivity determinations, errors in t and g are on the order of 5-10 % and most likely dominate the other sources of error. As mentioned previously, these errors do not affect the determination of $tan\delta$, tuning, or figure of merit.

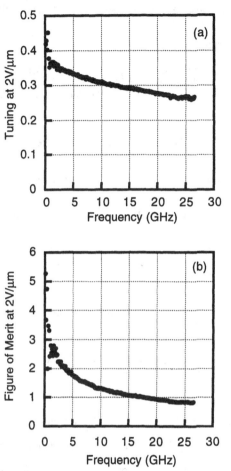

Fig. 9. Measured values for the (a) film tuning and (b) film figure of merit as a function of frequency at a bias field of 2 V/μm for a 0.4 μm film of $Ba_{0.5}Sr_{0.5}TiO_3$ at room temperature. The tuning and figure-of-merit are determined from $\Delta C_{film}(V)$ of the 25 μm gap calibration set.

CONCLUSIONS

We have demonstrated an experimental technique based on CPW transmission line measurements for the determination of the dielectric properties of thin film materials. Using multiline TRL calibrations of CPW transmission lines of different dimensions, we are able to extract the real and imaginary parts of the dielectric permittivity as a function of frequency from 1 to 40 GHz. By performing such measurements under the application of a bias voltage, we are able to determine the relevant dielectric quantities (dielectric loss, tuning, and figure of merit) for dielectric materials of interest for tunable microwave applications. We have demonstrated this technique with measurements of the dielectric properties of a 0.4 μm $Ba_{0.5}Sr_{0.5}TiO_3$ thin film at room temperatures. The ability to perform such measurements as a function of frequency and temperature should be valuable for the understanding and optimization of tunable thin film materials for microwave applications.

ACKNOWLEDGMENTS

We would like to acknowledge funding support from DARPA under the FAME program. We would also like to acknowledge considerable assistance from J. Morgan, M. Janezic and D. Williams, and useful discussions with I. Takeuchi, D. Rudman, and J. Baker-Jarvis.

REFERENCES

[1] A.T. Findikoglu, Q.X. Jia, I.H. Campbell, X.D. Wu, D. Reagor, C.B. Mombourquette, and D. McMurry, Appl. Phys. Lett 66, 3674 (1995).

[2] D.C. DeGroot, J.A. Beall, R.B. Marks, and D.A. Rudman, IEEE Trans. Applied Supercond. 5, 2272 (1995).

[3] Q.X. Jia, X.D. Wu, S.R. Foltyn, and P. Tiwari, Appl. Phys. Lett. 66, 2197 (1995).

[4] J.D. Baniecki, R.B. Laibowitz, T.M. Shaw, P.R. Duncombe, D.A. Neumayer, D.E. Kotecki, H. Shen, and Q.Y. Ma, Appl. Phys. Lett. 72, 498 (1998).

[5] S. Zafar, R.E. Jones, P. Chu, B. White, B. Jiang, D. Taylor, P. Surcher, and S. Gillepsie, Appl. Phys. Lett. 72, 2820 (1998).

[6] H. Chang, I. Takeuchi, and X.-D. Xiang, Appl. Phys. Lett. 74, 1165 (1999).
[7] F.A. Miranda, C.H. Mueller, C.D. Cubbage, K.B. Bhasin, R.K. Singh, S.D. Harkness, IEEE Trans. Appl. Supercond. 5, 3191 (1995).

[8] W. Chang, J.S. Horowitz, A.C. Carter, J.M. Pond, S.W. Kirchoefer, C.M. Gilmore, and D.B. Chrisey, Appl. Phys. Lett. 74, 1033 (1999).

[9] R.B. Marks, IEEE Trans. Microwave Theory Tech. 39, 1205 (1991).

[10] D.F. Williams and R.B. Marks, 38th ARFTG Conf. Digest, pp.68-81, March 1992.

[11] D.F. Williams and R.B. Marks, IEEE Trans. Microwave Guided Wave Lett. 1, 243 (1991).

[12] D.F. Williams and R.B. Marks, IEEE Trans. Microwave Guided Wave Lett. 1, 141 (1991).

[13] D.F. Williams and R.B. Marks, IEEE Trans. Microwave Guided Wave Lett. 3, 247 (1991).

[14] M.D. Janezic and D.F. Williams, IEEE International Microwave Symposium Digest, Vol.3, pp. 1343-1345, June, 1997.

[15] E. Carlsson and S. Gevorgian, IEEE Trans. Microwave Theory Tech. 47, 1544 (1999).

[†]Contribution of the U.S. government, not subject to U.S. copyright.

MICROWAVE DIELECTRIC TUNING AND LOSSES IN EPITAXIAL LIFT-OFF THIN FILMS OF STRONTIUM TITANATE

CHARLES T. ROGERS, MARK J. DALBERTH, AND JOHN C. PRICE
Department of Physics, University of Colorado, Boulder, CO 80309-0390

ABSTRACT

We report on the complex dielectric response function for $SrTiO_3$ (STO) thin films that have been removed from the original growth substrate via an epitaxial lift-off process. We measured the dielectric response as a function of frequency (10 kHz to above 2 GHz), as a function of temperature (4 K to 300 K), and as a function of applied electric field (between +/- 10 V/μm). The dielectric properties are well described by a model based on a dilute set of thermally activated two-level dipoles in a matrix of bulk STO. Activation energies for the dipoles are found to cluster around two central values near 50 meV and 100 meV. Attempt frequencies are generally around 10 THz. Assuming that the dipole moment of a typical two-level system is equal to that of a unit cell of polarized bulk STO, we find that a volume fraction of roughly 4% of a typical film is involved in the dipole response.

INTRODUCTION

Thin-film strontium titanate (STO), $SrTiO_3$, has been under extensive recent study due to its nonlinear dielectric properties. Single crystal STO has a number of unusual dielectric properties [1]. At 300 K, the dielectric response is unusually large and has a low frequency value of roughly 300. Upon cooling, the real portion of the dielectric response increases dramatically to values around 30,000 at 4 K. The response also becomes a function of the applied electric field. Below 40 K, static fields of roughly 1 V/μm can cause the small signal dielectric constant to decrease by factors of between 2 and 10 depending upon temperature of operation. A mean field model for the real part of the dielectric response, based upon soft phonon physics, has been proposed by Vendik [2] and is fairly successful at explaining the observed behavior. In single crystals, the dielectric loss, determined by the ratio of the imaginary part of the dielectric function to the real part (referred to as tanδ) can be as low as 2×10^{-4} in the microwave frequency region [3]. STO thin films with the observed dielectric properties of bulk crystals would be quite useful in a variety of microwave circuits [4]. The known compatibility of thin film STO with thin films of the high temperature superconductors (HTS) offers further interesting possibilities for very low loss and tunable microwave filters and other adjustable circuit elements based upon HTS components.

Early work on thin film STO concentrated on achieving the dielectric tuning found in bulk material. At this point, bulk tuning levels or even larger tuning is commonly achieved [5]. The major emphasis recently has been on reducing the dielectric losses in thin films. Typical tanδ values are at least two orders of magnitude worse than found in bulk STO. In this paper, we report on the microwave dielectric losses found in free-standing STO films. These films are removed from their growth substrates through an epitaxial lift-off process [6, 7]. We find that the total dielectric response, both the real and the imaginary portions, are simpler in these films than in more traditional STO films, still on the substrate. We provide experimental details on the growth and lift-off process; we show that a model of dilute dipolar defects in otherwise bulk-like STO is successful in describing most of the observed behavior.

EXPERIMENT

Thin-film Growth

All our STO films were deposited by laser ablation from single crystal targets, via KrF 248 nm, 10 nsec. optical pulses, focussed to a surface energy density on the target of 1.5 J/cm^2. The film growth process is substantially the same as is used in many laboratories during the growth of high temperature superconductor films. Briefly, we begin with single crystal substrates of (001) lanthanum aluminate (LAO) or (110) neodymium gallate (NGO). Both have served well as substrates for growth of high-quality STO in our lab [5, 8]. Substrates are mounted on a stainless steel heater block with silver paste. The substrate is then heated to 850 C in preparation for film growth. When preparing a STO film for epitaxial lift-off, we first grow a 'sacrificial' or release layer. This layer is chosen to have two properties. First, it must serve as a good epitaxial match for STO growth. Second, it must have distinct chemical differences from STO so that it may be chemically etched after growth to release the STO film. We have used both YBa$_2$Cu$_3$O$_{7-x}$ (YBCO) and PrBa$_2$Cu$_3$O$_{7-x}$ (PrBCO) as sacrificial layers. Our early efforts used 400 nm thick release layers. Most recently, we have prepared 50 nm release layers. We have found no systematic benefits to the use of either release material, or release layer thickness. Both are grown at typical YBCO deposition conditions that yield superconducting YBCO films with transition temperatures above 91 K (850C and 100 mTorr of molecular oxygen in our system).

After growth of the release layer, we immediately deposit STO films at 850 C and 600 mTorr oxygen, to a thickness of 500 nm. Deposition thickness is monitored *in situ* with HeNe laser ellipsometry and is regularly verified, post deposition, via profilometer. Finally, the chamber is back-filled with 500 Torr of pure oxygen and the film is cooled to room temperature over a 2 hour period. Ellipsometric monitoring of the bilayer film during the cool down process indicates that the underlying release layers fully oxygenate. For bilayers with YBCO release layers, we have verified via a mutual inductance probe that the YBCO films superconduct above 89 K.

X-ray diffraction on a subset of the films shows substrate peaks, (00l) peaks from the release layer (these peaks obscure the predominantly (001) STO peaks), and a very small contribution (roughly 0.04%) of STO (110) and (220) peaks.

Capacitor fabrication and lift-off process

After film growth, we first fabricate capacitor structures and then perform the lift-off process. Our capacitors are coplanar, inter-digital structures. Electrodes are 20 nm Ti followed by thicker 1 μm Au. Photolithographic patterning of the electrode structure is done using either NR8-1000 or NR8-3000 (Futurex, Inc.) photoresist. Metallization is performed by thermal evaporation followed by liftoff in acetone. After electrode fabrication, individual capacitors are diced from the wafer. A finished coplanar capacitor is shown in Figure 1a.

After dicing, the coplanar capacitors are ready for epitaxial lift-off. We place the capacitor chip in 100 ml of dilute (1%) HNO$_3$ in deionized water. Over a period of roughly 1 hour, the release layer is chemically removed and the STO film with its Au electrodes slides free of the substrate. A released STO capacitor is shown in Figure 1b.

The finished capacitors are surprisingly robust and can be 'scooped' up with a tweezers blade while in solution without significant damage (after some practice). However, efforts at

Figure 1. a) Coplanar inter-digital capacitor before epitaxial liftoff. The capacitor is shown electrically mounted on a ring resonator as described in the text. b) After lift-off, the released capacitor is seen lying next to the substrate. Notice that the released film is curved, indicating some strain relief. Capacitor is 1mm by 2mm with a 25 micron gap between electrodes.

removing these capacitors directly through the meniscus of water usually result in destruction of the capacitor. Apparently, the water surface tension causes the film to crumple.

Our present technique for handling the liftoff capacitors is to pipette them into a container of isopropyl alcohol. Once in alcohol, we use a pipette to deposit the capacitor on the surface of the microwave test circuit. The alcohol evaporates, leaving the intact capacitor behind without damage. Once dry, capacitors can be gently pushed into position and electrically connected with silver paste.

Dielectric measurement technique

Dielectric measurements are simultaneously acquired at low frequency via a capacitance meter (1 MHz down to 10 kHz), and at microwave frequencies via a microwave ring resonator technique. Figure 2a shows the logical design of our dielectric measurement apparatus. We fabricate sets of several capacitors simultaneously on each chip. A pair of capacitors is selected. They are then electrically mounted with silver paste, one on each side of the ring structure. The parallel capacitance of the pair is measured over the frequency range

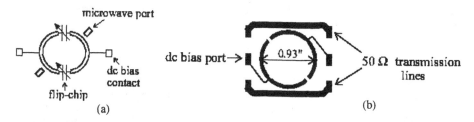

Figure 2. a) Schematic representation of our dielectric measurement apparatus. b) Actual layout of the ring resonator, dc bias leads, and 50 Ω transmission line microwave coupling ports. The microwave ports are terminated with 50 Ω loads during operation.

from 10 kHz to 1 MHz by connecting the two dc bias lines to coaxial cables running to the capacitance meter. These same lines are used for studying the influence of static electric field bias. These low frequency measurements are non-resonant and broad band. Microwave dielectric measurements are taken simultaneously by operating the ring at its microwave resonance frequencies. The details of the microwave design process are fully described in Ref. 8. Briefly, we concentrate on the behavior at the two lowest resonance frequencies. Each of these resonances has a voltage node at the position of the dc bias lines. Therefore, the low frequency lines have no impact on the resonance behavior. The easiest resonance to visualize has a half sine wave standing wave voltage on each arm of the ring. The resonance frequency is thus approximately determined by requiring that the wavelength is equal to the ring circumference, L_{eff}. The capacitors are located at the extremum positions of the standing wave; neither capacitor sees a voltage difference. Therefore, this resonance, which we refer to as the non-tuning mode, is insensitive to the capacitor properties. We use the non-tuning mode as a measure of microwave losses due to non-capacitor loss.

The second resonance has a more complicated voltage profile around the ring and is sensitive to the capacitor values. Straightforward microwave analysis shows that resonant frequency of this 'tuning mode' is determined by a combination of the capacitance, C, the electrical impedance of the ring arms, Z_0, and the ring circumference [8, 9]. Therefore, measurements of the resonant frequency can be used to determine the capacitance value. We find (here k_0 is the wave vector magnitude for the resonant wavelength):

$$C = \frac{1}{2\omega_0 Z_0 \tan\left(k_0 L_{eff}/4\right)} \qquad (1)$$

The sensitivity, S, of this technique for measuring the capacitance depends upon the relative change in resonance frequency with relative capacitance. For our rings, we find:

$$S \equiv \frac{C}{f_0}\frac{\partial f_0}{\partial C} = \left[1 + \frac{\left(k_0 L_{eff}/2\right)}{\sin\left(k_0 L_{eff}/2\right)}\right]^{-1} \qquad (2)$$

Finally, the natural width of the tuning resonance provides a measure of all losses, including the dielectric loss. Information from the non-tuning mode allows us to estimate losses other than those from the capacitors. However, capacitive losses dominate over much of the temperature, frequency, and electric field range. An upper bound for the dissipation factor, D, for the capacitor (ratio of loss impedance to capacitive impedance) can then be determined from the measured Q of the resonance (ratio of resonant frequency to 3 dB width of the resonance) by:

$$D \leq \frac{1}{2SQ} \qquad (3)$$

In the actual implementation, we use a microstrip design, described in detail in Ref. 8. The microwave substrate is two-side copper clad with a dielectric constant of 2.2 [10]. Our design has an L_{eff} of approximately 75 mm and a ring impedance of 67 Ω. These numbers are appropriate for an optimized sensitivity of roughly 0.3 for capacitors of 2 pF and resonance frequencies near 1 GHz.

Figure 2b is the actual physical design of the circuit. A transparency reproduction of Fig. 2b can be used as a photolithography mask for producing our microwave circuit. The entire ring resonator is mounted in a variable temperature 'dipping' cryostat. The apparatus allows for dielectric measurements over a frequency range from 10 kHz to 2 GHz and a temperature range from 4 K to 200 K. Dielectric response is determined from the measured coplanar capacitance via a conformal map [11].

RESULTS

Temperature dependent dielectric response

Figure 3 shows an example of the temperature dependence we find for the real and imaginary portions of the dielectric response at zero applied electric field. The sample shown is a 500 nm STO film lifted-off of a 400 nm PrBCO release layer on an NGO substrate, as described above. The curves correspond to different measurement frequencies. Three of the curves are measured by the low frequency capacitance meter at 10 kHz, 100 kHz, and 1 MHz. The fourth curve shows data from the microwave resonance technique. Thus, each point represents a different resonance frequency, though all are near 1 GHz.

In some regards, these results are similar to previous reports for STO films: The real part of the dielectric response increases with decreasing temperature, reaches a peak, and then decreases with continued decrease in temperature. The loss term also displays a pronounced peak. However, for the film in Figure 3, we see a pronounced shift in loss peak position and associated shifts in the real part of the dielectric function for the different measurement frequencies.

Dispersion of this type in the total dielectric response shows that the real and imaginary components of the dielectric function are related. It provides an opportunity to investigate the underlying dielectric physics. As an example relevant to our model presented below, Debye [12] has shown that peaks in the loss tangent mark positions where the measurement frequency and the characteristic response rate, $1/\tau$, of the system are equal i.e., $\omega\tau=1$. Therefore, the loss peaks offer a way to measure characteristic loss rates.

Figure 4 shows a plot of the natural logarithm of the measurement frequency plotted versus the inverse of the peak temperature. The straight line fit to the data indicates that the relaxation rates follow an Arrhenius behavior:

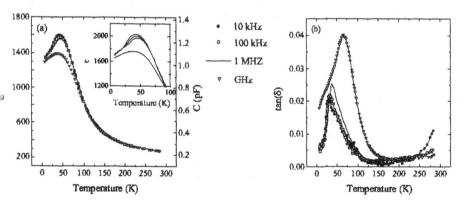

Figure 3. a) Real part of dielectric response and b) associated loss tangent as a function of temperature and measurement frequency for a 500 nm thick lift-off STO film. The inset to (a) shows the dispersion observed in the real part of the dielectric response.

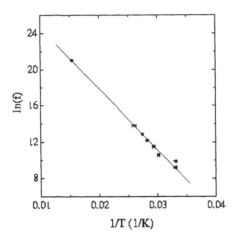

Figure 4. Arrhenius plot of inverse of loss peak temperature versus natural logarithm of the measurement frequency. The solid line fit is described in the text and yields an activation energy of $E_0/k_B = 671$ +/- 6 K and an attempt frequency $1/\tau_0 = 40$+/-10 THz.

$$\frac{1}{\tau} = \frac{1}{\tau_0}\exp(-E_0/k_B T) \qquad\qquad (4)$$

For this particular sample, the solid line is a fit indicating an activation energy of 671 K and an attempt frequency of 40 THz. We have found this type of thermally activated loss behavior in all of the lift-off STO films we have investigated. We generally find activation energies that cluster around two values, one near 600 K as for Figure 4, and a second near 1100 K. Fitted attempt frequencies have significantly larger error and are found in the range from 100 GHz up to 50 THz depending upon the sample.

While the peak positions do show Arrhenius behavior, our more detailed analysis of the temperature dependent peak shape [8] shows that the peaks likely arise from a distribution of activation energies around the central value. We find that the detailed distribution required is specific to each sample. In a very few cases, we have observed quite narrow loss peaks consistent with a single activation energy. However, typical films at present show broader loss peaks.

Electric field dependence of tuning and loss

Figure 5 shows the microwave measurements of the dielectric tuning and loss at 5 K for the same film discussed above. We find that the lift-off films generally display a factor of two tuning with an applied field of 2 V/micron. In the region above 1 V/micron, all of them show microwave loss tangents below 0.01. Our best lift-off STO films show dielectric loss tangents below 0.003. These results are by far the lowest dielectric loss we have observed on STO films of any type.

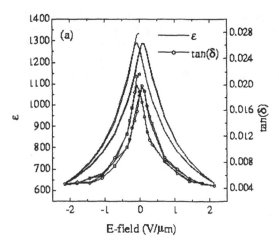

Figure 5. Microwave frequency measurements of dielectric tuning and loss in a 500 nm lift-off film of STO taken in the vicinity of 1 GHz.

Dielectric model

The overall simplicity of the observed dielectric dispersion has led us to consider a model for the dielectric function. Debye has developed a model for a dilute system of dipoles in a homogenous dielectric fluid [12]. The same model applies to dilute dipoles in a cubic solid, where the dielectric response is also homogeneous. In this model, the homogeneous system contributes a real-valued dielectric response of magnitude, ε_{STO}, which we identify as the behavior of bulk STO.

We model this dielectric response with Vendik's dielectric function, modified to include a static random electric field, E_R, arising from disorder in the film [2]. E_R causes the bulk dielectric response to peak at a finite temperature. This function is based on treating the STO as a one-dimensional lattice of titanium and oxygen ions and calculating in mean field theory the lowest frequency transverse optical phonon mode. The restoring force between the positively charged titanium and negatively charged oxygen ions, which is associated with this mode, determines the net ionic motion in the presence of an applied electric field. Thus, this phonon mode determines the dominant dielectric response at frequencies low compared with the mode frequency.

Vendik's model can be cast in several equivalent forms. The version we use relates the dielectric response to five unknown quantities. First, the electrostatic attraction between ions in the unit cell leads to a natural tendency for dimerization and creation of a dipole moment per unit cell. This tendency is counteracted in Vendik's theory by a combination of thermal vibration and ionic zero-point motion. The end result is a temperature-dependent spring constant between the titanium and oxygen ions. This spring constant is given by:

$$a_T(T) \approx 72b \frac{\hbar^2}{m_T (k_B \Theta_D)^2} k_B \left[\sqrt{\left(\frac{\Theta_D}{4}\right)^2 + T^2} - T_C \right] \qquad (5)$$

This form depends upon the known mass of the titanium ion, m_T, and upon three unknown parameters: The Curie temperature, T_C, where ionic attraction would exceed thermal effects (and lead to a ferroelectric phase transition), the Debye temperature, Θ_D, which characterizes the zero-point motion, and on a nonlinear force constant, b, which is the coefficient of a quartic term in the bare titanium-oxygen potential and limits the ultimate dipolar distortion.

Vendik uses these parameters along with the unknown effective charge, Z, transferred between titanium and oxygen (assumed equal but opposite in the theory) to define a nonlinear electric field scale

$$E_{NL} \equiv \left(\frac{4a_T^3}{27b}\right)^{1/2} \Big/ eZ \tag{6}$$

Vendik introduces an electric field parameter, ξ, which combines the applied electric field, E, and the random electric field, E_R, in quadrature:

$$\xi = \sqrt{E^2 + E_R^2} \tag{7}$$

Finally, we have:

$$\varepsilon_{STO}(T, E) \approx \frac{(eZ)^2}{a_T \varepsilon_0 v_{STO}} \frac{1}{\left\{\left[\left[\frac{\xi}{E_{NL}} + \sqrt{\left(\frac{\xi}{E_{NL}}\right)^2 + 1}\right]\right]^{2/3} + \left[\left[\frac{\xi}{E_{NL}} + \sqrt{\left(\frac{\xi}{E_{NL}}\right)^2 + 1}\right]\right]^{-2/3} - 1\right\}} \tag{8}$$

Here, v_{STO} is the unit cell volume of STO. This function depends upon five unknown parameters, namely, the Curie temperature, the Debye temperature, the nonlinear force constant, the effective charge, and the random field. Figure 6 shows a fit to the bulk data of Neville et al. [1], along with the fitting parameters. These results are representative of the parameters we find for bulk.

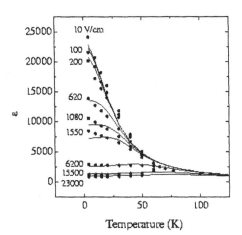

Figure 6. Fitting results using Eq. (8) to fit the data of Neville et al. Fitted parameters are: $Z=1.73+/-0.07$, $T_C=79+/-1$ K, $b=(3.1+/10.4)\times10^{20}$ N/m^3, $\Theta_D=328+/-7$ K, and $E_R=0$.

The dilute dipoles in Debye's picture contribute an additional dielectric response, modeled with a simple relaxation frequency dependence. The complete model dielectric function has real and imaginary parts given by:

$$\mathrm{Re}\,\varepsilon = \varepsilon_{STO}(T,E) + \varepsilon_D(T,E)\frac{1}{1+\omega^2\tau^2}$$

$$\mathrm{Im}\,\varepsilon = \varepsilon_D(T,E)\frac{\omega\tau}{1+\omega^2\tau^2}$$

(9a)

The relaxation rate, $1/\tau$, is assumed thermally activated, Eq. 4, as we observe. The dipolar dielectric function, ε_D, in this model is linearly related to the total density of dipoles in the system.

Unfortunately, the temperature and electric field dependence of the dipolar response is presently unknown. However, as a tentative simplifying guess, we have assigned them the same thermal and electric field behavior as is observed for unit cells of STO. In this picture, the dipolar defects have the same dipole moment and electric field sensitivity as is found in a unit cell of STO, but respond more slowly to applied electric fields. Under this assumption, the common factors of ε_{STO} can be collected and the dielectric function then depends upon the fraction of total STO volume, α, occupied by dipolar defects:

$$\mathrm{Re}\,\varepsilon = \varepsilon_{STO}(T,E)\left[1+\alpha\frac{1}{1+\omega^2\tau^2}\right]$$

$$\mathrm{Im}\,\varepsilon = \varepsilon_{STO}(T,E)\alpha\frac{\omega\tau}{1+\omega^2\tau^2}$$

(9b)

This model contains the essential physics that we believe is important in describing our data. However, as we discussed above, although we have occasionally found cases where the loss peak is well described by a single relaxation rate, far more commonly we find that a distribution of activation energies around a central value is required to understand the observed peak widths. Therefore, we have also investigated this model for the case where the volume fraction is distributed over a range of activation energies:

$$\mathrm{Re}\,\varepsilon = \varepsilon_{STO}(T,E)\left[1+\int\frac{D(E_0)dE_0}{1+\omega^2\tau^2(E_0)}\right]$$

$$\mathrm{Im}\,\varepsilon = \varepsilon_{STO}(T,E)\int\frac{\omega\tau(E_0)D(E_0)dE_0}{1+\omega^2\tau^2(E_0)}$$

(9c)

One simple distribution that allows for exact evaluation of the integrals is the uniform distribution over a given range of activation energies [13]. Figure 7 shows an example of using the uniform distribution model to reproduce the data of Figure 3. The model clearly yields the essential features of the frequency dependent dispersion in the real and the imaginary parts of the dielectric response. The major deviations arise largely from our ignorance of the actual distribution of thermal activation parameters. Note also that the STO parameters are similar to those found for bulk material.

273

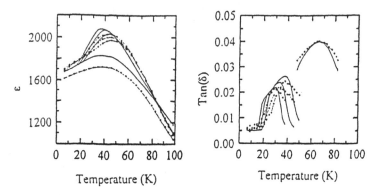

Figure 7. Results using Eq. (8) to model the data of Fig. 3. Parameters are: Z=1.73, T_C=96 K, b=8x10^{20} N/m^3, Θ_D=390 K, E_R=1x10^6V/m, E_0=50meV, $1/\tau_0$=40 THz, width of the distribution is 50 meV, total volume fraction of dipoles is 4%.

Finally, we note that this model provides an interesting cross check on the underlying physics: In the Vendik theory, the random field is added to mimic various defects in thin films. Our data sets typically require a random field of 5-10x10^5 V/m to explain the peak and eventual decrease in the dielectric response at low temperature. In our model, the volume fraction of dipoles provides a natural source of this random field: Once we are at low enough temperature or high enough frequency that the dipoles cannot follow the applied field, then they are reasonably thought of as static. Therefore, they can provide a static random field. Given the density of dipoles from our model, we can calculate the root mean squared dipolar field at a typical position, due to a single dipolar defect. We find [8]:

$$\sqrt{\left\langle E_{DIPOLE}^2 \right\rangle} = \frac{\sqrt{2\alpha}}{3}\left(\frac{p}{\varepsilon_0\varepsilon}\right)\frac{1}{\nu_{STO}} \tag{10}$$

Here, α is the volume fraction of dipoles, p is the dipole moment of a given defect, and ε is the dielectric constant of the homogenous STO material. Our modeling shows that the density is 4% of the STO unit cell density. Taking a dipole moment of 1x10^{-29} C-m, similar to the dipole moment of a BaTiO$_3$ ferroelectric unit cell, and ε=1000 yields a root mean square dipolar field of 10^6 V/m. This result is close to the observed random fields; it serves as further confirmation that dipolar defects are likely the dominant source of dielectric losses and deviations of our thin film STO dielectric properties from those observed for bulk single crystals of STO.

CONCLUSIONS

We have observed frequency and temperature dependent dispersion in both the real and imaginary parts of the dielectric response function of epitaxial lift-off STO films. The internal consistency between the real and imaginary parts shows that both arise from the underlying motion of charge within the sample. In other words, the losses are distinctly *dielectric* in nature. These observations rule out the possibility that the losses arise from unrelated magnetic or resistive effects. The dielectric dispersion displays distinct thermally activated behavior. This behavior is consistent with a hypothesized distribution of individual dipolar defects. We have investigated a simple Debye model for the dielectric response. This picture is an effective medium theory based upon a dilute system of dipoles within a matrix of bulk STO. This model succeeds in providing a semi-quantitative description of all the major observations. The picture suggests a 4% volume density of dipoles in the present films. The model also demonstrates that the dipoles are largely responsible for the observed bulk random field.

It is not clear how the strain relief upon lift-off actually leads to narrow activation energy distributions. However, the strain field is certainly coupled to lattice defect energies, again suggesting that a model of charged defects in the lattice is likely to be correct. Our results show that STO lift-off films are becoming high enough in quality to see single defect behavior. Further development of the film growth processes should lead to unambiguous identification of the defects involved in dielectric losses and will likely reduce the defect density into the range necessary for microwave applications.

ACKNOWLEDGMENTS

We gratefully acknowledge the support of the Office of Naval Research, Superconducting Electronics Program under grants N00014-97-1-0141, N00014-97-1-1054, and N00014-96-1-0907.

REFERENCES

1. R. C. Neville, B. Hoeneisen, and C. A. Mead, J. Appl. Phys. 43, 2124 (1972).
2. O. G. Vendik and S. P. Zubko, J. Appl. Phys. 82, 4475 (1997).
3. G. Rupprecht and R. O. Bell, Phys. Rev. 125 1915, (1962).
4. D. Galt, J. C. Price, J. A. Beall, R. H. Ono, Appl. Phys. Lett., 63, 3078 (1993).
5. M. J. Dalberth, R. E. Stauber, J. C. Price, and C. T. Rogers, Appl. Phys. Lett. 72, 507 (1998).
6. E. Yablonovitch, D. M. Hwang, T. J. Gmitter, L. T. Florez, and J. P. Harbison, Appl. Phys. Lett. 56, 2419 (1990).
7. M. M. Eddy, R. Hanson, M. R. Rao, B. Zuck, J. S. Speck, and E. J. Tarsa, Mater. Res. Soc. Symp. Proc., 474, 365 (1997).
8. *Dielectric loss of strontium titanate thin films*, Mark J. Dalberth, Ph. D. thesis, University of Colorado (1999).
9. D. Galt, T. Rivkina, and M. W. Cromar, Mater. Res. Soc. Symp. Proc., 493, 341 (1997).
10. Duroid 5880 obtained from Rogers Microwave Corporation.
11. H.-D. Wu, F. S. Barnes, D. Galt, J. C. Price, and J. A. Beall, SPIE International Symposium on Optoelectronic and Microwave Engineering, 2156, 131 (1994).
12. *Polar Molecules*, P. Debye, Dover Publications, Inc. New York, NY (1929).
13. *Dielectric Relaxation*, Vera V. Daniel, Academic Press, New York, NY (1967).

Local Optical Probes of Microwave Dielectric Dispersion in Ferroelectric Thin Films

Charles Hubert, Jeremy Levy[†]
Department of Physics and Astronomy, University of Pittsburgh, Pittsburgh, PA 15260

E. J. Cukauskas, Steven W. Kirchoefer, Jeffrey M. Pond, and William J. DeSisto
Naval Research Laboratory, Electronics Science and Technology Division
Washington, DC 20375

ABSTRACT

The soft-mode contribution to the dielectric response of $Ba_{0.5}Sr_{0.5}TiO_3$ ferroelectric thin films is measured locally at microwave frequencies using time-resolved confocal scanning optical microscopy. Optical measurements performed on an ensemble of nanometer-scale regions show a well-defined phase shift between the paraelectric and ferroelectric response at 2-4 GHz. Application of a static electric field produces large local variations in the phase of the ferroelectric response. These variations are attributed to the growth of a-axis ferroelectric nanodomains whose size-dependent relaxation frequencies lead to strong dielectric dispersion at mesoscopic scales. The in-plane paraelectric response is believed to arise from the surrounding c-axis matrix and non-polar regions, and shows negligible dispersion. These results provide a direct view of the soft-mode mechanisms of microwave dielectric loss in ferroelectric thin films.

INTRODUCTION

A central question arising in the physics of ferroelectrics is the determination of the relationship between polar structure and dynamic response. For example, ferroelectric domains influence the dielectric permittivity *via* the motion of domain walls[1]. An understanding of this relationship is important for improving the microwave properties of thin films of these materials for frequency-agile electronic applications.

Ferroelectric thin films exhibit a number of features that are reminiscent of bulk relaxors. In the $Ba_xSr_{1-x}TiO_3$ (BST) system, a sharp first-order ferroelectric phase transition observed for single crystals is broadened by several hundred degrees when thin films are prepared. This broadening is not, for the most part, due to chemical disorder between Ba and Sr, since these effects are observed for $BaTiO_3$ thin films as well[2]. More likely candidates of the disorder are Oxygen vacancies, point defects and inhomogeneous strain from the grain boundaries. For BST thin films, a coexistence of ferroelectric and paraelectric behavior has been observed over nanometer-scale regions[3]; images of ferroelectric nanodomains with sizes as small as 50 Å have been obtained using near-field optical techniques[4]. Apart from a low-frequency cutoff that is likely related to finite-size effects, the temperature and frequency dependence of ferroelectric thin films bears a striking resemblance to relaxor ferroelectrics.

[†] To whom correspondance should be addressed.

In this paper, we report results from time-resolved confocal scanning optical microscopy (TRCSOM) experiments that measure the local soft-mode contribution to the dielectric response of ferroelectric thin films at microwave frequencies. Both the ferroelectric and paraelectric response are measured with diffraction-limited (~500 nm) spatial resolution and picosecond temporal resolution. The spatially inhomogeneous dielectric dispersion observed for the soft mode provides a direct experimental link between nanopolar structure and fast (sub-nanosecond) dynamical response.

EXPERIMENTAL DETAILS

$Ba_{0.5}Sr_{0.5}TiO_3$ thin films are grown on MgO substrates by off-axis co-sputtering from $SrTiO_3$ and $BaTiO_3$ targets[5]. The films are c-axis oriented and polycrystalline, with a high degree of in-plane order. The electrical properties of the films are also characterized at microwave frequencies 1-20 GHz. Silver/gold interdigitated electrodes are deposited on the sample, and s-parameter measurements are performed and analyzed using a series capacitor and resistor equivalent circuit model. The microwave measurements yield quality factors $Q=\varepsilon_1/\varepsilon_2 \sim$ 50-70 for the samples investigated here[5].

Figure 1 shows schematically the experimental arrangement for the TRCSOM measurements. Ultrashort (~120 fs) light pulses from a mode-locked Ti:Sapphire laser (repetition frequency $\omega_1/2\pi \sim$76 MHz) clock a phase-locked oscillator that produces a high harmonic n of the repetition rate of the laser: $\omega_n = n\omega_1$, 2 GHz $< \omega_n / 2\pi < 4$ GHz ($26 \leq n \leq 53$) A combined dc and ac voltage is applied to an interdigitated capacitor $V(t) = V_{dc} + V_{ac}\cos\omega_n t$. The laser is focused between the capacitor fingers, where the electric field is fairly uniform and in-plane oriented, so that $E(t) \approx V(t)/d$, where $d\sim$5-10 μm is the finger spacing. The ferroelectric and paraelectric response to the microwave field is measured *via* the electro-optic effect[3]. An electrical delay line varies the time t between the electrical "pump" and the optical

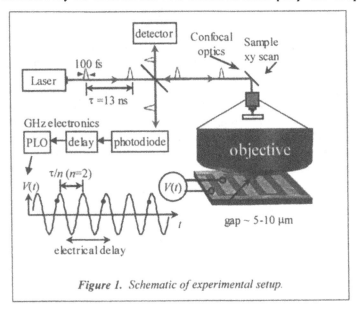

Figure 1. Schematic of experimental setup.

278

"probe", allowing the light to interrogate the sample response at different phases of the microwave driving field. All of the experiments described here are performed at room temperature (T=294 K). A more detailed description of the experimental technique is given in Ref. 6.

EXPERIMENTAL RESULTS

Representative time scans, acquired at a single location on the sample for several dc bias fields E_{dc}, are shown in Figure 2. The measured signal $S(t)$ is periodic and well described by Fourier components at the microwave driving frequency ω_n and the second harmonic $2\omega_n$:

$$S_{fit}(t) = S + \text{Re}[\mathbf{F}\exp(-i\omega_n t) + \mathbf{P}\exp(-2i\omega_n t)]$$

(1a)

$$= S + F_1\cos(\omega_n t) + F_2\sin(\omega_n t) + \\ P_1\cos(2\omega_n t) + P_2\sin(2\omega_n t)$$

(1b)

where $\mathbf{F} = F_1 + iF_2$ and $\mathbf{P} = P_1 + iP_2$. The coefficients $(S, F_{....})$, obtained by a linear least-squares fit, are listed in Table 1. Both the magnitude $|\mathbf{F}| \equiv \sqrt{F_1^2 + F_2^2}$ and phase $\delta_F \equiv \tan^{-1} F_2 / F_1$ of the ferroelectric response depend on the value and history of E_{dc}. The paraelectric magnitude $|\mathbf{P}| \equiv \sqrt{P_1^2 + P_2^2}$ and phase $\delta_P \equiv \tan^{-1} P_2 / P_1$, by contrast, is nearly constant, independent of the spatial location (see Table 1); therefore, on the physical grounds that the paraelectric response has negligible loss compared to the paraelectric mode, we adjust "zero delay" so that the average paraelectric phase is identically zero: $\overline{\delta}_P \equiv 0$.

Figure 3 shows the results from an ensemble of 60 different locations on the film ($j = 1, 2, ..., 60$). At each location, time-resolved traces similar to those shown in Figure 2 are acquired at four sequentially applied values of the dc electric field, $E_{dc} = (0^+, +E, 0^-, -E)$, E=30 kV/cm. The resulting complex phasors $\{\mathbf{F}_j = F_{1j} + iF_{2j}\}$ and $\{\mathbf{P}_j = P_{1j} + iP_{2j}\}$ are presented in (a,c)), which shows a well-defined phase angle (modulo π) between the ferroelectric and paraelectric response $\overline{\delta}_F - \overline{\delta}_P \equiv \delta_0 \approx 1.6$ rad. The phase distribution for δ_F is bimodal because the linear electro-optic coefficient changes sign with the polarization direction. By assumption $\overline{\delta}_P = 0$, which implies that ferroelectric response, on average, *lags* the driving field by δ_0, giving rise to dielectric loss. The magnitude of the ferroelectric response shows a large amount of variation for $E_{dc} = 0$ (see Table 2). Application of a nonzero bias field, shown in Figure 3(b,d), demonstrates a central result: large field-induced local dispersion in the ferroelectric response. By comparison, the paraelectric response (Figure 3(f,h)) is unaffected. In Figure 3, three representative points are also shown so that comparisons can be made of the response at individual locations as a function of E_{dc}. The motion of a given $\mathbf{F}_j$ in the complex plane is not linear with field, nor is it symmetric about the origin; however, it is reproducible. By contrast, the values of $\mathbf{P}_j$ do not change with E_{dc}.

Figure 4 plots the difference in the ferroelectric (a,b) and paraelectric (c,d) response for the extreme and zero values of E_{dc}. Figure 4(a) shows that for the most points the change in F at the extreme values is large and lies on the upper-half plane. The zero-field hysteresis Figure 4(b) is much smaller, showing correlations along the same direction δ_0.

Table 2 lists a number of statistical quantities related to the extracted coefficients, including the average coefficients $\{\overline{F}_1, \overline{F}_2, \overline{P}_1, \overline{P}_2\}$, phase angles $\overline{\delta}_F \equiv \tan^{-1} \overline{F}_2 / \overline{F}_1$, $\overline{\delta}_P \equiv \tan^{-1} \overline{P}_2 / \overline{P}_1$ and

RMS values along the principal axes for the four values of E_{dc}. Note that $\bar{\delta}_F$ *increases* with field—opposite to what is predicted by Gurevich and Tagantsev for bulk materials [7], but in agreement with the microwave loss tangent measurements[5].

Qualitatively similar results are observed for an annealed sample grown and processed in the same manner, and for a number of samples grown and processed by other techniques.

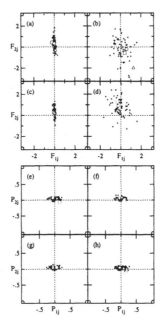

Figure 3. *Plot of fitting parameters $\{F_{1j}, F_{2j}, P_{1j}, P_{2j}\}$ for an ensemble of non-overlapping locations $j=1,...,60$ at various biases E_{dc}. (a,e) $E_{dc}=0^+$ kV/cm, (b,f) $E_{dc}=+30$ kV/cm, (c,g) $E_{dc}=0^-$ kV/cm, (d,h) $E_{dc}=-30$ kV/cm.*

	S	F_1	F_2	P_1	P_2	$\tan^{-1} F_2/F_1$	$\tan^{-1} P_2/P_1$	
$E_{dc}=0^+$	-0.253	-0.114	0.267	0.23	0.086	-1.167	0.358	*Table 1. Extracted fitting*
$E_{dc}=30$ kV/cm	-0.269	0.273	-0.143	0.287	0.086	-0.483	0.291	*coefficients for data shown in*
$E_{dc}=0^-$	-0.11	-0.093	0.584	0.121	0.046	-1.413	0.363	*Figure 2.*
$E_{dc}=-30$ kV/cm	-0.128	-0.516	1.695	0.186	0.066	-1.275	0.341	

Average	S	F_1	F_2	P_1	P_2	δ_F	δ_P
$E_{dc}=0^+$	0.256	-0.006	0.294	0.005	0.043	1.59	1.46
$E_{dc}=30$ kV/cm	0.230	-0.129	0.799	0.016	0.032	1.73	1.11
$E_{dc}=0^-$	0.273	-0.008	0.299	-0.003	0.045	1.60	1.64
$E_{dc}=-30$ kV/cm	0.250	0.136	-0.261	0.005	0.047	2.05	1.47
RMS	S	F_1	F_2	P_1	P_2	δ_F	δ_P
$E_{dc}=0^+$	0.256	-0.006	0.294	0.005	0.043	1.59	1.46
$E_{dc}=30$ kV/cm	0.230	-0.129	0.799	0.016	0.032	1.73	1.11
$E_{dc}=0^-$	0.273	-0.008	0.299	-0.003	0.045	1.60	1.64
$E_{dc}=-30$ kV/cm	0.250	0.136	-0.261	0.005	0.047	2.05	1.47

Table 2. Statistical information for fitting coefficients S, F_1, F_2, etc.

DISCUSSION

Any explanation of the above results must address a number of findings: (1) A well-defined phase between the ferroelectric and paraelectric response when $E_{dc}=0$, (2) field-induced broadening of the ferroelectric phase (Figure 3(b,d)), (3) field-induced increase and broadening of the magnitude of the ferroelectric phase response, (4) no correlation (other than qualitative) between data taken at adjacent or distant locations (>500 nm). One likely explanation is that the static field induces the growth (and shrinkage) of a-axis oriented nanodomains from the c-axis matrix. The field-dependent dispersion might arise from a size-dependent relaxation τ^8 coupled to an underdamped mode[9]. Relaxation alone cannot describe the observed results, since the dispersion for a pure relaxation $\chi_R(\omega) \propto (1-i\omega\tau)^{-1}$ is restricted to the upper right quadrant of the complex plane, whereas the observed $\{F_j\}$ in an applied field is much broader.

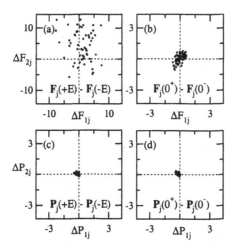

Figure 4. *Plot of difference in fitting coefficients $\{F_{1j}, F_{2j}, P_{1j}, P_{2j}\}$ for various dc biases E_{dc}, where E=30 kV/cm. Note the change in scale for (a).*

The physical manifestation of this underdamped mode is not clear, but might be associated with domain wall motion. Arlt et al., have developed a theory in which microwave dielectric relaxation in single crystal and ceramic bulk ferroelectrics is explained by the emission of shear waves from 90° domain walls[10]. However, this theory is not straightfowardly mapped onto the situation for thin films, in which both strain and disorder play pivotal roles.

It is possible to estimate the effect of E_{dc} on the growth of a-axis nanodomains from the RMS fluctuations along the δ_0 direction. The factor of 1.4 increase implies that the average nanodomain volume has increased by a factor $(1.4)^2 \sim 2$. The increase in volume size leads to strong dispersion, assuming that the response is sharply resonant and that the resonance frequency depends on domain size. The 5x increase in the phase dispersion implies a $Q \sim 4$ for the resonant mode in this system.

Further work is clearly required to explore the temperature and frequency dependence of this mesoscopic dielectric dispersion. The growth of nanodomains is a central feature of a number of theories of relaxor phenomena, and it is possible that similar behavior is occuring over a more restricted frequency range in these ferroelectric thin films. Additional information may also be obtained from high-spatial probes such as aperatureless near-field scanning optical microscopy[4], which have the potential to combine the time resolution demonstrated here with near-atomic spatial resolution.

ACKNOWLEDGEMENTS

The authors thank O. Tikhomirov for helpful discussions. This work was supported by the Office of Naval Research (N00173-98-1-G011) and the National Science Foundation (DMR-9701725).

REFERENCES

[1] M. E. Lines and A. M. Glass, *Principles and Applications of Ferroelectrics and Related Materials* (Clarendon, Oxford, 1977).

[2] B. H. Hoerman, G. M. Ford, L. D. Kaufmann, and Y. B. W. Wessels, Appl. Phys. Lett. **73,** 2248-50 (1998).

[3] C. Hubert, J. Levy, A. C. Carter, W. Chang, S. W. Kiechoefer, J. S. Horwitz, and D. B. Chrisey, Appl. Phys. Lett. **71,** 3353-5 (1997).

[4] C. Hubert and J. Levy, Appl. Phys. Lett. **73,** 3229-31 (1998).

[5] E. J. Cukauskas, S. W. Kirchoefer, W. J. DeSisto, and J. M. Pond, Appl. Phys. Lett. **74,** 4034 (1999).

[6] C. Hubert and J. Levy, Rev. Sci. Instr. **70,** 3684-7 (1999).

[7] V. L. Gurevich and A. K. Tagantsev, Adv. Phys. **40,** 719-67 (1991).

[8] M. P. McNeal, J. Sei-Joo, and R. E. Newnham, J. Appl. Phys. **83,** 3288-97 (1998).

[9] A. Sawada and M. Horioka, Jpn. J. Appl. Phys. Suppl. **31,** 390-2 (1985).

[10] G. Arlt, U. Bottger, and S. Witte, Annalen der Physik **3,** 578-88 (1994); G. Arlt, U. Bottger, and S. Witte, Appl. Phys. Lett. **63,** 602-4 (1993).

PERFORMANCE OF A FERROELECTRIC TUNABLE PRE-SELECT FILTER/LOW NOISE AMPLIFIER HYBRID CIRCUIT

GURU SUBRAMANYAM*, FELIX A. MIRANDA**, ROBERT R. ROMANOFSKY**, FRED VAN KEULS**, AND CHONGLIN CHEN***
*Department of Electrical &Computer Engg., University of Dayton, Dayton, OH 45469
**NASA Glenn Research Center, 21000 Brookpark Road, Cleveland, OH 44135
***Texas Center for Superconductivity, University of Houston, Houston, TX 77204

ABSTRACT

In this paper we discuss the performance of a proof-of-concept of a tunable band pass filter (BPF)/Low Noise Amplifier (LNA) hybrid circuit for a possible gain-compensated down-converter targeted for the next generation of K-band satellite communication systems. Electrical tunability of the filter is obtained through the nonlinear electric field dependence of the relative dielectric constant of a ferroelectric thin-film such as strontium titanate ($SrTiO_3$) or barium strontium titanate ($Ba_xSr_{1-x}TiO_3$). Experimental results show that the BPFs are tunable by more than 5%, with a bipolar biasing scheme employed. The BPF/LNA tunable hybrid circuit was used to study the effect of tuning on the hybrid circuit's performance especially on the amplifier's noise-figure and the gain.

INTRODUCTION

Tunable microwave components such as resonators, filters, and phase shifting elements based on the nonlinear electric field dependence of ferroelectric thin films have been recently demonstrated [1-4]. $SrTiO_3$ (STO), is suitable for low temperature applications below 77K, where one can reduce its relative dielectric constant by more than a factor of 5 by applying a dc electric field [4]. $Ba_xSr_{1-x}TiO_3$ (BSTO), on the other hand, is a room temperature tunable ferroelectric, as one could reduce its relative dielectric constant by more than a factor of 3 at or near room temperature [5]. Tunabilities higher than 15% have been reported in tunable filters and resonators using both micro-strip and coplanar wave-guide transmission lines [3,6]. Recently, we have reported on K-band tunable band-pass filters using ferroelectric thin-films based on the conductor /ferroelectric/ dielectric two-layered micro-strip structure [7]. In this work, this type of ferroelectric tunable filter is integrated with a commercial K-band LNA to study the effect of tunability on the hybrid circuit's associated gain and noise-figure, for both room temperature as well as cryogenic applications. The performance of this hybrid circuit may be relevant for the implementation of a gain-compensated down converter for the next generation of satellite communication systems.

DESIGN

Two different filters were used in this study. One is a K-band, two-pole bandpass filter designed for 4% bandwidth, the second is a K-band, narrow-band (<1-% bandwidth) 3 pole band-pass filter. The filters were designed with edge coupled half wavelength resonators for a center frequency of 19GHz, corresponding to a dielectric constant of the tunable ferroelectric at approximately 1600, achievable under an applied dc bias. The design of the tunable filters

283

has been described elsewhere [7]. At present, the relative dielectric constant of high quality laser ablated thin films of STO (ε_{rSTO}) is typically tunable between 4000 and 300 at temperatures below 77K[4] whereas BSTO has tunabilities up to 65% and (ε_{rBSTO}) as high as 4660 at room temperature [5].

EXPERIMENTAL

The commercial LNA used was the Alpha Industries Inc.'s AA022N1-00, a 20-24 GHz GaAs MMIC LNA, designed for single bias supply operation ($V_D=V_G=4.5V$), with a typical noise figure of 2.5 dB and an associated gain of 22 dB both @23 GHz. The LNA is a three-stage reactively matched chip using a 0.25 μm low noise pseudo-morphic HEMT (PHEMT) technology. The LNA was characterized at room temperature and low temperature applications for integration with BSTO and STO ferroelectric tunable filters. An automated noise-figure measurement system, consisting of a HP8970 Noise Figure Meter, an ATN tuner amplifier and software, and a custom-made on wafer cryogenic probe station, was used for amplitude of noise and the small signal gain measurements. In this set-up, a known noise source is introduced periodically into the system gain. A noise source with known excess noise ratio (ENR) at the frequencies of interest, is pulsed to turn off and on to produce two noise levels.

The samples with 300-nm BSTO thin film on LaAlO₃ (LAO) substrates were obtained from the University of Houston. The samples with STO thin-film of 1000 nm were obtained from Superconductor Core Technologies (SCT Inc.), Boulder, CO. Both sets of samples were deposited by laser ablation. The Au/BSTO/LAO (2 pole BPF) and Au/STO/LAO (3 pole BPF) circuits were fabricated using standard positive photo-resist lithography. The filters were studied under a full bipolar bias (FBB) configuration where alternate sections (including the input and output lines) were biased positive and negative (see figure 1).

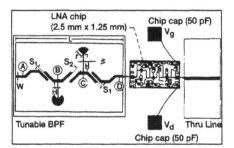

Figure 1. Schematic diagram showing a packaged BPF/LNA hybrid circuit. The filter's dimensions are: L=6.8 mm; W=86.25 μm; S_1 = 100 μm; S_2 = 300 μm; H=1.33 mm, w=12.5 μm, and r=200 μm. The overall dimensions of the packaged circuit are 1.25 mm x 1 cm.

The BSTO ferroelectric based filters were studied at room temperature and the STO samples over a wide temperature range (300 K down to 50K) under vacuum, to eliminate any possibility of arcing at high bias voltages. Bias voltages were applied up to ± 400V, with minimal power consumption, because of the dielectric nature of the ferroelectric film. The filters were characterized separately by swept frequency scattering parameter measurements performed using a HP-8510C automatic network analyzer (ANA). After characterizing the filters, each filter was packaged with an LNA, to form the hybrid circuit on a grounded, gold-plated alumina carrier. A layout of the packaged circuit is shown in figure 1.

RESULTS AND DISCUSSIONS

The noise-figure and gain measurements performed on the LNA at room temperature and at cryogenic temperatures are shown in figure 2a and 2b respectively. As shown in figure 2a, the associated gain of the LNA is greater than 20 dB from 15- 18 GHz. The noise-figure for the LNA is generally below 2 dB at room temperature in the frequency range of 15 to 18 GHz. At 80 K, the gain improves by several dB and the noise-figure falls below 0.5 dB between 16 and 20 GHz.

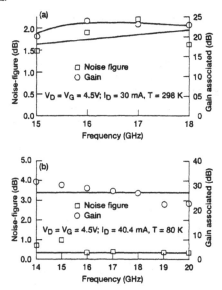

Figure 2. The associated gain and noise-figure for one of the on-chip probed LNA at (a) room temperature and (b) 80 K. The bias conditions for the LNA were $V_D = V_G = 4.5$ V. The solid-lines are trend lines plotted by the ATN software.

The swept frequency scattering parameter measurements performed on one of the gold/BSTO/LAO based K-band microstrip BPF is shown in figure 3. The measurements were performed at room temperature with a FBB from 0 V to ±400V. The minimum insertion loss of the filter is about 5.75 dB, at a bias voltage of ±400V. The higher insertion loss in the BPF is primarily due to the higher loss tangent (tan δ) in BSTO thin films in the range of 0.01-0.1 at room temperature [5]. The filter is tunable from a center frequency of 16.55 GHz at 0 V, to a center frequency of 17.4 GHz at ±400V, a frequency tunability of 5% at room temperature.

The frequency response of the packaged hybrid circuit was tested for various applied bias voltages of the filter, with the drain and gate voltages of the LNA biased at 4.5V. The

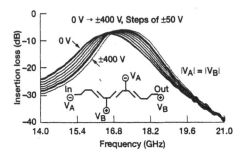

Figure 3. The room temperature frequency tunability of a BSTO based BPF under full bipolar bias (FBB), with bias voltages from 0V to ±400 V.

hybrid circuit exhibited frequency tunability as well as improvement in gain with applied bias, as shown in figure 4. The passband bandwidth of the hybrid circuit is larger (more than 10%) compared to the passband of the filter itself (7%) primarily due to the wire-bonds and packaging of the hybrid circuit. Shorter wire-bonds could eliminate this problem. The hybrid circuit was frequency tunable by approximately 3.5% at room temperature. The non-

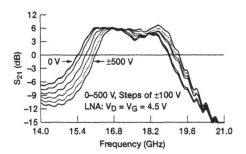

Figure 4. The room temperature frequency response of a BPF/LNA hybrid circuit with the filter under a FBB bias of 0V to ±400 V, and the LNA biased at $V_D = V_G = 4.5$ V.

de-embedded associated gain of the hybrid circuit is around 6.2 dB in the passband of the filter. The hybrid circuit was also tested in the noise-figure setup with two different bias conditions for the LNA of $V_D=V_G=4.5$V, and $V_D=2.5$V, $V_G=5$V. The noise figure and the associated gain of the hybrid circuit for $V_D=V_G=4.5$V is shown in figure 5.a, with no applied bias to the filter. The highest associated gain of the circuit is ~13 dB at 17 GHz, and the corresponding noise figure of approximately 13 dB. Since the noise figure of the LNA is around 2 dB, and the filter's noise figure (equal to insertion loss) is ~ 6 dB, the additional 5 dB loss most likely arises from the integration of the two components (i.e., matching losses) comprising the hybrid circuit. These losses are excessive, and may be improved with modifications to the hybrid

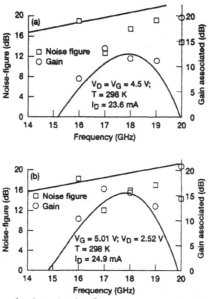

Figure 5. The associated gain and noise-figure measurements of one of the hybrid circuits with the filter unbiased, and the LNA biased at (a) $V_D = V_G = 4.5$ V. (b) $V_D = 2.5$ V, and $V_G = 5$ V. The solid lines show the trend, plotted by the ATN software.

circuit packaging. When the LNA bias conditions were changed to $V_D = 5$ V and $V_G = 2.5$ V (corresponding to lowest noise figure at room temperature), the gain of the hybrid circuit improved to 15 dB at 17 GHz, and the corresponding noise figure of 12 dB(see figure 5.b). Further studies are underway to reduce the noise figure and improve the associated gain of the circuit.

The second filter studied, the narrowband (<1-% bandwidth) 3 pole STO ferroelectric tunable filter was integrated with the hybrid circuit similar to the setup of figure 1. The 3-pole filter exhibited a narrow passband between 19.687 and 19.862 GHz when the filter was not biased. The filter's insertion loss was ~8 dB at room temperature. The hybrid circuit exhibited a gain of 11.5 dB at the center frequency of the narrow passband (19.787 GHz) for $V_D = V_G$ =4.5 V for the LNA (see figure 6). As the hybrid circuit was cooled to 150K, the gain improved to 15.15 dB for the same bias condition. Further cooling down to 85K reduced the associated gain to less than 10 dB (filter under zero-bias) and approximately 12 dB (filter under FBB of ±200 V). Due to the large gain of the LNA at K-band, some undesirable responses were also observed in the frequency range of measurements. Measurements in the noise figure setup are currently underway for this hybrid circuit. The challenge for this hybrid circuit is to match the optimum input impedance of the LNA possibly through different biasing arrangements for the filter, to obtain the maximum possible associated gain and lowest possible noise-figure. Work is currently underway to improve the packaging as well as matching the impedances required at the input of the LNA.

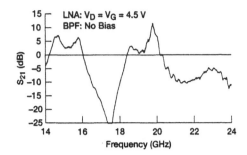

Figure 6. The frequency response of a hybrid circuit with a narrow-band 3-pole filter (center frequency of 19.7 GHz) replacing the 2 pole filters. The LNA was biased at $V_D = 2.5$ V, and $V_G = 5$ V. The filter was unbiased.

CONCLUSIONS

Ferroelectric tunable bandpass filter/LNA hybrid circuits were packaged and tested for possible applications in a Ku and K-band receiver down-converter. A commercial LNA was chosen for this application to show the proof of concept. The effect of tunability of the filter on the hybrid circuit's associated gain and the noise-figure was studied. To our knowledge, this is the first study of a ferroelectric tunable filter integrated with an LNA. The performance of the hybrid circuit shows that the electrical tunability of the filters could be advantageously used for improvements in the hybrid circuit's associated gain as well as the noise-figure.

ACKNOWLEDGEMENTS

The author G.S. acknowledges the NASA/OAI summer Faculty Research Fellowship program, and continuing support from NASA Glenn Research Center. The author F.W.V acknowledges the National Research Council associateship. The authors thank fabrication support offered by Mr. Nick Varaljay, Mr. Bruce Viergutz, and Ms. Liz McQuaid.

REFERENCES

1. D. G. Vendik, L.T. Ter-Martiosyan, A. I. Dedyk, S.F. Karmanenko, and R.A. Chakalov, Ferroelectrics, vol. 144, pp 33-43 (1993).
2. A.M. Hermann, R.M. Yandrofski, J.F. Scott, A. Nazirpour, D. Galt, J.C. Price, J. Cuchario, and R.K. Ahrenkiel, J. of Superconductivity, vol. 7, no.2, pp 463-469 (1994).
3. A.T. Findikoglu, X. Jia, X.D. Wu, G.J. Chen, T. Venkatesan, and D.W. Reagor, Appl. Phys. Lett., vol. 68, no.12, pp (1996).
4. D.M. Dalbert, R.E. Stauber, J.C. Price, C.T. Rogers, and D. Galt, *Appl. Phys. Lett.*, vol. 72, no.4, pp 507-509 (1998).
5. C. M. Carlson, T. V. Rivkin, P. A. Parrilla, J. D. Perkins, D. S. Ginley, A. B. Kozyrev, V. N. Oschadchy, and A. S. Pavlov, submitted to Appl. Phys. Lett, 9/23/99.
6. S.S. Gevorgian, et al., IEEE trans. Appl. Superconductivity, vol. 7, pp 2458 (1997).
7. G. Subramanyam, F.W. Van Keuls, and F.A. Miranda, IEEE Microwave and Guided wave lett., vol. 8, no.2, pp 78-80 (1998).

NEAR-FIELD IMAGING OF THE MICROWAVE DIELECTRIC PROPERTIES OF SINGLE-CRYSTAL PbTiO$_3$ AND THIN-FILM Sr$_{1-x}$Ba$_x$TiO$_3$

Y.G. WANG^a, M.E. REEVES^{a,b}, W. CHANG^b, J.S. HORWITZ^b and W. KIM^b

a Department of Physics, The George Washington University, Washington, DC, 20052
b Naval Research Laboratory, Washington, DC, 20375

ABSTRACT

A PbTiO$_3$ crystal and Sr$_{1-x}$Ba$_x$TiO$_3$ films have been studied by near-field scanning microwave microscopy (SMM). In the PbTiO$_3$ crystal, dielectric properties and topography are obtained simultaneously. In Sr$_{1-x}$Ba$_x$TiO$_3$ films, local variations and sample-to-sample differences are observed. To quantitatively determine local dielectric permittivity and loss, we also carry out theoretical calculations on dependence of resonant frequency and quality factor on dielectric constants of bulk samples and thin films. Good agreements between experimental and theoretical results are obtained.

I. INTRODUCTION

With the rapid progress in communication technology and growing demand for high permittivity materials, microscopic characterization of dielectric materials in GHz frequency range has become more important. So far, several types of near-field scanning microwave microscopies (SMM) have been developed, including $\lambda/4$ resonator, [1] coaxial line, [2] ring resonator [3], and submicron resolutions have been achieved. However, it is still a challenge to quantitatively characterize the microwave properties on a microscopic scale because the Coulomb force between the tip and sample depends not only on the material properties, but also on geometric factors of tip, sample and experimental setup.

In this paper, we report both our experimental and theoretical approaches to characterize dielectric properties by SMM using a $\lambda/4$ resonator [1]. We obtain the dependence of the local dielectric permittivity on the measured resonant frequency, f_0, of the cavity under various conditions. In particular, we find both experimentally and theoretically that the measured df_0/dz is a linear function of the material dielectric constant, ϵ, when ϵ is high enough. Topographic features also contribute to the measured signal. We are able to distinguish the topographic contribution to the electrical response by adjusting the position of the tip accurately in measurements of a PbTiO$_3$ crystal.

II. EXPERIMENTAL AND THEORETICAL APPROACHES

The SMM used in our study mainly consists of a $\lambda/4$ coaxial resonator [1] with a resonant frequency about 1.75 GHz and an HP 8753D network analyzer. The $\lambda/4$ resonator is shown in the inset of Fig.1. A bottom metal plate is attached to the cavity so that the electric field is strongest at the end of the central conductor. A tungsten STM tip protrudes from the central conductor of the cavity and comes close to samples under study. Thus near-field microscopy in the submicron range can be realized. To accurately control the position of the tip in three dimensions, piezoelectric actuators are used to position the sample.

For comparison with experiments, we perform finite-element calculations on the electromagnetic fields inside the cavity and of the tip-sample assembly. When calculating the field near the tip, a static approximation is used considering that tip size is much smaller than the wavelength (17 cm at 1.75 GHz). The tip, consisting of a cylinder and a cone with a spherical end, is assumed to hold a constant potential. The cylindrical symmetry further simplifies the solution so that the electric field near tip can be obtained by solving the Poisson's equation $\nabla^2\phi=0$ in two-dimensions. To be consistent with experiments, a metal ground plane is added below the sample. The changes of resonant frequency and Q are then obtained using perturbation theory. That is, we assume that the small contribution of the fields generated within the sample are capacitively coupled to the cavity.

III. RESULTS AND DISCUSSION

First, we calculated the electric and magnetic fields inside the cavity. It is found the electric field is strongest at the open end of the central conductor where the STM tip is incorporated. On the contrary, the magnetic field is zero here. A bottom metal plate closes the cavity, effectively concentrating the electric field near the tip, reducing the resonant frequency. This result agrees with our experimental observation.

The calculated resonant frequency is further decreased by incorporating the tip into the cavity. This has been confirmed experimentally by using tips of various length. We also calculate the sensitivity to changing dielectric properties and the spatial resolution for various tip sizes. It is found that decreasing the tip radius decreases the sensitivity of the system but increases the spatial resolution.

Calculations of changes in f_0 and Q with distance z between tip and samples with different dielectric constants show excellent consistency to the corresponding experimental observations. It is interesting to note that the derivative of resonant frequency with tip-sample distance, df_0/dz, increases linearly with the dielectric constant, ϵ, when $\epsilon >10$. Both the theoretical calculations and the experimental measurements exhibit the tendency shown in Figure 1. This is quite important because most materials interesting for microwave applications have dielectric constants within this linear range. By our method, we are able to determine the dielectric properties more accurately irrespective of the effects of sample size and shape. In general, the resonant frequency is expected to change with the size of the sample under study due to the long-range character of Coulomb interactions. We have tried to measure f_0 and df_0/dz for the same kind of samples of different size and the same sample in the center and close to its edge. There are distinct change in f_0 but negligible variations in df_0/dz.

We also calculate f_0 as a function of the dielectric constant and thickness for thin films on a given substrate (see Fig.2). In this case, f_0 is no longer a linear function of of the dielectric constant of films. Nevertheless, the dielectric constant of the film can be determined as long as df_0/dz is measured and the film thickness is given.

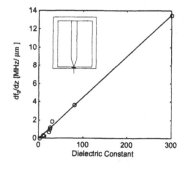

Fig. 1. Dependence of df_0/dz on dielectric constant of samples. The straight line is the theoretical calculation and the circles are experimental results measured on bulk crystals of different dielectric constants. The inset shows the coaxial $\lambda/4$ resonator used in our experiment

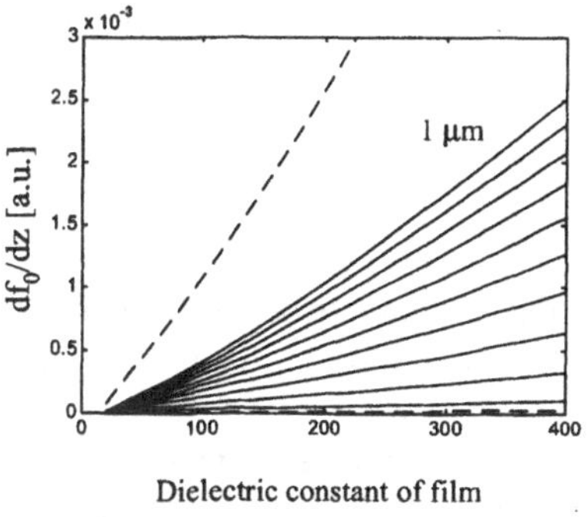

Fig. 2. Calculated df_0/dz as a function of dielectric constant for films grown on a substrate with $\epsilon=10$. The curves represent films with thicknesses ranging from 0 (bottom dashed line) to 1 μm (labeled line) in 0.1 μm steps. The top line (dashed) is for a film thickness of 0.5mm and represents the bulk value.

In the literature, there have been reports of nanometer-scale spatial resolution [1,3] and of 1 in 10^5 property sensitivity [1] in near-field microwave microscopies. However, high spatial resolution and high sensitivity have not been obtained simultaneously. Our calculation shows that for a cylindrical impurity of $\epsilon=120$ imbedded in a surrounding of $\epsilon=80$, the shift in the resonant frequency decreases with the size of the impurity and goes below the detection limit of 10^{-5} when the impurity is smaller than $1\mu m$. When the size of impurity is fixed, the shift of f_0 increases with contrast in permittivities as expected.

To determine material properties accurately, it is highly desired to obtain the topographic features and the dielectric properties simultaneously so that the contributions of topography can be excluded. [4,5] We have successfully obtained both in the measurement of a PbTiO$_3$ crystal by moving the tip vertically at various horizontal positions and measuring df$_0$/dz of the sample. Fig.3 shows the dependence of f$_0$ on tip-sample distance when the tip is above an a domain (circles) and a c domain (stars), respectively (df$_0$/dz drops to zero at touching).

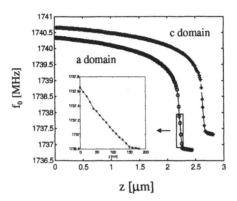

Fig. 3. Changes of f$_0$ with tip-sample distance for a (circles) and c (stars) domains in a PbTiO$_3$ crystal. The inset shows the dependence of f$_0$ on z close to the sample.

Fig.4 shows the simultaneously obtained f$_0$, Q and topography of the PbTiO$_3$ crystal. The data show that the crystal consists of parallel a and c domains. The a domains have a higher dielectric constant (ϵ_a=105) and a lower loss tangent (tanδ_a=0.04) than the c domains where ϵ_c and tanδ_c are 35 and 0.08, respectively. [6]. As a result, f$_0$ is lower but Q is higher in the a domains. Because the domain wall forms an angle about 45° to the crystal surface, the peak of f$_0$ shifts to the left. The observed topography agrees with AFM results [7] and the bending angle between a and c domains is 3.8^o, consistent with the theoretical prediction of 3.65^o. [7,8] At the edge of the area under study, there is a small region, denoted by F in Fig.4(b), with much lower Q than the surroundings. This is due to defects in the crystal introduced during the growth process. In fact, we have found several such regions of various sizes and shapes.

We have measured Sr$_{1-x}$Ba$_x$TiO$_3$ films grown on MgO and LaAlO$_3$ substrates by pulsed laser deposition (PLD) process. [9] The image of as-deposited sample contains islands about 5μm in size. SEM and ANSOM measurements rule out surface roughness or grain size, since these features are smaller than 10 nm and 200 nm, respectively. [10] After annealing the sample at 1100°C for 6 hours, the film becomes much more uniform and the spatial variation in the permittivities at different regions is much smaller, as seen by comparing the scales in Fig.5(a) and (b).

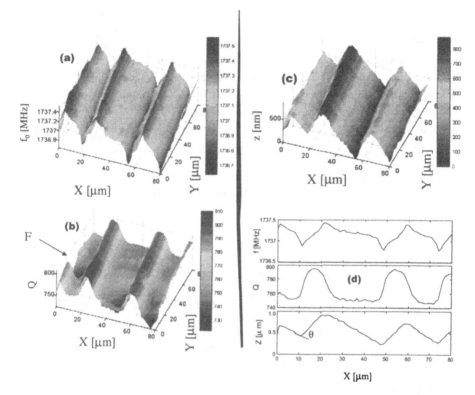

FIG. 4. Scanning-near-field images of a PbTiO$_3$ crystal showing (a) resonant frequency, (b) quality factor, (c) topography, and (d) a cross-section profile at y=70μm. The angle θ denotes the bending at the a-c domain wall.

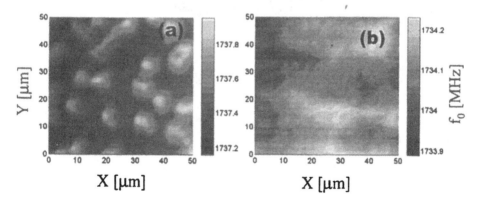

Fig. 5. Local variations of dielectric constants of Sr$_{1-x}$Ba$_x$TiO$_3$ films for (a) an as-deposited film and (b) a film annealed at 1100°C for 6 hours. The gray scale represents the measured resonant frequency.

IV. CONCLUSIONS

In summary, we have calculated the dependence of f_0 on material properties and parameters of the experimental setup. A linear dependence of df_0/dz on dielectric constant is found both experimentally and theoretically. In a $PbTiO_3$ crystal, the dielectric properties and topography are obtained simultaneously. For $Sr_{1-x}Ba_xTiO_3$ films, local variations for different processing conditions are observed. Future work will be focus on the accurate determination of loss in dielectric films.

Acknowledgements: The work is supported by DARPA. The authors would like to thank Dr. F. Rachford and Dr. J. Byers for helpful discussions.

[1] C. Gao, X.D. Xiang, Rev.Sci.Instrum., **69**, 3846 (1998). Y. Lu, T. Wei, F. Duewer, Y. Lu, N.B. Ming, P.G. Schultz and X.D. Xiang, Science, **276**, 2004 (1997).

[2] D.E. Steinhauser, C.P. Vlahacos, F.C. Wellstood, S.M. Anlage, C. Canedy, R. Ramesh, A. Stanishevsky and J. Melngailis, Appl.Phys.Lett., **75**, 3180 (1999).

[3] Y. Cho, S. Kazuta and K. Matsuura, Appl.Phys.Lett., **75**, 2883 (1999).

[4] F. Duewer, C. Gao, I. Takeuchi, X.D. Xiang, Appl.Phys.Lett., **74**, 2696 (1999).

[5] C.P. Vlahacos, D.E. Steinhauser, S.K. Dutta, B.J. Feenstra, S.M. Anlage and F.C. Wellstood, Appl.Phys.Lett., **72**, 1778 (1997).

[6] A.V. Turik, Phys.Stat.Sol., **B94**, 525 (1979).

[7] Y.G. Wang, J. Dec and W.Kleemann, J.Appl.Phys., **84**, 6795 (1998).

[8] J.P. Remeika and A.M. Glass, Mater.Res.Bull. **5**, 37 (1970).

[9] W. Chang, J.S. Horwitz, A.C. Carter, J.M. Pond, S.W. Kirchoefer, C.M. Gilmore and D.B. Chrisey, Appl.Phys.Lett., **74**, 1033 (1999).

[10] C. Hubert and J. Levy, Appl.Phys.Lett., **73**, 3229 (1998).

AUTHOR INDEX

SUBJECT INDEX

Printed in the United States
By Bookmasters